Operations Management and Industrial Engineering

Operations Management and Industrial Engineering

Edited by Alden Bennett

CLANRYE
INTERNATIONAL
www.clanryeinternational.com

Clanrye International,
750 Third Avenue, 9th Floor,
New York, NY 10017, USA

ISBN: 978-1-64726-664-6

Cataloging-in-Publication Data

Operations management and industrial engineering / edited by Alden Bennett.
p. cm.
Includes bibliographical references and index.
ISBN 978-1-64726-664-6
1. Production management. 2. Industrial management. 3. Operations research.
3. Industrial engineering. 4. Management science. I. Bennett, Alden.
TS155 .O64 2023
658.5--dc23

For information on all Clanrye International publications
visit our website at www.clanryeinternational.com

CLANRYE
INTERNATIONAL

Contents

Preface

Operations management is a field of management that deals with the design and control of the production processes, and redesigning of business operations for producing goods or services. Industrial engineering is a branch of engineering management involved in the optimization of complex processes and systems. It achieves optimization of processes and systems by creating, improving and implementing an integrated system of resources, which include labor, money, knowledge, information, and machinery. There are various sub-disciplines of industrial engineering, namely, methods engineering, financial engineering, energy engineering, and operations research and optimization. Operations management and industrial engineering deal with the application of finance, analytics, engineering, logistics, and structured problem-solving techniques for meeting the goals of contemporary and globalized supply chains. This book brings forth some of the most innovative concepts and elucidates the unexplored aspects of operations management and industrial engineering. With state-of-the-art inputs by acclaimed experts of this field, it targets students and professionals.

This book has been the outcome of endless efforts put in by authors and researchers on various issues and topics within the field. The book is a comprehensive collection of significant researches that are addressed in a variety of chapters. It will surely enhance the knowledge of the field among readers across the globe.

It gives us an immense pleasure to thank our researchers and authors for their efforts to submit their piece of writing before the deadlines. Finally in the end, I would like to thank my family and colleagues who have been a great source of inspiration and support.

Editor

1

Managing Environmental and Operational Risks for Sustainable Cotton Production Logistics: System Dynamics Modelling for a Textile Company

M. Ali Ülkü [1,2,3,*], Melek Akgün [4], Uday Venkatadri [2,3], Claver Diallo [2,3] and Simranjeet S. Chadha [2,3]

1 Rowe School of Business, Dalhousie University, Halifax, NS B3H 4R2, Canada
2 Centre for Research in Sustainable Supply Chain Analytics, Dalhousie University, Halifax, NS B3H 4R2, Canada; uday.venkatadri@dal.ca (U.V.); claver.diallo@dal.ca (C.D.); chadha@dal.ca (S.S.C.)
3 Department of Industrial Engineering, Dalhousie University, Halifax, NS B3H 4R2, Canada
4 Sakarya Business School, Sakarya University, Sakarya 54050, Turkey; melekakgun@sakarya.edu.tr
* Correspondence: ulku@dal.ca

Abstract: Effective management of cotton production logistics (CPL) against volatile environmental conditions while maintaining product quality and yield at acceptable costs has become challenging due to increasing global population and consumption and climate change. In CPL, the harvesting, processing, and storage of cotton are all linked, prone to various environmental risks (e.g., flooding) and operational risks (e.g., excess spraying of pesticides). Thus, it is crucial for a resilient and sustainable supply chain management to prioritize risks and chart suitable risk response strategies. For a CPL, we employ a system dynamics (SD) approach to investigate the likelihoods of environmental and operational risks and their impacts in four dimensions: variable costs, fixed costs, quality performance, and yield. Using the case of a textile company in Turkey, we demonstrate an end-to-end framework for mitigating CPL risks. SD simulation results show that increases in seed prices and machine and equipment breakdowns are the risks that most affect the unit cost, whereas pests and plant diseases most hurt cotton harvest yield. Via scenario analyses, we demonstrate that a proper risk response strategy, compared to doing nothing, may reduce variance in cotton quality by about 35% at the expense of about an 11% increase in unit cost variability.

Keywords: cost management; system dynamics; risk management; cotton production logistics; scenario analysis; supply chain

---

1. Introduction

Environmentally unsustainable production methods risk future capacities: agriculture is no exception. The most widespread non-food crop, cotton, has been a staple product instrumental for the economy, especially in developing countries. Competition in many sectors, including textiles, currently occurs between supply chains (SCs) rather than between companies. Companies cannot achieve a competitive advantage by isolating themselves from their SC partners. Pressing market and social demands and an increasing need for resiliency against supply disruptions at the global scale oblige SC partners to coordinate production, distribution, and customer service effectively.

The concept of supply chain management (SCM) has gradually gained importance since the 1980s when companies discovered the benefits of cooperation. SCM deals with the management of material, information, and financial flows in a network of suppliers, producers, distributors, and customers. Coordination and integration of these flows between and within enterprises are of critical importance for effective SCM. An SC's ability to offer end-customer goods and services of the required quality at a low cost depends on its success in utilizing process efficiencies and technological innovations. With price increasingly dependent on market conditions, cost management is key to efficiency and profitability. In this respect, cost management strategies have become pivotal in SCM (e.g., [1]).

Regardless of their type, size, or field of activity, all enterprises face risks. SCM deals with a complex structure exposed to various risks—from small-scale risks such as delivery delays to more significant risks such as deterioration of the entire chain [2]. In a general sense, "risk" is the probability of not achieving the desired result. Therefore, it is important to develop an effective risk management policy in SCM. Where the term "supply chain risk management" (SCRM) is relatively recent, the concept of risk management has quite a long history. While the growth of global activities may significantly reduce a SC's costs, it makes it more susceptible to risks because of increased scope and interdependencies. Companies that face increasing risks are more likely to establish partnerships and consciously participate in SC cooperation [3]. In so doing, cost–benefit trade-offs need to be carefully evaluated before adopting SCRM strategies [4].

A profitable and the most widespread non-food crop, cotton is the primary material for almost half of all textiles. More than 250 million people worldwide, mostly in developing countries, depend on cotton production for their income. The use of pesticides, high water consumption, and the conversion of natural habitat to agricultural land (deforestation) count as the most influential factors for cotton's environmental footprint. To maintain future capacity, required are more sustainable cotton production and storage methods, whereby risks are proactively managed. However, this remains an exigent and challenging task [5].

Figure 1 displays a visual of the problem we study in this paper. We segment into two main stages a textile supply chain (TSC), which uses cotton as a raw material: cotton production and textile manufacturing. Our scope is cotton production logistics (CPL) that includes seeding, growing, harvesting, and then pre-processing/storage of cotton stocks. Both the environmental risks such as drought and the operational risks such as labor shortage determine the overall quality, yield, and cost of the CPL.

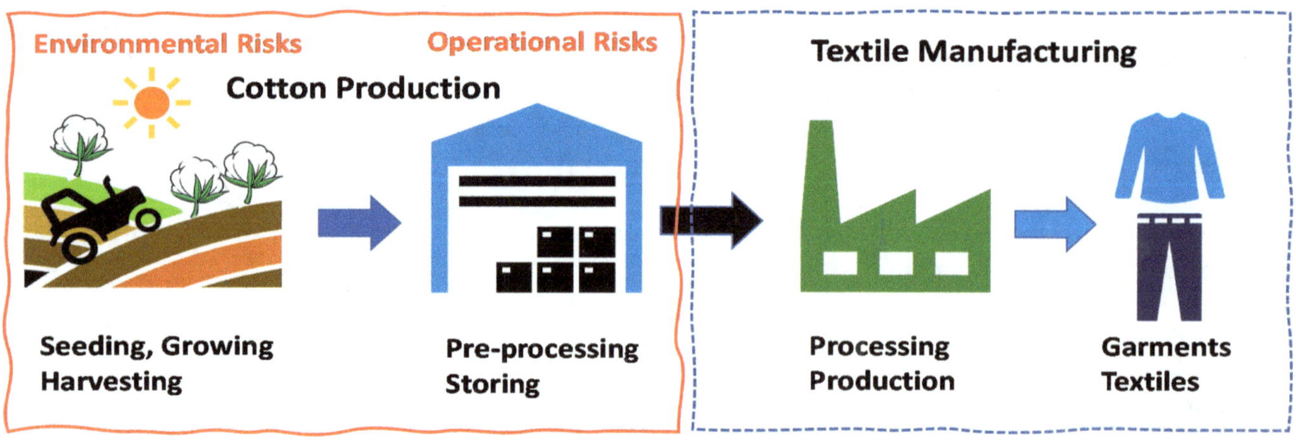

Figure 1. Visualization of the problem scope.

This paper contributes to the body of literature on supply chain risk management whereby system dynamics (SD) is employed to model complicated and confounding parameters and decision variables. To that end, as our salient contributions, (i) we analyze, via an SD modelling approach, the effects of responses to risks that may occur during CPL operations, (ii) we employ an integrated textile SC example on two essential performance criteria: cost and quality, and (iii) we provide by a case study of an end-to-end framework in developing risk response strategies.

The remainder of this paper is organized as follows: In Section 2, we provide a detailed literature review. Section 3 introduces the substructure of the study and describes building blocks of the system dynamics model (SDM) to assess and prioritize CPL risks. Drawing on Turkey as a case study, we discuss risk assessment and prioritization in Section 4 and then describe several risk response scenarios in Section 5. Finally, in Section 6, we summarize inferences obtained from the study and provide closing remarks.

2. Literature Review

The study of SCRM emerged after the 1990s and increased rapidly after 2000. SCs are largely exposed to very similar risks: according to [6], among others, 34% of SC interruptions result from processes within the company itself, 15% from suppliers, 13% from customers, 4% from nature, 4% from governments, and 6% from the chain relationships. There are a variety of risks that may cause interruptions in SC operations. For a disaster relief SC, Peng et al. [7] use the SD model by simulating interrupted road and communication networks damaged by an earthquake. SCRs can be categorized into three types: internal, chain, and external. Internal risks result from in-house operations or from the operations themselves (c.f., [8]). These include accidents, machine and equipment (M&E) breakdown, the failure of information technology system, human error, quality, and errors in management decisions such as the determination of the lot size, secure stock levels, financial problems, and delivery plans. Chain risks are those outside the institution but inside the SC. These result from insufficient coordination and the lack of information exchange between SC actors. Safety, the suitability of the material, time of supply, delivery problems, strike, lock-out, etc., are supplier-related risks, whereas issues related to variable demand, payments, and order processing are customer-related risks. External risks result from an interaction between the SC and the environment, such as accidents, severe weather conditions, laws and regulations, natural disasters, and wars. Bradley [9] analyzed the differences between frequent and rare risks that cause an interruption in the SC and suggested a new risk measurement and prioritization method that explains the features of rare risks.

There also exist studies on cost minimization and profit maximization in the SCRM literature. For example, to mitigate demand uncertainty in SC planning, Gupta and Maranas [10] suggested a stochastic programming–based planning model that reacts to the demand over time. Yu and Goh [11] investigated the visibility of SC and the effects of SCRs on SC performance. Wieland [12] suggested a model that ensures the selection of SC strategy based on risk probability and impact. The literature also includes studies investigating the effects of cooperation and integration, which have significant roles in the success of SCRM. Braunscheidel and Suresh [13] highlighted the impact of market tendency and learning tendency on SC agility. They also found that internal integration, external integration with the main suppliers and customers, and flexibility positively affect the SC's agility. Chen et al. [14] explored the relationship between three different types of risks—supply, demand, and process. In another study, from the perspectives of uncertainty, variability, and trust, Chen et al. [15] showed that supply risk could be mitigated by building trust in the buyer-supplier relationship in addition to information knowledge sharing. Glock and Ries [16] studied risk due to lead time variability for an SC multiple-supplier, single-buyer integrated inventory problem. Kull et al. [17] analyzed the risk of failure that results from a second-level supplier using stock and source dependence theories and reported that the supplier's excessive stock sometimes increases the risk of supply rather than decreasing it. Chen and Lee [18] suggested a multi-product, multi-stage, and multi-period model to fulfil more than one aim in multi-stage supply networks with uncertain demand and product price.

Depending on the context, SCRM is tailored according to the definition of risk used and the type of risk assessment tools available for decision-making. To that end, ample discussion of SCRs exists in the literature. For example, Kern et al. [19] developed a model for managing risks related to the supplier by attributing the risk definition, risk assessment, and risk-reducing processes to risk performance in a detailed empirical study. The effects of constantly redefining risk, its assessment, and its reduction were also modelled. Tummala and Schoenherr [20] studied SCRM processes and suggested techniques for risk assessment and management. Hallikas et al. [21] explained the risk analysis concept from both producer and supplier perspectives and investigated two approaches as computer-aided cause–effect analysis and internal audit for the risk analysis using case study and qualitative information. Trkman and McCormack [22] researched new techniques for estimating and defining supply risk. Their approach of assessing and classifying suppliers is based on suppliers' qualities, performance, and characteristics of the suppliers and modified according to the specific environmental conditions.

The degree of sensitivity of the SC to risks is a significant determinant for risk management strategies. Foroughi et al. [23] focused on the risks in the supplier part and investigated the reasons behind the increase in the SC sensitivity and investigated the processes of defining and evaluating threats and charting precautions to be taken in an SCRM process. Using probability theory, Grötsch et al. [24] identified conditions enabling proactive SCRM practices: The bankruptcy of past suppliers was chosen to indicate the organization's level of sensitivity. It was determined that a management control system, a logical cognitive style, and a good supplier–buyer relationship positively affect the proactive management of the bankruptcy risk of the supplier. Wagner and Neshat [25] defined SC sensitivity and comparatively measured it for companies in different categories using graph theory. In this empirical study, the authors found a negative relationship between SC sensitivity and performance but a positive relationship between SC sensitivity and risk management. After investigating the main reasons for the sensitivity and risks of SCs in the general sense, Thun and Hoenig [26] defined the possibilities and potential effects on the SC. The results were investigated separately as internal and external risks in the probability–effect matrix. Therefore, the effect of the SCRM on performance was tested. The performance difference between the companies that apply reactive and proactive risk management was shown.

SD, pioneered by Forrester [27], is a powerful modelling and simulation methodology for obtaining insights about complex systems, especially for long-term decision-making in dynamic industrial management problems. SD has been a modelling approach used to study systems' structure and behaviour with dynamic qualities that consist of interactive feedback cycles. SD modelling (SDM) aims to simulate the complex relationships within a system over a certain period and then identify possible worst-case scenarios. Furthermore, SDM helps analyze how the system reacts to external effects within a given period. The SD approach emerged from the need to understand better and analyze industrial processes and the complex structure of these processes by top decision-makers and stakeholders. With a heightened interest recently, the SD approach has been used to investigate various social, economic, and environmental systems in which the integrative idea and feedback cycles are critical to understanding the relationship between variables ([28]). The system dynamics model (SDM) can be a tool for analyzing the behaviours of complex socio-economic systems to show how organizations and politics affect actions [29] and as a powerful means for analyzing manufacturing systems [30]. Kim and Park [31], for example, successfully employed SD to investigate otherwise highly intractable inventory decisions for SC coordination. Although alternative SCRM tools exist, SD as an analytical method has been widely preferred by researchers for studying various operational environments (see, for example, [32–36]).

In the face of mounting challenges around sustainably matching global demand and supply, and the ever-changing landscape of the business world with environmental imperatives due to climate change, consumer pressures for sustainability, and rampant disruptions such as the COVID-19 pandemic, there is much room for research for managing risks in supply chains. To that end, this paper contributes to SCRM literature by study CPL risk mitigation via an SDM approach.

3. Research and Case Study Background

Cotton is the main source of natural fibre for textile supply chains (TSCs) and, as such, is a crop plant with high economic value. Moreover, cottonseed is a good source for vegetable oil for human consumption and vegetable protein in animal feed. In Turkey, cotton production is precious.

According to the Global Agricultural Information Network Report [37], "The textile industry continues to be one of the leading sectors in the Turkish economy ... Approximately 36,000 companies operate in the sector, providing about 515,000 jobs in garment production alone. Exports of ready-to-wear items in 2018 were [USD] 17.6 billion, and textiles were [USD] 8.4 billion, both up about four percent compared to a year before ... buyers ... prefer to work with Turkey due to geographical proximity, short response time, and good quality."

With 4.3% of the total cotton production in the world, Turkey ranks first in Europe and sixth in the world. With a total of 742,000 hectares, Turkey ranks eighth in the world in terms of planting area. Turkey also has the highest fiber yield among countries with a cotton planting area above 300,000 hectares. With cotton shares making up 62% of total fiber production in Turkey (compared to 48% globally), Turkey's textile industry generates 34% of the total export income, highlighting the value of cotton production in this developing economy [38]. Yield, quality, and cost of cotton affect the subsequent members of a TSC and its production process. Further, the risks in cotton production that affect the cost, quality, and yield also affect yarn weavers and cloth producers, distributors, and retailers. Cotton planting areas in the world are limited. Fiber production uses plant, animal, and artificial sources as alternative inputs. As cotton fibre consumption increases with the rise in population, there remains a risk of significant cotton fibre deficit worldwide.

The aim of this study, using an SD approach, is to analyze and evaluate the effects of the risks in CPL based on three primary performance criteria: yield, quality, and cost. The risk management process comprises the following steps: risk definition, measurement, assessment, evaluation, response, monitoring, and control. This process provides a successful framework in defining potential risk factors and their probability of materialization [20]. Risk management defines the risks, then measures and evaluates them; from here, it is necessary to determine the response/reaction to hedge against prioritized risks. Risk assessment is the joint function of the probability of materialization of the risk and the effect that will be created should it occur. The criteria used for the likelihood of the materialization of risks are generally digitized quantitative expressions. While the impact of the risk can be financial criteria, they can also be based on qualitative definitions such as the loss of business prestige. These criteria can be used in the conceptual modelling and evaluation of SCRs. The probability of risk materialization may be estimated using archival data, risk logs, expert opinion, or statistical data. Caution should be exercised as some of the risk management strategies may affect other risks negatively [39]. Therefore, it is also essential to understand the variability and mutual relations between the risks. Such an approach can be applied by taking threats and their sources into consideration (e.g., [2,40]), as well as their spheres of influence.

To quantify risk scores for risk prioritizations, we convert and multiply effect and probability values. Calculating risk scores in this way is beneficial in prioritizing risks and determining risk responses. However, we believe that determining which area the risk will affect, should it occur, and its level of effect (by correlating with a criterion by the field of effect in the enterprise) will yield more effective results both in risk analysis and prioritization. We note that it is a formidable task to implement a formal risk management process over the SC as a whole; various integration strategies are required for this.

3.1. Case Study and Related Parameters

In our case study, our focal company is 01TEX (a fictional name to preserve confidentiality), a textile enterprise established in 1989 in the Province of Adana, Turkey. 01TEX is a vertically integrated cotton production logistics: it rents cotton cropping lands on which cotton is produced and used as raw material for production in its indoor facility area of 30,000 m^2. It is among the biggest textile companies in Turkey, with eight spinning factories and one denim factory with weaving, knitting, and dyeing capabilities. The company produces the raw material—its primary input—on land that it hires each year. Therefore, the transfer-selling price is dependent on the unit production cost and the quality of the cotton produced. The quality of the cotton obtained, and its price based on its cost affect the textile production stages' performance. The cotton bought is used in yarn production, and the yarn is used in fabric production. 01TEX does not sell any yarn or fabric; the fabric produced is used as an input for ready-to-wear clothing that caters to both domestic and international markets. The company management oversees a major part of the SC, that is, from the production of cotton to delivering the apparels to the end consumer. Nevertheless, problems that occur during the cotton cultivation and the mobility of the cotton market prices can affect 01TEX's performance. Therefore, it is necessary to estimate the management policies at the cultivation and storage stages of cotton and their results.

To preserve confidentiality, the parameter values used in this study are approximations. Cotton production costs have variable and fixed-cost (FC) components. The bulk of the costs resulting from activities related to cotton production starts in September and October with soil preparation. Then, the soil is processed once with gobble (disc harrow). The planting is performed using a sowing machine after processing the earth four times with the bed-forming machine. The planting process occurs between March and April, and thinning is performed between May and June following the planting. This process is labor-intensive. Hoeing is then carried out five times using manual labor or a hoeing machine between June and July following the thinning. Middle breaking is also performed using the bed-forming machine between June and July. With planting, the base dressing is irrigated once in June and July, twice together with the fertilization process, and four times between June and August. Labor is used in activities such as planting, thinning, hoeing, middle breaking, fertilization, disinfestation, irrigation, and harvesting in cotton production. The average use of labor is 10 hours/acre. The average machine use in cotton production is estimated as 1.5 h/acre. The seed use per acre in the enterprise is 2.60 kg on average, and the average kilogram cost of cotton seed is USD 9. Compound fertilizer is used as base dressing. The base dressing cost of enterprises per acre is USD 19 on average, and the top-dressing cost is USD 17, resulting in a total average cost of USD 36.

The adverse effects of diseases and pests in cotton-growing are closely related to climate conditions, such as high humidity and temperature, and loss prevention methods. Therefore, five rounds of disinfestation are required each year. The average pesticide cost is USD 58/acre, and the average irrigation costs are USD 38.40/acre. It is assumed that the enterprise plants cotton over 2500 acres per year, and it is expected to harvest 400 kg/acre. As shown in Table 1, the targeted or standard unit cost of cotton is determined to be USD 1.10/kg. According to the International Cotton Advisory Committee [41], the cost of cotton production is continuously increasing and is estimated at USD 1.5/kg of lint. However, the risks that occur during the planting, growing, harvesting, or cotton cause deviations from targeted levels.

Table 1. Cotton Production Costs for 01TEX Case.

	Measure	**Quantity Used**	**Unit Cost (USD)**	**Cost USD/acre**
		Variable Costs		
Seed	kg/acre	2.6	9	23.4
Fertilizer	kg/acre	300	0.12	36
Lime	ton/acre	0.35	22	7.7
Labor	h/acre	10	10	100
Electricity	kwh/acre	1.5	0.42	0.63
Pesticide	times/acre	5	11.6	58
Water	ton/acre	48	0.8	38.4
Transportation Fuel	acre	1	70	70
Annual land rent	acre	1	24	24
	Total Variable Cost			**USD 366.13/acre**
		Fixed Costs		
M&E Depreciation	acre/year	2500	30	75.000
Overhead	acre/year	2500	44	110,000
	Total Fixed Cost (for 2500 acre)			185,000
	Total Cost			USD 440.13/acre
	Unit Cost (400 kg/acre yield)			**USD 1.10/kg**

In defining the cotton quality, six basic standard criteria are used: micronaire, fibre strength, fibre length, length uniformity, colour grade, and extraneous matter. The quality index of cotton is created based on the evaluation of these criteria as a whole. 01TEX defines its suitable quality index level for production levels as 4 (based on a quality range of 1 to 5). However, the risks incurred during cotton planting, growing, harvesting, or storage led to deviations from the targeted quality level. 01TEX management decided to implement suitable risk management strategies by determining the risk incidents that lead to deviations in targeted cost, yield, and quality metrics. A structured risk management system is developed to this end.

3.2. *Risk Identifications and Measurement*

The risk identification process is the first and most crucial stage of the risk management process. It requires preparing an extensive list of risks organized by category for business units, corporate functions, and investment projects. This stage helps form the risk profile of the company. Each risk identified requires prioritization levels to draw the attention of management. This prioritization is performed by making a risk assessment [42]. Methods used in the risk identification process vary according to the enterprise's recorded risk data, risk types, or the risk management culture in the enterprise.

Risk incidents likely to occur during CPL of O1TEX were classified as operational risks that are relatively under the control of the enterprise (the process and the chain risks, c.f., [8]), and environmental risks that may be encountered as a result of natural conditions outside the control of the enterprise. To identify those risk factors, 01TEX stroke a risk team consisting of cross-functional production managers, finance, purchasing, marketing, logistics, and cost analysis units chosen among the company management, and experts employed in professional chambers outside the enterprise. Recall Figure 1: Because 01TEX owns the farming and pre-processing/storage functions, it controls operational decisions for its CPL. Using a Delphi method (e.g., Soni et al. [43]), an interactive forecasting method that relies on a panel of experts, 01 TEX identified the risks, their likelihoods, and the ranges of impacts on the performance measures. These data directly provided to us were used as the base case for the risk simulations.

It is necessary to estimate the probability of materialization and the effects of the risks defined as critical. Risk exposure value is calculated by multiplying the likelihood of a risk with its impact. The most common way of presenting risks is the probability–effect matrix. The likelihood of occurrence may be high, intermediate, or low level. Similarly, the effects or results may be high, average or low. This process is called "risk mapping." A high, intermediate, and low 3 × 3 matrix is suitable for many organizations' needs, but a 5 × 5 matrix (or 7 × 7) may be more suitable for others. Besides, it is critical to define what is meant by "high", "intermediate" and "low" in terms of the probability and effect [44]. Additionally, the type and impact of risk over time add another layer of complexity to the analysis. For example, a risk with a negligible financial impact in the short term may have important effects on the reputation of the enterprise in the long term. Therefore, identified risks were investigated in four dimensions in terms of their effects over time. The risks that occur during cotton production have various levels of effects on fixed cost (FC), variable cost (VCs), yield, and cotton quality. See Table 2 for identified risks, likelihood of materialization, possible impact areas, and their effects.

01TEX used a 5-point Likert type scale to measure the likelihood of the risks identified. The likelihood categories were defined as:

1 Rare, risks that are less than 0–10% likely to occur.
2 Unlikely, risks that are 10–35% likely to occur.
3 Possible, risks that are 35–60% likely to occur.
4 Likely, risks that are 60–85% likely to occur.
5 Almost sure or frequently, risks that are 85–90% likely to materialize.

Table 2. Cotton production logistics (CPL) risks, likelihoods, and impacts for 01TEX case.

Type	CPL Codes and Risks	Risk Likelihood	Risk Impact Range (min–max)			
			↑ VC USD/kg	↑ FC 1000 USD/y	↓ Yield (kg/acre)	↓ Quality (Index)
Environmental Risk	ER1: Flood	4	-	-	10–30	0.05–0.10
	ER2: Pests and diseases	4	-	-	6–10	0.09–0.12
	ER3: Drought	3	2–4	6–16	3–6	0.06–0.08
	ER4: Increasing seed prices	3	4–7	-	-	-
	ER5: Deficient rainfall	4	-	-	6–8	0.10–0.14
	ER6: Excess rainfall	4	-	-	3–7	0.05–0.08
	ER7: Excess hot weather	3	-	-	4–6	0.07–0.10
	ER8: Excess cold weather	2	-	-	3–6	0.05–0.08
	ER9: Service interruption	2	-	-	1–2	-
	ER10: Seed quality problems	2	-	-	4–8	0.07–0.09
	ER11: Deficient workforce	5	-	-	3–4	-
Operational Risk	OR1: M&E breakdown *	4	0.2–0.5	10–20	-	-
	OR2: Lack of skilled workers	3	-	-	-	0.02–0.03
	OR3: Fire in warehouse *	2	-	20–40	-	0.09–0.12
	OR4: Fire in field	2	4–6	4–5	3–6	0.10–0.15
	OR5: Work accident	2	-	40–50	-	-
	OR6: Excess spraying	3	-	-	6–8	0.10–0.14
	OR7: Deficient spraying	4	-	-	3–7	0.05–0.08
	OR8: Pesticide residue	3	-	-	4–6	0.07–0.10
	OR9: Lack of humidity control in warehouses *	4	-	-	3–6	0.05–0.08
	OR10: Deficient fertilization	4	-	-	5–7	0.09–0.15

* These risks also negatively affect the cotton stock in the warehouse if it occurs.

The Delphi panel members responded to 21 questions, one for each risk factor above in Table 2, such as "Risk factor ... (e.g., excess rainfall) is ... (e.g., possible) to occur and impact our company's performance." Then, the average of responses were taken and rounded to the nearest numerical scale. O1TEX's Delphi panel included ten members: three from top management (e.g., a board member, a unit director), two experts from outside of the company, two from middle management (e.g., a foreman), and three from the shop floor and field workers. The members responded to all of the questions, and all the members had equal weight when calculating the average response to each risk factor.

In Table 2, the likelihood of the occurrence of drought in a particular year is estimated as scale (category) 3 (i.e., 35% at minimum, 60% at maximum). Note that in our SD simulation, for each year, the probability of drought is randomly chosen between 0.35 and 0.60. If this risk materializes, it leads to a decrease of a minimum of 3 kg and a maximum of 6 kg in cotton yield, a decrease of a minimum 0.09 and a maximum 0.12 points in cotton quality, an increase of a minimum of USD 2 and a maximum of USD 4/kg in VCs, and a minimum of USD 6000 and a maximum of USD 16,000 in FCs. It is necessary to create a correct risk reaction by evaluating the effects of the risks in more than one dimension and taking their interactions into account. Compared to traditional risk assessments, our process has a more complex structure in calculating risk scores; therefore, we employ an SD approach for risk assessment.

3.3. System Dynamics Model and Initial Conditions

The stages of creating an SD model are as follows (c.f., [27,44]):

1. Structural analysis to understand the elements that make up the system and the relations between them. This stage benchmarks the system and finds system properties and structures at the macro scale.
2. Causal loop diagram to define the logical structure of the system. The modelling elements can be divided into level variables, rate variables, and auxiliary variables.
3. Analysis of the quantitative relations between variables by creating equations for them in mathematical form. It also reflects the curvilinear relations between variables. The creation of such relational dynamic equations facilitates computer simulation of a complex system.
4. Simulation of these flow charts and equations to adjust the control variables to investigate the effects of different factors in the system, followed by debugging.

The SD causal loop diagram consists of rate, grade or level, and auxiliary variables. Causal loop diagrams are used in the determination of feedback cycles regardless of the types of variables. Causal transformation schemes fulfil two essential functions in SD studies. First, they are used in the determination of the causal relations between the variables as the model is being developed. Second, causal loop diagrams simplify the demonstration of a model.

Figure 2 illustrates the relationships between yield, cotton quality and costs, and the negative impacts on cost and quality if those risks materialize. The logical relations of the interaction between the risks were also created. For example, for each year, the cases of excessive and deficient rainfall, drought and flood risks, or excessively hot and cold temperatures are estimated. As an example of another logical condition, we allow pesticide residue risk in our model to emerge only when disinfestation pesticides are sprayed excessively on the cotton farm.

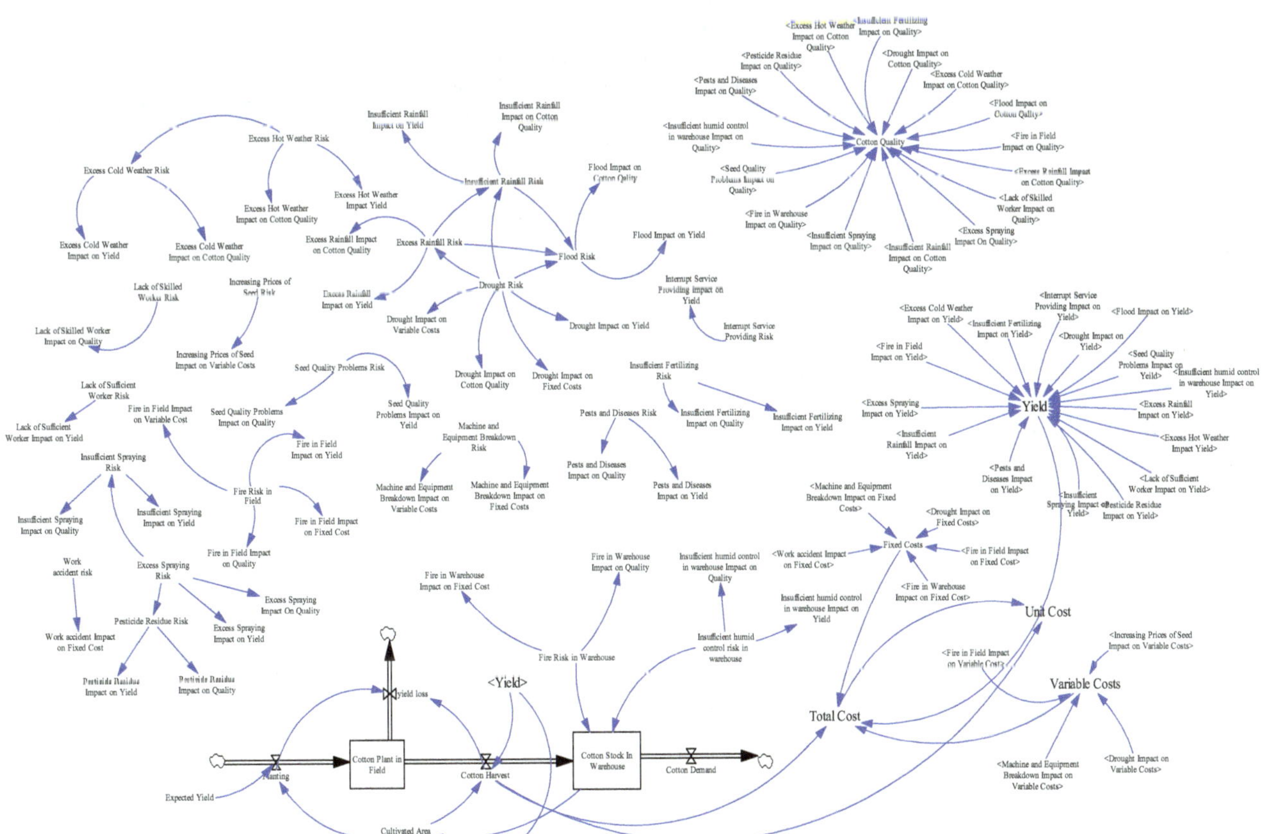

Figure 2. Causal loop diagram of CPL risks for the 01TEX case.

Drawing on the causal loop diagrams, we develop relational equations (see Section 3.4). In this study, the total simulation horizon is 20 years. The following assumptions were employed in describing the scenarios: the planting area was 2500 acres, and the expected average cotton yield was 400 kg/acre. The VCs of cotton production expected to occur in the risk-free environment under normal conditions are USD 366.13/acre, and the annual FCs are USD 185,000/2500 acres. The expected cotton quality index is 4. Various risk elements and the responses given to these risks affect cotton quality at multiple levels. The annual cotton planting varies between 800,000 kg and 900,000 kg according to the textile production facility's capacity. For each year, 01TEX desires to hold in its warehouse an initial stock of 50,000 kg of cotton.

3.4. Model Variables and Relational Equations

SD models contain three types of variables: flow variables which determine the change in stocks (e.g., products that enter and exit the warehouse); stock variables which represent accumulations in the system, and auxiliary variables which represent the required steps to determine the flow variables using stock variables (e.g., orders). See Table 3 for model variable notations, their definitions, and measurement units.

Table 3. Nomenclature.

Notation	Definition
t	Index for year, $t \in [0, T]$
i	Index for risk type, $i \in [ER1, \ldots, ER11] \cup [OR1, \ldots, OR10]$
A	The size of cotton planting area rented for the next T years (in acres)
A_t	The cultivated area in year t (in acres)
D_t	Demand for cotton in year t (in kgs)
F_t	Total FCs materialized during year t (in dollars)
H_t	Cotton harvested in year t (in kgs)
Q_t	Cotton quality index at year t $Q_t \in \{1, 2, 3, 4, 5\}$, a scalar value
V_t	Total VCs materialized during year t (in USD/kg)
W_t	Cotton stocked at warehouse at the start of year t
Y_t	Cotton yield realized during year t (in kg/acre)
$F_{i,t}$	Impact of risk type i on FCs, during year t
$I_{i,t}$	Impact of risk type i on initial stock, during year t
$R_{i,t}$	Probability of risk type i materializing during year t, $0 \le R_{i,t} \le 1$
$Q_{i,t}$	Impact of risk type i on quality level, during year t
$V_{i,t}$	Impact of risk type i on VC, during year t
$Y_{i,t}$	Impact of risk type i on cotton yield, during year t
$\overline{F}$	Overall expected FC
$\overline{Q}$	Target quality index, set at the desired level of 4 for 01TEX
$\overline{V}$	Overall expected VC
$\overline{Y}$	Target yield, set at 400 kg/y for 01TEX
TC_t	Total cost spent during year t, in dollars
UC_t	Unit cost realized at year t, in USD/kg

For the next 20 years, 01TEX plans on a demand requirement of between 800,000 and 900,000 kg of cotton/year. In deciding on the size of the cotton planting area to be rented for the next year, the amount of cotton already available in the warehouse is deducted from the demand requirement of the year first. Then, the area of planting is determined with the expectation that the yield will be 400 kgs of cotton from each acre seeded. The fixed-asset investment was made to plant approximately 2500 acres annually, and the FCs are planned accordingly.

$$W_t = W_{t-1} + H_t - D_t - \sum_{\forall i} R_{i,t} I_{i,t} \text{ where } H_t = Y_t A_t,\ A_t = A - (W_{t-1}/\overline{Y}), \text{ and} \quad (1)$$

$$Y_t = \overline{Y} - \sum_{\forall i} R_{i,t} Y_{i,t} \quad (2)$$

with initial conditions for O1TEX taken as W_0 = 50,000 kgs, A = 2500 acres, and $\overline{Y}$ = 400 kgs/acre.

Environmental and operational risks may also be effective on the yield per acre, in addition to their effects on cotton quality and costs. The expected cotton yield per acre is 400 kg. The decrease in yield leads to the increase in unit cost in addition to the cotton cost per kg in the relevant planting period, in addition to reducing the amount of cotton produced. That relationship is formulated in Equation (2) above.

The amount of cotton in the warehouse is affected by specific risk incidents. The quality index, variable production costs, FCs, and the risks that may be effective on yield are presented in Table 4 with their probabilities. O1TEX set its target quality index value ($\overline{Q}$) to 4, but the incidents that reduce that index value over time are more likely to occur if no risk-mitigation action is taken. Then, Q_t the realized quality index at year t, can be calculated as in Equation (3) below.

$$Q_t = \overline{Q} - \sum_{\forall i} R_{i,t} Q_{i,t} = 1, \quad (3)$$

Table 4. Simulated total impact of the risks for 01TEX case.

Risk Code	CPS Risks	Risk Impact			
		Variable Cost USD/kg/y	Fixed Cost USD/year	Yield (kg/acre)	Cotton Quality Index
ER1	Flood	-	-	23.907	0.0739
ER2	Pests and diseases	-	-	108.176	1.4057
ER3	Drought	31.585	98,209.06	51.013	0.6932
ER4	Increasing seed prices	35.679	-	-	-
ER5	Deficient rainfall	-	-	12.290	0.2168
ER6	Excess rainfall	-	-	33.079	0.4983
ER7	Excess hot weather	-	-	53.983	0.9913
ER8	Excess cold weather	-	-	8.434	0.1101
ER9	Service interruption	-	-	6.732	-
ER10	Seed quality problems	-	-	32.424	0.3875
ER11	Deficient workforce	-	-	58.985	-
OR1	M&E breakdown	4.966	182,592.1	-	-
OR2	Lack of skilled workers	-	-	-	0.1820
OR3	Fire in warehouse	-	114,754.3	-	2.1347
OR4	Fire in field	5.443	4875.57	3.122	0.1397
OR5	Work accident	-	177,814.3	-	-
OR6	Excess spraying	-	-	50.028	0.8536
OR7	Deficient spraying	-	-	29.552	0.4756
OR8	Pesticide residue	-	-	19.172	0.3399
OR9	Lack of humidity control in warehouse	-	-	56.527	0.8764
OR10	Deficient fertilization	-	-	76.186	1.6143

VCs occur per acre planned. In Table 1, VCs per acre were calculated to be USD 366.13/acre. Environmental and operational risks may increase VCs over the next 20 years, as long as no risk reduction work is carried out. We calculate V_t, the realized VC in year t, by adding the effects of all the risks to the targeted VC $\overline{V}$. Equation (4) reflects this relationship.

$$V_t = \overline{V} + \sum_{\forall i} R_{i,t} V_{i,t}, \quad (4)$$

FCs result from the organization structure formed to make 2500 acres of plantation per year. These consist of elements such as machine and equipment (M&E) investment, planning, and control. The expected value of FCs is USD 185,000/year. Incidents that increase FCs may occur as long as environmental and operational risk-reducing actions are not performed in 20 years. Similar to Equation

(4), in Equation (5), the materialized FCs accrued during the year t, denoted by F_t, is the sum of the targeted FCs and those additional factors due to environmental and operational risks.

$$F_t = \overline{F} + \sum_{\forall i} R_{i,t} F_{i,t}, \tag{5}$$

Finally, in Equation (6), we express the yearly unit costs as functions of the fixed and VCs, the size of the planting area, and the harvest amount.

$$UC_t = TC_t / H_t \;\; \text{where}\; TC_t = F_t + V_t A_t, \tag{6}$$

For our SD model development, we used Vensim PLE as the software for coding and analysis. Equations (1)–(6) were used to reveal the relationships between performance measures of interest. What would have happened if no risk mitigation action is taken? With the initial conditions defined above, we ran our simulation model for the planning period, 20 years. Figure 3 displays the set of those outcomes for each of the key variables: (a) cotton harvesting yield, (b) FCs, (c) VCs, (d) unit costs, and finally (e) cotton quality. Those simulated variables demonstrate the fluctuations in the ranges given in Table 2.

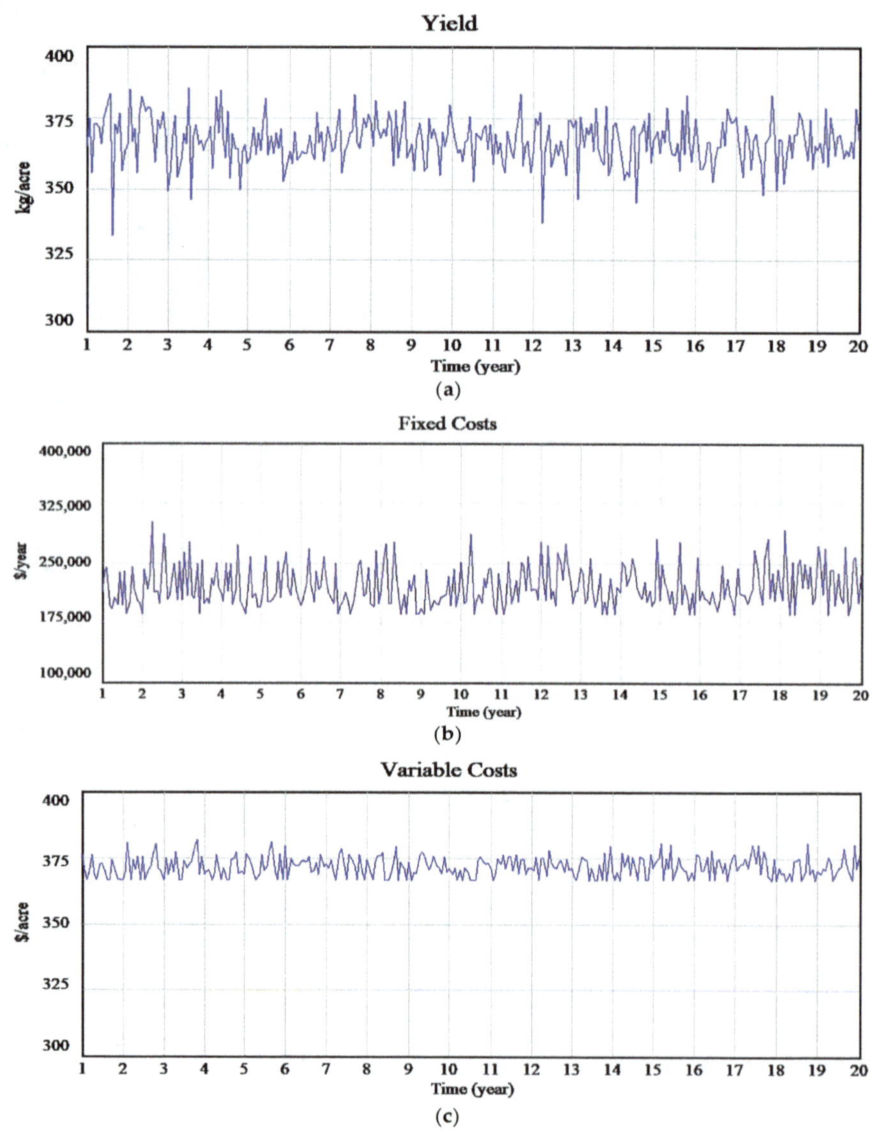

Figure 3. *Cont.*

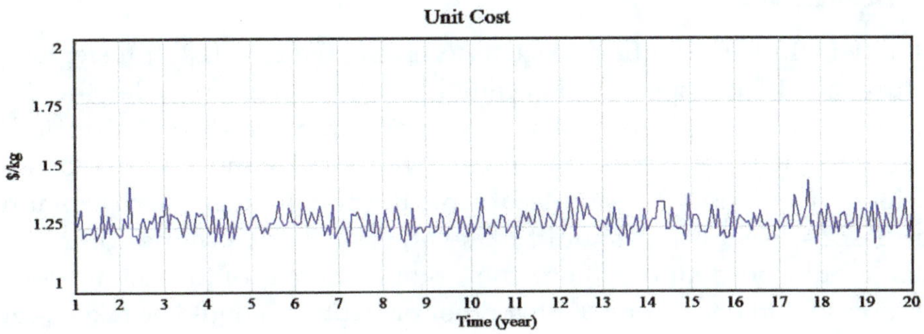

(d)

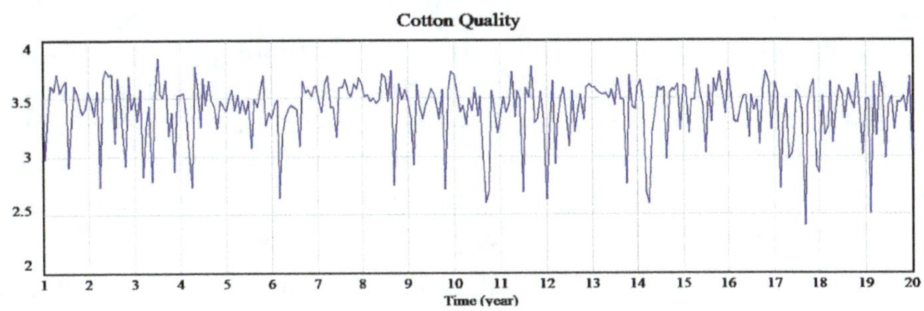

(e)

Figure 3. Simulation results for key variables when no risk mitigation action is taken.

Next, we turn our attention to identifying the most influential risk factors for 01TEX, to chart possible scenarios to be used in risk mitigation strategies.

4. Risk Assessments and Risk Prioritization

When the risk analysis (definition and estimation) is completed, it is necessary to compare the enterprise's risks based on its risk appetite, opportunities for recourse, and exposure level. Risk assessment is about the importance of the enterprise's risks, and whether a suitable method or alleviation option is available should a risk arise. Risk prioritization is defining risk management priorities by comparing the level of risk with previously described target risk levels and tolerance thresholds. The risk is about financial effect and probability and subjective criteria such as health and security effect, reputation effect, sensitivity, and the initial rate [42].

Table 4 presents quality, yield, FCs, VCs, and unit cost results of operational and environmental risks under initial conditions. Additionally, presented are the effects on quality, yield, FCs, and VCs with no risk-reducing action over the 20 years ahead. These results show the impact of risk types on cotton quality, costs, and total yield. The risk of an increase in seed prices (i.e., ER4) most a ects the rise in VCs. Drought and fire in planting areas are the next most significant risks.

The first three risks that have a negative effect on FCs are M&E breakdowns, work accidents, and fire that may break out in the cotton warehouses. The first three risks that may cause a negative effect on cotton yield are pests and diseases, inadequate fertilization, and deficiency in the number of workers. Lastly, the three types of risks that most negatively affect cotton quality are fires that break out in the warehouse, insufficient fertilization, and pests and diseases.

Of course, it is necessary that the risks that affect the identified performance values be investigated and evaluated by management. However, risk priorities precede planning risk responses. Determining risk responses against risk incidents, with the possibility of occurrence within the next 20 years, is among the risk management targets of 01TEX management. The cost and benefits of the risk response to 01TEX are now investigated under various scenarios.

5. Risk Response Scenarios

Risk control may include precautions such as avoiding the risk, risk transfer, and risk financing (e.g., hedging, insurance). Risk response, also called risk treatment, includes the decision to avoid, reduce, transfer, or not accept specific risks. All information produced historically is of critical importance in defining the enterprise's tolerance towards risks and risk response. It is essential for resource efficiency to identify the limits during the initial stage of risk planning. All risks have underlying reasons. Risk teams should make an effort to determine common risk responses. First, evident risks should be defined; then, the organization should review possible responses; lastly, the organization should decide on whether other precautions should be taken [45]. Risk response strategies are generally divided into four categories:

1. Accept (A): This risk is accepted and can be tolerated without performing any other procedure. Although the risk level is high, the capability to make decisions about certain risks may be restricted, or the cost of any activation may be disproportional to the potential benefit obtained. The risk is accepted if it is not possible or practical to overcome the risk, or no response can be guaranteed as a result of the importance of the risk. We note here that Figure 3 displays the case when "Accept" is the response strategy used as a "base," the benchmark scenario for 01TEX.
2. Mitigate (M): Reduce or mitigate the effects of the risk. While the activities that cause the risk to continue within the organization, an action (check) should be performed to bring that risk to an acceptable level. Risk mitigation reduces an acceptable threshold probability and/or effect of an adverse risk incident. Taking early action to mitigate the likelihood and/or impact of a risk is generally more effective than repairing the damage after it emerges. Risk mitigation may require resources or time; thus, risk mitigation offers a trade-off between doing nothing against its cost.
3. Transfer (T): The best response to certain risks may be to transfer them. This may be performed by conventional insurance or third-party payment. This option fits well for mitigating financial risks or the risk of the assets. Risk transfer includes finding another party that wants to be responsible for the management and undertaking responsibility when the risk occurs. The goal is to make sure the risk is possessed and managed by the best party. Risk transfer generally requires the payment of a premium; this trade-off should be factored in formulating the strategy.
4. Avoid (V): Certain risks may be reduced to acceptable levels only by ending the activity. It is impossible to avoid all risks, and this approach may sometimes be cost- or time-inhibitive.

The typical performance result of the data, apart from the quality index, is unit cost because it is found by dividing the sum of fixed and VCs by the planted area times the yield in the relevant year (i.e., the amount of harvest). In this case, considering that the yield is also effective together with fixed and VCs, the quality index of unit costs was used as a secondary performance criterion. Both cotton quality and unit cotton cost are important for 01TEX.

Having identified the most influential risk types, we now develop the risk scenarios for various risk responses for each risk (see Table 5). The rationale behind these scenarios is twofold. First, O1TEX wanted to mitigate these risks; therefore, it omitted the "avoid" option as a risk response. Second, we developed these scenarios by the risk likelihood and the scope of O1TEX in mitigating them. The "base" risk response strategy accepts all the risks and therefore corresponds to a "do nothing" scenario. A pure exhaustive enumeration (combinatorial) approach would call for an additional 19,862 (=3^9-1) cases. To eschew this computational burden, but more so for clarity, we adopted a sequential approach in formulating scenarios. To that end, depending on base scenario results, we assigned a risk response strategy to the most significant risk factors that hurt the criteria at hand. For Scenario 1, for example, pests and diseases, M&E breakdown, and fire in the warehouse called for a mitigation strategy. However, a risk transfer strategy (e.g., insurance) for increases in seed prices were deemed appropriate. We similarly formulated the remaining scenarios. In the following, these scenarios, which also serve as sensitivity analyses, are described in detail.

Table 5. Risk response strategies of the scenarios (A: Accept, T: Transfer, M: Mitigate).

Source of Risk	Base	S1	S1.1	S1.2	S1.3	S1.4	S2	S2.1	S2.2	S2.3	S3	S3.1	S3.2
Pests and diseases	A	M	A	M	M	M	M	M	M	M	M	M	M
Incr. in seed prices	A	T	T	A	A	A	A	A	A	A	A	A	A
M&E breakdown	A	M	M	M	A	M	M	M	M	M	M	M	M
Fire in warehouse	A	M	M	M	M	A	M	M	M	M	M	M	M
Drought	A	A	A	A	A	A	M	A	A	A	A	A	A
Work accident	A	A	A	A	A	A	M	M	A	M	M	M	M
Deficient fertilization	A	A	A	A	A	M	M	M	M	A	M	M	M
Deficient workforce	A	A	A	A	A	A	A	A	A	A	M	A	M
Fire in field	A	A	A	A	A	A	A	A	A	A	M	M	A

Scenario 1 (S1)

The results of the strategies applied for reducing the highest priority risks is examined in S1. Risks of primary importance create adverse effects on the pests and diseases risk yield and quality index. This risk is at the first rank on yield and third level of importance on the quality index. Here, management plans to start a disinfestation program using a new method with an annual FC of USD 15,000 to reduce this risk. It was learnt that the effects of risk on yield and quality may be reduced by 50% when this option is applied. The risk that affects VCs most is the risk of an increase in seed prices. A response that can be developed here is risk transfer. For this, there is the possibility for long-term price lock-in with a future contract. It is estimated that this reaction will increase by USD 5000 in annual FCs; in this case, the increase in seed prices will impact. The risk of fire in the cotton warehouses has a negative effect at the first rank on cotton quality and third rank on FCs. An investment that may lead to a FC increase of USD 10,000 per year should be made to reduce this risk. In this case, the effect of a fire in a warehouse on FCs decreases to the level of USD 5000–10,000. Its impact on the quality index will decrease to the score level of 0.045–0.60 by decreasing at a rate of 50%, and its effect on the cotton stock in the warehouse to the level of 0.05–0.15 kg.

The risks resulting from M&E breakdowns increase FCs at the top level. An FC investment of USD 2000 should be made together with annual preventive maintenance to mitigate this risk. In this case, the increase in FCs created by M&E breakdowns will decrease to USD 1000–2000, and the VC will decrease to USD 0.01–0.02. S1 was created by applying all of these strategies. The unit cost and quality index results were investigated should some of the risks in the sub-scenarios of S1 be accepted. The results of the failure to get no reaction against the risk of diseases and pests were investigated in sub-scenario S1.1, the acceptance of the risk of increase in seed prices in S1.2, acceptance of the risk of fire in the warehouse in S1.3, and the risk exposure results of the unit cost and quality index in case the M&E breakdown risk is accepted were investigated in S1.4.

Scenario 2 (S2)

Among the risks of second-degree priority, the response plan against the risk of drought suggests increasing the amount of irrigation. This risk response will cause an additional VC of USD 9.60 and will reduce the effect of the risk on VCs by 50%. An outsourced consulting service with an annual FC of USD 5000 will be received as a precaution of risk mitigation to reduce workplace accidents. In this case, there is a decrease of 80% in risk exposure. The need for fertilizers will be determined more accurately through regular soil analyses by bearing a VC of USD 4/acre against the risk of insufficient fertilization. This precaution will reduce the yield and quality risk by 1/3. S2 was created by applying all of these strategies. The results of accepting the risk of drought were investigated in S2.1, work accidents in S2.2, and deficient fertilization among the sub-scenarios of S2.

Scenario 3 (S3)

The risks with the lowest priority are the lack of sufficient workers and fire risk in the field. A wage increase may mitigate the risk of labor shortage, at the expense of a USD 5 increase in VCs. This will cause a decrease of 50% in the risk impact. There is the option to get an outsourced service to mitigate the risk of fire in the field. This option is likely to reduce risk impact at a rate of 75% while causing an increase of USD 10,000 in the annual FCs. S3 was created by applying both strategies. The results of accepting the risk of the lack of sufficient workers were investigated in S3.1, and the results of accepting the risk of fire in the cotton field were investigated in S3.2.

Table 6 displays scenario simulations over the next 20 years of planning horizon for the targeted quality index (4) and unit cost (USD 1.1/kg). Initially, all the risks negatively affect 10.99 score quality indices in 20 years. As such, the cumulative deviation from unit costs amounts to USD 3.06 per kg. The results of the mitigation and transfer responses (S1) taken against the risks with a first-level priority reduced the deviation in the quality index to the level of 8.22. This means an improvement of 25.17% in terms of quality. However, the quality improvement at this level took the amount of deviation in unit costs to USD 3.26 by increasing at the level of 6.3%. The results of accepting the pests and disease risks were investigated in S1.1 when compared to S1. An improvement occurred in the quality index deviation in this scenario compared to the initial state, but this improvement remained below the results of S1. The increase in unit costs was higher when compared to S1. According to this result, it is not advisable to accept the risk of pests and diseases. The results when the risk of an increase in seed prices is accepted are investigated in S1.2. No change occurred in S1.2 compared to S1, since the risk of an increase in seed prices does not affect the quality index. The deviation in unit costs increased at a rate of 6.0%, according to the initial scenario. Accordingly, it is not recommended to apply the transfer strategy against the risk of an increase in seed prices. This risk should be accepted. There is no precaution against this risk in subsequent scenarios, and this risk remains as it is in the initial scenario.

Table 6. Total risk variances (over 20 years) of risk response scenarios.

Variance Type	Base	S1	S1.1	S1.2	S1.3	S1.4	S2	S2.1	S2.2	S2.3	S3	S3.1	S3.2
Quality	10.99	8.22	9.06	8.22	10.15	8.22	6.80	7.15	7.15	8.00	7.04	7.04	7.15
Unit Cost	3.06	3.25	3.31	3.25	3.12	3.43	3.75	3.39	3.45	3.30	3.78	3.60	3.57

The results of accepting the risk of fire in the warehouse were investigated in S1.3. While the quality index deviation increases at a rate of 7.6% in the simulation results of this scenario, unit cost deviation increased at a rate of 1.9%. While the increase in the deviation of costs is quite low in this option, the improvement in the level of quality occurred was quite low when compared to S1.2. At this point, a choice should be made between quality improvement and unit costs. The application of the risk mitigation strategy against the risk of fire in the warehouse was applied. The results of accepting the M&E breakdown risk are investigated in S1.4. This scenario includes the simulation results of the pests and diseases and risk of fire in the warehouse mitigation strategies. Accordingly, the same degree of deviation occurred in S1.4 with the quality index level in S1 and S1.2. However, the rate of increase in the variation in unit costs was 11.9%. Therefore, the mitigation strategy should be applied and the risk should not be accepted against the risk of M&E breakdown. In S1, the risks with the highest priority were examined. Amongst S1 and its sub-scenarios, the most suitable risk mitigation suite is S1.2. S1.2 includes mitigation strategies against pests and diseases, fire in the warehouse, and M&E

breakdowns. The risk of an increase in seed prices is accepted, with no risk response. S1.2 data were primarily regarded as the starting point in the design of S2.

Mitigation strategies were applied against drought, work accidents, and insufficient fertilizing risks in S2. Here, the deviation in the quality index decreased to the lowest level. According to the base scenario, the quality index deviation decreased at a rate of 38.1%. However, the deviation in unit costs increased 22.3%. If company management is satisfied with the increase in the level of quality achieved in return for the costs borne, the mitigation strategies in this scenario can be applied. The results of the option of accepting the drought risk were investigated in S2.1. According to the simulation results, the quality index here occurred with a lower deviation at the rate of 34.9% when compared to the base scenario. The deviation in unit costs increased by 10.5% when compared to the base scenario. This option was found suitable by the management in that it ensures a balance between cost and quality level, and it was decided to accept the risk of drought. The acceptance of the risk of work accidents in S2.2 did not create a difference on the quality index when compared to S2.1, but it led to a deviation of 12.7% more in costs when compared to the main scenario. This means that a mitigation strategy should be applied against work accidents, and the risk should not be accepted. A reduction of 27.2% occurred in the quality index when compared to the initial conditions as a result of accepting the insufficient fertilizing risk in S2.3, and the deviation increase in unit costs occurred at a rate of 7.7%. While the rise in deviation in unit costs is low compared to other options, the decrease in the quality index deviation is relatively low, especially compared to S2.1. Therefore, S2.1 was taken as the basis for subsequent scenarios.

In Scenario 3, quality index variance decreased 35.91% compared to the base scenario due to the precautions taken against the shortage of workers and fire risks in the cotton field. While it is a good result in terms of the level of quality, an increase of 23.4% in the deviation in unit costs is not regarded to be very suitable in terms of the balance between quality and costs. If the risk is accepted against the lack of sufficient workers among these risks, the quality level does not change compared to S3. Still, the increase in the deviation in costs occurs at a 17.4% rate compared to the base scenario. Finally, the quality index deviation decreased 34.9%, and the deviation increase in unit costs became 16.5% when the risk of fire in the field was accepted.

Based on how much deviation from the level of quality and unit costs over the planning horizon is regarded as tolerable by management, 01TEX can decide its risk response, as shown in Table 6. The scenario that ensures the most appropriate balance between quality and unit cost is S2.1. The increase in seed prices, drought, lack of sufficient workers, and fire risks in the field are among the first four risks in the risk prioritization that should be accepted, and mitigation strategies should be activated accordingly.

6. Concluding Remarks

This study suggests a framework to identify, assess, and prioritize risks that occur during cotton production and proposes risk responses that affect the whole textile SC. The framework uses a simulation approach with the SDM to effectively assess cotton production risks. The model was primarily simulated to define the effects of identified risks on four performance criteria. The results show that it is not possible to mitigate all risks with high priority in terms of their impact on yield, quality, and variable and fixed costs and that a risk response that enhances quality levels and cotton yield, need also increase the costs. Therefore, assessing a single-dimension risk effect and risk response planning may not prove useful. To proactively manage and prevent this disadvantage, the effects of risks as well as their probability of realization and the business performance they affect should be well-identified, and risk response strategies should be planned accordingly.

Managing risks in supply chains fosters resilience and robustness in SC planning and execution (e.g., El Baz and Ruel [46]). Especially in the face of massive disruptions such as COVID-19, SC managers have once more realized the vital role of SCRM. Exacerbated with climate change, uncertainties mounting in SCs both from supply and demand sides call for more holistic and integrated approaches to SCRM. In this paper, we demonstrated the impact of some uncertainties in CPL of a textile SC via a case study from Turkey. However, this is all but one application using SDM, wherein the data are based on approximations and the experts' input. Naturally, various other approaches can be integrated into forecasting models for SCs. Depending on the type of the product, industry, market conditions, the length of the SC, and its vulnerability to environmental conditions, one model may fit better than others. For example, structural equation modelling could be another alternative to the Delphi method if the company has resources to invest in improving its forecasting process. With the advent of enhanced technology and pressing environmental and social pressures for sustainability, there is much room for research in developing analytical models that aid in the design and operation of sustainable and resilient SCs (e.g., Araz et al. [47], Ülkü and Engau [48]).

Although the operational and social risks were the focus in this paper, the issues around social risks in SCs are rampantly emerging. Especially for those industries such as agriculture and textiles that are still heavily labor-intensive in some countries, the inclusion of social risks for a holistic SCRM is required. We considered the lack of skilled labor as a parameter in this study; ultimately, this relates to the quality of the working conditions and fair wages, among others (c.f., Bhavsar et al. [49]). The COVID-19 pandemic has exacerbated the lack of skilled labor because of heightened restrictions on international workers' visas and health conditions. This again nuances the need to incorporate into SCRM some metrics related to social sustainability.

Naturally, the numerical findings and insights in this study are confined to the parameters for and the context of the 01TEX case. Nevertheless, it provides an end-to-end framework in assessing risks and formulating risk response strategies and highlights trade-offs in management objectives. An approach focusing solely on cost minimization may differ from one that places the highest cotton quality first. To that end, one limitation, and thereby future research, is to develop a more sophisticated decision method that simultaneously factors in weights or preferences over multiple decision criteria and risk response strategies. Another limitation is the accuracy of the input data. Although the Delphi method in assessing various uncertainties has its merits from practical perspectives, combining expert judgements with quantitative data, such as in structural equation modelling, may improve the consistency and accuracy of forecasts.

This study has focused on the supply side whereby the quality, yield, and cost of cotton were decision criteria in formulating the risk response strategies. Amalgamating CPL with the downstream textile manufacturing (recall Figure 1) and retailing would enlarge the SC's scope by including the demand side, enabling price, rather than cost, as a decision criterion. Such a multi-stage problem setting within which multiple SC partners act with conflicting interests would be worthy of future research. Because most SCs are currently globalized, including those in the textile industry, it would be interesting to study via SDM the management of and mitigation strategies for globally sustainable cotton production, especially in the context of social sustainability. There is an emerging concern around modern slavery in cotton production and historical child labor use in the cotton fields (e.g., Stevenson and Cole [50]). Finally, developing a robust optimization model to trade-off risks with risk mitigation costs warrants a new research extension.

Author Contributions: Conceptualization, M.A.Ü. and M.A.; methodology, M.A.; software, M.A.; validation, U.V., C.D., and S.S.C.; formal analysis, M.A.; resources, M.A.Ü.; data curation, M.A. and S.S.C.; writing—original draft preparation, M.A., M.A.Ü., and U.V.; writing—review and editing, U.V, C.D., and M.A.Ü.; visualization, M.A.; supervision, M.A.Ü.; project administration, M.A.Ü. All authors have read and agreed to the published version of the manuscript.

References

1. Hofmann, E.; Bosshard, J. Supply chain management and activity-based costing. *Int. J. Phys. Distrib. Logist. Manag.* **2017**, *47*, 712–735. [CrossRef]
2. Waters, D. *Supply Chain Risk Management, Vulnerability, and Resilience in Logistics*; Replica Press: New Delhi, India, 2011.

3. Zeng, B.; Yen, B.P.-C. Rethinking the role of partnerships in global supply chains: A risk-based perspective. *Int. J. Prod. Econ.* **2017**, *185*, 52–62. [CrossRef]
4. Wang, X.; Tiwari, P.; Chen, X. Communicating supply chain risks and mitigation strategies: A comprehensive framework. *Prod. Plan. Control* **2017**, *28*, 1023–1036. [CrossRef]
5. World Wide Fund for Nature. 2020. Available online: https://www.worldwildlife.org/industries/cotton (accessed on 20 July 2020).
6. Hendricks, K.B.; Singhal, V.R. An Empirical Analysis of the Effect of Supply Chain Disruptions on Long-Run Stock Price Performance and Equity Risk of the Firm. *Prod. Oper. Manag.* **2005**, *14*, 35–52. [CrossRef]
7. Peng, M.; Peng, Y.; Chen, H. Post-seismic supply chain risk management: A system dynamics disruption analysis approach for inventory and logistics planning. *Comput. Oper. Res.* **2014**, *42*, 14–24. [CrossRef]
8. Fan, Y.; Stevenson, M. A review of supply chain risk management: Definition, theory, and research agenda. *Int. J. Phys. Distrib. Logist. Manag.* **2018**, *48*, 205–230. [CrossRef]
9. Bradley, J.R. An improved method for managing catastrophic supply chain disruptions. *Bus. Horiz.* **2014**, *57*, 483–495. [CrossRef]
10. Gupta, A.; Maranas, C.D. Managing demand uncertainty in supply chain planning. *Comput. Chem. Eng.* **2003**, *27*, 1219–1227. [CrossRef]
11. Yu, M.; Goh, M. A multi-objective approach to supply chain visibility and risk. *Eur. J. Oper. Res.* **2014**, *233*, 125–130. [CrossRef]
12. Wieland, A. Selecting the right supply chain based on risks. *J. Manuf. Technol. Manag.* **2013**, *24*, 652–668. [CrossRef]
13. Braunscheidel, M.J.; Suresh, N.C. The organizational antecedents of a firm's supply chain agility for risk mitigation and response. *J. Oper. Manag.* **2009**, *27*, 119–140. [CrossRef]
14. Chen, J.; Sohal, A.S.; Prajogo, D.I. Supply chain operational risk mitigation: A collaborative approach. *Int. J. Prod. Res.* **2013**, *51*, 2186–2199. [CrossRef]
15. Chen, J.; Sohal, A.S.; Prajogo, D.I. Supply risk mitigation: A multi-theoretical perspective. *Prod. Plan. Control* **2016**, *27*, 853–863. [CrossRef]
16. Glock, C.H.; Ries, J.M. Reducing lead time risk through multiple sourcing: The case of stochastic demand and variable lead time. *Int. J. Prod. Res.* **2013**, *51*, 43–56. [CrossRef]
17. Kull, T.; Closs, D.J.; Kull, T.; Closs, D.J. The risk of second-tier supplier failures in serial supply chains: Implications for order policies and distributor autonomy. *Eur. J. Oper. Res.* **2008**, *186*, 1158–1174. [CrossRef]
18. Chen, C.; Lee, W. Multi-objective optimization of multi-echelon supply chain networks with uncertain product demands and prices. *Comput. Chem. Eng.* **2004**, *28*, 1131–1144. [CrossRef]
19. Kern, D.; Moser, R.; Hartmann, E.; Moder, M. Supply risk management: Model development and empirical analysis. *Int. J. Phys. Distrib. Logist. Manag.* **2012**, *42*, 60–82. [CrossRef]
20. Tummala, R.; Schoenherr, T. Assessing and managing risks using the Supply Chain Risk Management Process (SCRMP). *Supply Chain Manag. Int. J.* **2011**, *16*, 474–483. [CrossRef]
21. Hallikas, J.; Virolainen, V.; Tuominen, M. Risk analysis and assessment in network environments: A dyadic case study. *Int. J. Prod. Econ.* **2002**, *78*, 45–55. [CrossRef]
22. Trkman, P.; McCormack, K. Supply chain risk in turbulent environments—A conceptual model for managing supply chain network risk. *Int. J. Prod. Econ.* **2009**, *119*, 247–258. [CrossRef]
23. Foroughi, A.; Albin, M.; Kocakulah, M. Perspectives on global supply chain supply-side risk management. In Proceedings of the 2006 Technology Management for the Global Future-PICMET IEEE Conference, Istanbul, Turkey, 8–13 July 2006; Volume 6, pp. 2732–2740.
24. Grötsch, V.; Blome, C.; Schleper, M.C. Antecedents of proactive supply chain risk management: A contingency theory perspective. *Int. J. Prod. Res.* **2013**, *51*, 2842–2867. [CrossRef]
25. Wagner, S.M.; Neshat, N. A comparison of supply chain vulnerability indices for different categories of firms. *Int. J. Prod. Res.* **2012**, *50*, 2877–2891. [CrossRef]
26. Thun, J.H.; Hoenig, D. An empirical analysis of supply chain risk management in the German automotive industry. *Int. J. Prod. Econ.* **2011**, *131*, 242–249. [CrossRef]
27. Forrester, J.W. *Industrial Dynamics*; MIT Press: Cambridge, MA, USA, 1961.
28. Fagundes, M.V.C.; Teles, E.O.; De Melo, S.A.V.; Freires, F.G.M. Decision-making models and support systems for supply chain risk: Literature mapping and future research agenda. *Eur. Res. Manag. Bus. Econ.* **2020**, *26*, 63–70. [CrossRef]

29. Sterman, J.D. *Business Dynamics: Systems Thinking and Modeling for a Complex World*; McGraw-Hill: New York, NY, USA, 2000.
30. Adane, T.F.; Bianchi, M.F.; Archenti, A.; Nicolescu, M. Application of system dynamics for analysis of performance of manufacturing systems. *J. Manuf. Syst.* **2019**, *53*, 212–233. [CrossRef]
31. Kim, B.; Park, C. Coordinating decisions by supply chain partners in a vendor-managed inventory relationship. *J. Manuf. Syst.* **2010**, *29*, 71–80. [CrossRef]
32. Baines, T.S.; Harrison, D.K. An opportunity for system dynamics in manufacturing system modelling. *Prod. Plan. Control* **1999**, *10*, 542–552. [CrossRef]
33. Tako, A.A.; Robinson, S. The application of discrete event simulation and system dynamics in the logistics and supply chain context. *Decis. Support Syst.* **2012**, *52*, 802–815. [CrossRef]
34. Ghadge, A.; Dani, S.; Chester, M.; Kalawsky, R. A systems approach for modelling supply chain risks. *Supply Chain Manag. Int. J.* **2013**, *18*, 523–538. [CrossRef]
35. Li, C.; Ren, J.; Wang, H. A system dynamics simulation model of chemical supply chain transportation risk management systems. *Comput. Chem. Eng.* **2016**, *89*, 71–83. [CrossRef]
36. Mehrjoo, M.; Pasek, Z.J. Risk assessment for the supply chain of fast fashion apparel industry: A system dynamics framework. *Int. J. Prod. Res.* **2015**, *54*, 28–48. [CrossRef]
37. Global Agricultural Information Network, USDA. Turkey, Cotton and Product Annual, GAIN Report# TR9007. 2019. Available online: https://www.fas.usda.gov/data (accessed on 3 May 2020).
38. Basal, H.; Karademir, E.; Goren, H.K.; Sezener, V.; Dogan, M.N.; Gencsoylu, I.; Erdogan, O. Cotton production in Turkey and Europe. In *Cotton Production*; Jabran, K., Chauhan, B.S., Eds.; John Wiley & Sons Ltd.: Hoboken, NJ, USA, 2019; pp. 297–321.
39. Chopra, S.; Sodhi, M.S. Managing risk to avoid supply-chain breakdown. *Sloan Manag. Rev.* **2004**, *46*, 53–61.
40. DuHadway, S.; Carnovale, S.; Hazen, B. Understanding risk management for intentional supply chain disruptions: Risk detection, risk mitigation, and risk recovery. *Ann. Oper. Res.* **2019**, *283*, 179–198. [CrossRef]
41. International Cotton Advisory Committee. Did You Know? *2019*. Available online: https://www.icac.org/LearningCorner/LearningCorner?CategoryId=1&MenuId=14 (accessed on 20 May 2020).
42. Akgün, M. The Operational Risk Assessments in Manufacturing Industry. In *Global Business Expansion: Concepts, Methodologies, Tools, and Applications*; IGI Global: Hershey, PA, USA, 2018; pp. 653–675.
43. Soni, U.; Jain, V.; Kumar, S. Measuring supply chain resilience using a deterministic modeling approach. *Comput. Ind. Eng.* **2014**, *74*, 11–25. [CrossRef]
44. Forrester, J.W. System dynamics, systems thinking, and soft OR. *Syst. Dyn. Rev.* **1994**, *10*, 245–256. [CrossRef]
45. Hopkin, P. *Fundamentals of Risk Management: Understanding, Evaluating, and Implementing Effective Risk Management*; IRM Kogan Page: London, UK, 2012.
46. El Baz, J.; Ruel, S. Can supply chain risk management practices mitigate the disruption impacts on supply chains' resilience and robustness? Evidence from an empirical survey in a COVID-19 outbreak era. *Int. J. Prod. Econ.* **2020**, 107972. [CrossRef]
47. Araz, O.M.; Choi, T.-M.; Olson, D.L.; Salman, F.S. Role of Analytics for Operational Risk Management in the Era of Big Data. *Decis. Sci.* **2020**. [CrossRef]
48. Ülkü, M.A.; Engau, A. Sustainable supply chain analytics. In *Industry, Innovation and Infrastructure. Encyclopedia of the UN Sustainable Development Goals*; Leal Filho, W., Azul, A.M., Brandli, L., Lange Salvia, A., Wall, T., Eds.; Springer: Berlin, Germany, 2021. [CrossRef]
49. Bhavsar, A.; Diallo, C.; Ülkü, M.A. Towards sustainable development: Optimal pricing and sales strategies for retailing fair trade products. *J. Clean. Prod.* **2020**, 124990. [CrossRef]
50. Stevenson, M.; Cole, R. Modern slavery in supply chains: A secondary data analysis of detection, remediation and disclosure. *Supply Chain Manag. Int. J.* **2018**, *12*, 81–99. [CrossRef]

2

Machine Learning Methods for Quality Prediction in Production

Sidharth Sankhye and Guiping Hu *

Department of Industrial and Manufacturing Systems Engineering, Iowa State University, Ames, IA 50011, USA; ssankhye@iastate.edu

* Correspondence: gphu@iastate.edu

Abstract: The rising popularity of smart factories and Industry 4.0 has made it possible to collect large amounts of data from production stages. Thus, supervised machine learning methods such as classification can viably predict product compliance quality using manufacturing data collected during production. Elimination of uncertainty via accurate prediction provides significant benefits at any stage in a supply chain. Thus, early knowledge of product batch quality can save costs associated with recalls, packaging, and transportation. While there has been thorough research on predicting the quality of specific manufacturing processes, the adoption of classification methods to predict the overall compliance of production batches has not been extensively investigated. This paper aims to design machine learning based classification methods for quality compliance and validate the models via case study of a multi-model appliance production line. The proposed classification model could achieve an accuracy of 0.99 and Cohen's Kappa of 0.91 for the compliance quality of unit batches. Thus, the proposed method would enable implementation of a predictive model for compliance quality. The case study also highlights the importance of feature construction and dataset knowledge in training classification models.

Keywords: machine learning; quality; classification

---

1. Introduction

Production and operational efficiency are essential for the competitiveness of manufacturing companies, especially effective quality control. Reworks, reverse logistics, and production plan disruptions due to defective products are some of the problems caused for an organization's logistics by lack of quality. In the recent decades, modern manufacturing quality departments have shifted their focus from reactive to proactive methods. There has been significant progress in defect prevention via process improvements. However, as the requirement for quality increases, the cost of preventing defects could start to outweigh potential savings [1]. Current quality control strategies in manufacturing focus on continuous improvement to identify root causes of defects and implementing preventive measures. However, defects such as a mislabeled product, or a product assembled with missing parts have significantly greater penalties for escaping as compared to the costs incurred correcting such defects. In such situations, using a prediction model to preemptively contain and rework batches of defective product could provide greater return on investment as compared to efforts aimed at eliminating the causes of the defect. Currently used qualitative methods such as root cause analysis require significant involvement from skilled personnel to subjectively evaluate the effects of various causes. Thus, there is potential for root cause analysis to be supplemented with insight on factors that attribute to the defects by examination of the predictive model.

Presently, research has successfully demonstrated the use of machine learning techniques to predict the effects of process parameters in various manufacturing operations [2]. However, there have been relatively fewer applications that aim to predict the compliance quality of finished goods from data pertaining to a sequence of operations, on a production line. The research by Kerdprasop et al. could predict quality defects from multistage semiconductor manufacturing with Cohen's Kappa of 0.57 [3], while the work by Kao et al. and Arif et al. in the same context could predict defects with Cohen's Kappa of 0.98 [4] and 0.04 [5], respectively. However, these papers did not use classification algorithms like neural networks or bagged and boosted ensembles. Furthermore, the datasets used for training and evaluation were relatively balanced in class distribution, thus not needing the methods to address class imbalance. Defect rates in modern manufacturing tend to be low, resulting in datasets where the predicted class has significant imbalance. This paper aims to address these areas of improvement in the current research by addressing various practical issues arising from the use of machine learning in multistage manufacturing and training prediction models with comparable performance.

Data mining is the practice of using data to discover patterns and relations between attributes. With the abundance of modern manufacturing and process data, it has become feasible for the application of data mining in various use-cases such as inductively discovering factors that affect the quality of a product [6] or effectiveness of digital marketing [7]. Data mining methods can reveal information that may often be overlooked with a hypothesis-driven approach, such as the use of statistics. There are a number of data mining and knowledge discovery techniques such as clustering and association rule mining. Clustering analysis is typically adopted for preprocessing and analysis of large datasets to find clusters of similar datapoints [8]. Association rule mining is used to discover interesting relations between attributes in a dataset [9]. Both of these techniques do not have a clearly defined response attribute. Machine learning shares similarities with data mining such as being a subset of artificial intelligence, but it differs in its objective. Machine learning is used to aid in the completion of specific tasks rather than extract knowledge from data [10]. Classification is an example of machine learning since its methods are used to create classification models which are used for the prediction of discrete response variables [11]. Predicting the compliance quality of incoming parts from a manufacturing line is the problem addressed in this paper and it is an example of the use-case for classification in manufacturing industries [12].

Modern classification methods such as the ensemble models benefit from improvements in the computing capacity of modern systems [13], supporting the widespread use of machine learning methods. This is because the frequently used ensemble classifiers such as random forest and adaptive boosting often combine weak learners like simple decision trees, Naive Bayes, and k-nearest neighbor into an ensemble to combine the strengths of individual learners. Another challenge often faced in real applications is the classification of imbalanced datasets [14]. This is because most classification algorithms tend to learn models that are biased in favor of the majority class. Compliance quality in manufacturing plants is a good example of class imbalance since defective product units are in the heavy minority as compared to compliant units [15]. This study aims to shed some light on classification for imbalanced datasets since the focal problem is the classification of manufacturing quality non-compliance. Special attention has been devoted in the method design to address the imbalanced nature of such datasets.

Imbalanced classification has become an increasingly prominent obstacle in practical settings such as credit card fraud detection [16], fault detection in machinery [17], and diagnosis of cerebrovascular disease [18]. In these applications, the importance of correctly classifying the minority outweighs that of the majority class. When a dataset has a significant imbalance, it poses a challenge for classification of the minority class since a prediction model tends to favor correctly classifying the majority class. Lack of data points of the minority class makes it challenging to train an accurate model [14]. Current research has yielded solutions in the forms of feature engineering and learning algorithm-based approaches to improve the accuracy of minority class prediction [18].

Feature engineering methods to address imbalance datasets include the general sampling techniques and methods that are dataset specific such as high-level feature construction and selection [19]. Sampling methods focus on reducing the class imbalance by oversampling the minority, under sampling the majority class, or generating new data points of the minority class from existing data points via SMOTE [20]. However, the use of random under sampling results in loss of information regarding the majority class whereas random oversampling can result in overfitting the model to training data. SMOTE attempts to reduce overfitting by generating new data points via interpolation of nearest neighbor minority class. Improving the performance of classifiers for the minority class have yielded methods such as cost-sensitive boosting [21] and other ensemble methods that incorporate sampling within their training loops [22,23].

However, the above-mentioned methods to address imbalance do not resolve issues wherein the available features cannot be used to train an acceptable model, regardless of the currently available classification algorithms. Similar conditions can be encountered when attempting to predict the compliance quality of finished product from process data on an assembly line. Data pertaining to the manufacturing processes varies significantly in different factories and requires feature engineering before it can be used to train a classification model. An example of such feature engineering is the construction of explanatory features based on a thorough understanding of the dataset and its relevant subject matter [24]. Thus, feature construction based on dataset knowledge can provide substantially improved results. This served as one of the major motivations for application of the method described in this paper and its demonstration in the case study.

The knowledge required to construct additional features to improve classification models may require domain expertise and solutions discovered in this manner would tend to be specific for the problem addressed. This is because most practical applications of classification entail significant efforts in data consolidation, pre-processing, and understanding the use-case [18,25]. The method described in this paper pertains to quality prediction in manufacturing. As such, domain knowledge is useful when constructing features from the raw dataset and drawing conclusions from the results of training the classification models.

The rest of this paper is organized as follows. In Section 2, the proposed method for predicting manufacturing quality compliance is presented. Descriptions have been detailed for the classification methods, explanation of common features for manufacturing quality prediction and evaluation method for model performance comparison. Section 3 focuses on describing a case study to establish the context in which the proposed method is applied and the problems that it will address. Section 4 presents the results of this application in the case study and discusses possible improvements in the method specific to that case. The paper is concluded in Section 5 with a summary of findings and future research directions.

2. Materials and Method

Despite research pertaining to the application of machine learning methods to predict the response of specific processes such as quality in manufacturing [12] or arrival time for containerships [26], there is still unexplored potential for its use in predicting the final outcome of a chain of linked processes. Thus, the method proposed in this paper is intended for use in prediction of quality in manufacturing via classification but can be generalized for use in cases such as prediction of pass-or-fail compliance quality for multi-stage operations. A large dataset containing previous manufactured units with features corresponding to factors describing the unit or process parameters is required to train the model. Datasets with more features will result in a better classification model obtained via this method.

Figure 1 provides an overview of the method proposed for addressing the problem described in this paper. This method used established best practices in order to satisfactorily train and evaluate a classification model for its specialized application in the prediction of manufacturing compliance quality. Thus, applying domain knowledge when constructing features, selecting classification algorithms, or tuning hyperparameters results in superior model performance. As a first step, the data were split into training and independent test sets for evaluation. Despite the problem that would arise from sampling variance, it was beneficial to prevent overfitting the model on the training data and obtain an unbiased estimate of how the model would perform in its capacity [27] to predict manufacturing quality non-compliance. Feature selection and construction was then performed on the training dataset. Tenfold Cross Validation was then used to evaluate multiple models trained from the training dataset and tune hyperparameters of the training algorithms. While it would be beneficial to use Leave-one-out Cross Validation when tuning the hyperparameters of the classification model, it would be too computationally intensive given the number of data points available [28]. Using metrics explained later in this section, the trained models were then evaluated with the independent test dataset.

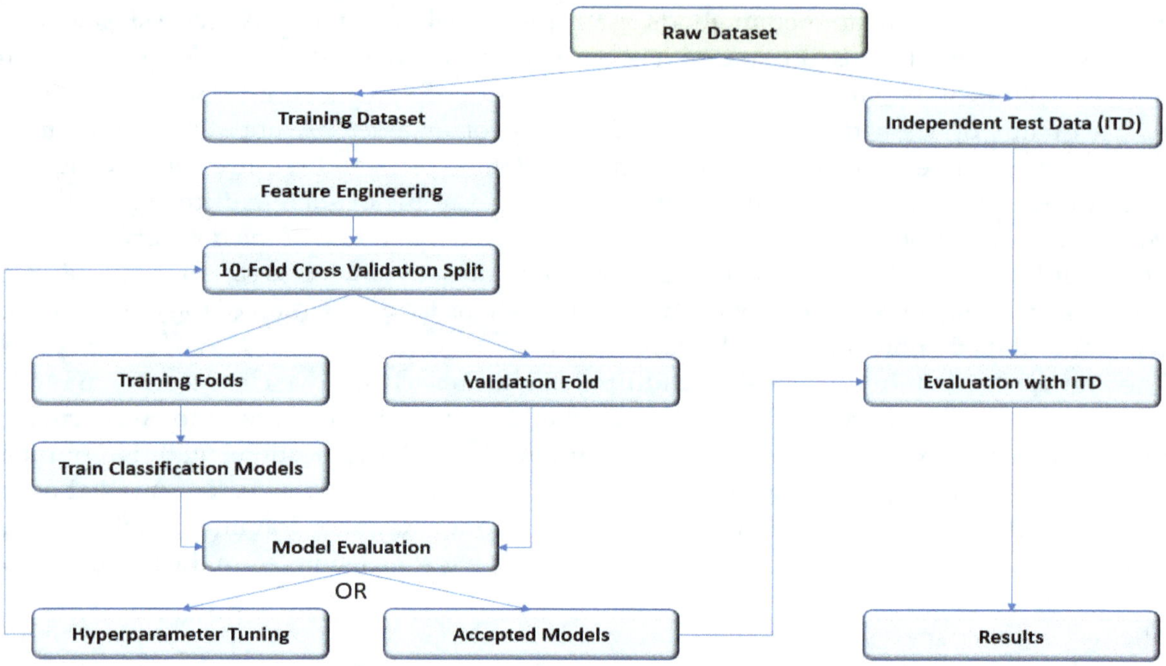

Figure 1. Method overview.

2.1. *Classification Methods*

Classification is used to predict the response of a discreet variable by considering its relations with other variables in the dataset [11]. Since the pass-or-fail compliance determined during final inspection in manufacturing is a discreet variable, classification methods can be used to predict such an outcome. There are multiple classification methods that can be used ranging from basics like decision trees, support vector machines and Naive Bayes to complex ensemble methods which can be broadly categorized into bagging or boosting. Since most methods have specific strengths and rely on assumptions about the nature of the dataset, the methods proposed in this paper were generalized enough for most cases pertaining to prediction of compliance quality in a manufacturing plant. Thus, the following classifiers could be used in combination to that effect.

2.1.1. Random Forest

This popular classification method is an improved form of bagging that utilizes decision trees as the weak learners [22,29]. The main advantage of random forest as compared to bagged ensembles is that it attempts to reduce bias by learning trees from a subset of features sampled from the dataset. The trained model can also be improved by tuning the number of features to sample, maximum decision tree depth and minimum instances per node. As such, its popularity and general effectiveness allowed for a good starting point in creating a collection of prediction models.

2.1.2. XGBoost

XGBoost is a relatively recent development in inductive learning [30]. This learner has been used extensively with agreement that it tends to be more computationally efficient and generally applicable for a wide range of dataset types, from common classification problems [31] to pattern recognition in time-dependent features [32].

This algorithm builds upon the gradient boosting algorithm by introducing a regularization parameter, which reduces the individual regression tree's sensitivity towards outliers in the dataset. As such, this algorithm will result in a model that has less variance than that learned from gradient boosting alone. The use of this learner in combination with random forest provided two models that were learned from the principles of boosted and bagged ensembles of decision trees, respectively. Additional methods such as support vector machines [33] using various kernels or artificial neural networks [34] can also be used for comparison if the results of these learners do not prove satisfactory.

2.2. *Feature Engineering*

Raw data collected from various sources can rarely be used directly to train a classification model. Thus, feature engineering refers to the steps taken to prepare a dataset for classification. This entailed cleaning the dataset to ensure consistency of data type in each feature and removal of redundant variables. Additionally, features could be constructed to support the algorithm used to train classification models in order to obtain improved results. This was achieved by transforming and combining features in the available data to obtain new variables. In the context of manufacturing quality on a mixed-model assembly line, the following features could be derived from available production data and provide a basis for the classification model to be built from.

2.2.1. Suspicious Unit Batches

Since quality defects tend to be relatively rare in modern manufacturing and the prevalence of random inspection, only using identified defective units as a class variable will result in a poor prediction model. Thus, a new class variable accounting for random inspection for prediction was derived instead. This variable was constructed from raw data pertaining to quality defects found during unit inspection, and production line data containing serial numbers of units in the sequential order at which they flowed through production and inspection. Defining the size of a suspicious batch of units based on identified defects required consideration for the nature of randomness in inspection and desired confidence in product compliance quality.

If a factory uses production lines to manufacture goods, the progress of work-in-progress (WIP) through its production stages can be traced more consistently than a job-shop or manufacturing cell layout. As such, if the final product inspection is random and a defective unit is found, then based on the required confidence level, a certain number of units before and after the defective unit in production sequence would have to be considered under suspicion of the same defect. Thus, it would also be a good idea to segregate and inspect the other units centered around the defective unit.

The exact number of units quarantined for such a reason could vary depending on the factory's throughput, inspection strategy and best practices. However, if a 100% inspection strategy is used then the construction of this class variable would not be necessary since there is no need to account for random inspection.

2.2.2. Proximity to a Model Changeover

In modern multi-model production lines, model changeovers have been known to cause quality defects in the assembled products [35]. Thus, a feature constructed to represent how close a unit was to a previous model changeover in its production run can result in learning a significantly improved classification model. This variable can be presented as:

- An absolute measure (s) of the unit's sequence number (n) relative to the last model change in the sequence numbers (n_0) assigned to each unit in the daily production logs.

$$s = n - n_0, \tag{1}$$

- A continuous variable (x) that measures the relative position of the unit (s) with respect to the sequence at which the model change happened (s_0) and the next such model change (s_1).

$$x = (s - s_0)/(s_1 - s_0), \tag{2}$$

2.2.3. Model Color Change

This feature was proposed on the basis of general guidelines established when investigating recurrent quality issues under suspicion of human error. A model change is much more apparent when there are stark differences between the appearance of the two models and thus should potentially reduce the likelihood of assembly errors [36] caused by the model changeover. This feature can be derived from the model change feature discussed earlier but would only apply if there was a change in color between the two models. A possible way in which this feature could be constructed was as a binary variable (c) that describes whether the previous model in the production logs was of a different color such that $c = 0$, if model before changeover was of same color and $c = 1$ if the model before changeover was of different color.

Similarly, with knowledge of the use-case, new features corresponding to transformed production data could be derived following the same principle as that mentioned above. For example, brand, assembly complexity based on bill-of-material, general appearance, or packaging methods could also be used to construct new features. Many of the features discussed earlier are generalized starting points based on best practices when troubleshooting for quality defects. Thus, expertise in the subject matter of the manufacturing processes and factors contributing to compliance quality will result in constructed features that can be used to train better classification models.

2.3. Evaluation Method

Choosing correct metrics when evaluating the performance of a classification model is crucial since each metric places varying emphasis on the overall accuracy, precision, recall, or agreement between model and ground truth for different class values. Thus, the chosen metric needs to closely evaluate model performance based on what is required from the user. In this paper, evaluation of the models was completed via Cohen's Kappa since it is often used as a way to measure the ability of a model to predict binary classes with heavy imbalances [37].

Despite the fact that a confusion matrix can comprehensively explain the performance of a model by itself, a single performance measure could make the evaluation faster. Additionally, in the context of manufacturing plants, a single metric can be easier to present in cases where cross-functional cooperation is required. Since the problem of manufacturing compliance quality being addressed in this paper can be presented as a binary class variable, the confusion matrix is much simpler, as shown in Figure 2. The class value of 'yes' corresponds to potentially defective units while 'no' represents otherwise.

		Reference	
		no	yes
Prediction	no	TN	FN
	yes	FP	TP

Figure 2. Confusion matrix for a binary class with negative (no) and positive (yes) values. TN denotes true negative, FN denotes false negative, FP denotes false positive, and TP denotes true positive predictions.

Accuracy is a performance measure that can be used to evaluate classification models, using a test dataset.

$$Accuracy = \frac{TN + TP}{TN + TP + FP + FN} \tag{3}$$

It is a sum of all correct predictions divided by the total number of predictions made. This metric can be a conclusive measure of performance in most datasets where the class has attributes with mostly similar representation.

However, in cases where there is a heavy class imbalance, accuracy can be misleading as a performance measure since it does not penalize misclassification of the minority. For example, a dataset with 10% and 90% binary class distribution can allow for the high accuracy of 90% simply by allowing the classifier to predict all as the majority class. Thus, there is no guarantee with regards to the quality of the classification model, especially for the minority class when relying solely on accuracy as a performance measure.

Another popular performance measure in recent literature regarding classification is the F-measure or F_1 score. It is a harmonic mean of precision and recall for the prediction of a particular value [38]. In this case, we can consider the F_1 score for predicting the class attribute as 'yes' as shown in Figure 2.

$$F_1 = 2 \times \frac{precision \times recall}{precision + recall} \tag{4}$$

$$precision = \frac{TP}{TP + FP} \tag{5}$$

$$recall = \frac{TP}{TP + FN} \tag{6}$$

However, a drawback of relying on the F_1 measure is the fact that it does not provide a good measure of the model's overall performance. The F_1 score depends upon which value of the class variable is being considered for the evaluation [39]. In the case of Figure 2, the F_1 score for class value 'yes' could be very different as compared to that of class value 'no'. This would not be a problem in cases where the cost of misclassification for one value trivializes the other but in other situations, the F_1 score might not be sufficient as a performance measure by itself.

As such, Cohen's Kappa (κ) is another metric that is commonly used to evaluate the performance of classification models [37]. In the context of classification models, it is a measure of the agreement between the predictor and reference while accounting for the agreement that occurs due to chance [40].

It is a function which penalizes chanced agreement (p_e) from the observed accuracy (p_0).

$$\kappa = \frac{p_0 - p_e}{1 - p_e} \tag{7}$$

$$p_0 = \frac{TN + TP}{N} \tag{8}$$

N is the total number of observations being used to evaluate the model.

$$N = TP + TN + FP + FN \tag{9}$$

$$p_e = p_N + p_P \tag{10}$$

p_N and p_P are the probabilities of agreement between predictor and reference by chance for negative (no) and positive (yes) class values.

$$p_N = \frac{(TN + FN)(TN + FP)}{N^2} \tag{11}$$

$$p_P = \frac{(TP + FP)(TP + FN)}{N^2} \tag{12}$$

The advantage of using Cohen's Kappa as opposed to F_1 score was that it can summarize the performance of the model with regards to the prediction of both classes in one measure whereas the use of F_1 could require evaluation for two class values. Additionally, it will account for agreement between raters due to chance. Combined with the fact that Cohen's Kappa can be generalized for use in situations where the class can take on more than two values, it has gained popularity for use in more complex classification problems as a performance measure.

Thus, for the problem addressed in the use-case, this paper proposed the use of both Cohen's Kappa as the singular performance metric, and the confusion matrix to consider some of the finer details about model performance regarding particular class attributes.

3. Case Study

This case study focuses on an appliance manufacturing factory that has recently undergone some major changes to one of its assembly lines. These changes include the addition of vision systems and scanners to keep track of all its produced units by their serial numbers. Additionally, there were improvements in the organization of quality assurance data regarding inspection results, returns and issue descriptions. However, there was also a rise in the number of quality defects being reported during random inspection pertaining to parts being missing or assembled wrong in the manufactured units. Despite the efforts and suggestions of various employees, it took a few months to investigate the issue and during this time, a huge amount of money had to be spent to recall units suspected to be defective back to the factory for inspection and release.

This presented an opportunity for the newly available data to be used to predict compliance quality of production runs or individual units before they even arrive at final inspection. Supervised learning could be used to train a classification model in advance. The trained model could then be used to predict the quality of units passing through a production line in real-time and contain suspected units before they escape the factory. Applications of machine learning in studies regarding quality in various manufacturing operations had yielded significant success in semiconductor manufacturing [4,41] and additive manufacturing [12].

Since most of the production lines in the factory had mixed model production with almost negligible changeover time, the entire assembly and inspection process had been streamlined to produce nearly 800 units of product daily over 3 work shifts without pause. The rushed pace of production made it difficult to recover from errors in production and thus it had become the quality engineering department's main directive to fix and contain quality issues proactively rather than reactively.

Depending on the results of this study, a prediction model based on real-time production data could be potentially useful, as a tool to alert inspection teams to prioritize batches likely to be defective, or even as a way to learn the causes of some of the main quality issues. However, since the information systems had not undergone expansion at the time of data collection, there was a lack of detailed data available about the assembly processes. Although, if the preliminary study could deliver a feasible prediction system or useful learnings, then the general method used could be adapted to similar effect with a dataset containing more features pertaining to process and machine data.

3.1. Data Sources

The data that were used in this paper are masked quality and production data from a large-scale appliance manufacturing plant. The manufacturing dataset is structured to contain unit serials, models and production batch sequence identifiers for every unit that passes through the production line. The random customer acceptance inspection (RCAI) dataset contains the unit serial, model number, and some identifiers for the issues found. These data are then merged with the production dataset in the form of a variable that is defined by whether the unit was caught defective in the RCAI.

Table 1 describes the variables in the available merged dataset which contains 75,636 observations of eight variables, wherein the field SuspiciousUnit is the proposed class variable:

Table 1. A table describing the meaning of fields available in the raw dataset.

Variable Name	Description
Sequence Number	This is the serial number of each unit and serves as a unique sequential identifier in the production line.
Model	This is the model number of the unit passing through the production line.
Week	This is the week number at which the unit entered final inspection.
Color	Describes the color of the unit. This tends to get recorded as a derivative of the model number.
BrandGroup	Describes the make of the unit. Derived from Model
PlatformGroup	Platforms are used to group models with similar, discernible features. For example, a mountain bicycle and stunt bike can be easily distinguished
RCAIDefect	This denotes a unit found defective in RCAI for Wrong/Missing parts as 1, and 0 otherwise.
SuspiciousUnit	This binary variable denotes all of the approximately 50 units before and after the RCAI defective unit which are marked for inspection of similar issues.

Additionally, Table 2 provides a summary of the distribution of two possible class variables in the raw dataset, in order to depict the difference in class imbalance.

Table 2. Variable distribution of the two possible class variables.

RCAIDefect Value	Datapoints	SuspiciousUnit Value	Datapoints
0	75,584	0	70,501
1	52	1	5135

3.2. Process Overview

The quality inspections conducted at this plant fall into two categories:

1. Final Product Inspection (FPI). This is a very rapid inspection that checks for several predefined functional and aesthetic issues for the unit. Every unit undergoes this inspection. However, a major limitation of this inspection is the fact that due to its speed, only issues that are defined on the checklist are checked for and issues are rarely documented. Instead, this inspection tends to serve as a chance for last minute correction to visibly faulty or incomplete units.
2. Random Customer Acceptance Inspection (RCAI). From a batch, about 3% of units are selected for a more thorough inspection by a specialist. If an issue is detected at this stage, the entire batch of product needs to be placed on warehouse hold in order to check for this issue. As such, defects found at this level tend to be costly for the plant.

One of the main reasons why data from RCAI are used rather than FPI is due to the big difference in discipline followed when it comes to documenting defects in RCAI, since it is a more thorough inspection. Another distinction that can be made between FPI and RCAI is the fact the results of the FPI depend on what is already well known as a problem area for the product model, whereas the RCAI tends to discover new issues instead. As such, it can be assumed that the inspection checklist used in the FPI is derived indirectly from the outcome of the RCAI. The data obtained from FPI could be used invaluably in data mining to improve upon first pass quality of the product. However, since issues found at RCAI tend to be much more monetarily damaging for the organization and the limitations of the FPI data, this paper will focus on the RCAI dataset instead.

Additionally, compliance quality failures during the RCAI tend to be costly because of the fact that the RCAI is a 3% random sampling inspection that is performed just a few steps before the product is packaged and dispatched to the distribution centers. As such, if defects are found at this stage, it is likely for other units in the batch to also be affected. However, due to the long time required to perform RCAI and how quickly the rest of the units are shipped off to the distribution centers, it is also very likely for some suspected units to escape the factory. This makes it very costly for a company to track down, recall and perform 100% inspection on all the suspected units. Thus, it could be very beneficial for a classification model to be used in order to predict certain failures before the product even reaches the final stages of the production process. Additionally, an inductively learned model might be able to explain how different variables interact and affect the outcome of product compliance quality in regard to assembly of wrong parts.

In order to formulate an effective approach to this problem, Figure 3 helps to visualize the key steps that correspond to each variable in the dataset on the production line. On the mixed model production line, units of a particular model are represented by color and, as depicted in the figure, there is a negligible changeover time between different models on a production that runs without pause. Data are collected at key points in the production line via barcode scanners and vision systems. Most of this data pertains to serial numbers and model numbers of units that pass through the production line at various stages or are redirected to testing and rework loops. Each production run of a particular model can vary widely in the number of units, from a few units of an experimental batch to several thousand that continue production over multiple days. Despite this, at no point is the production line stopped due to a model changeover.

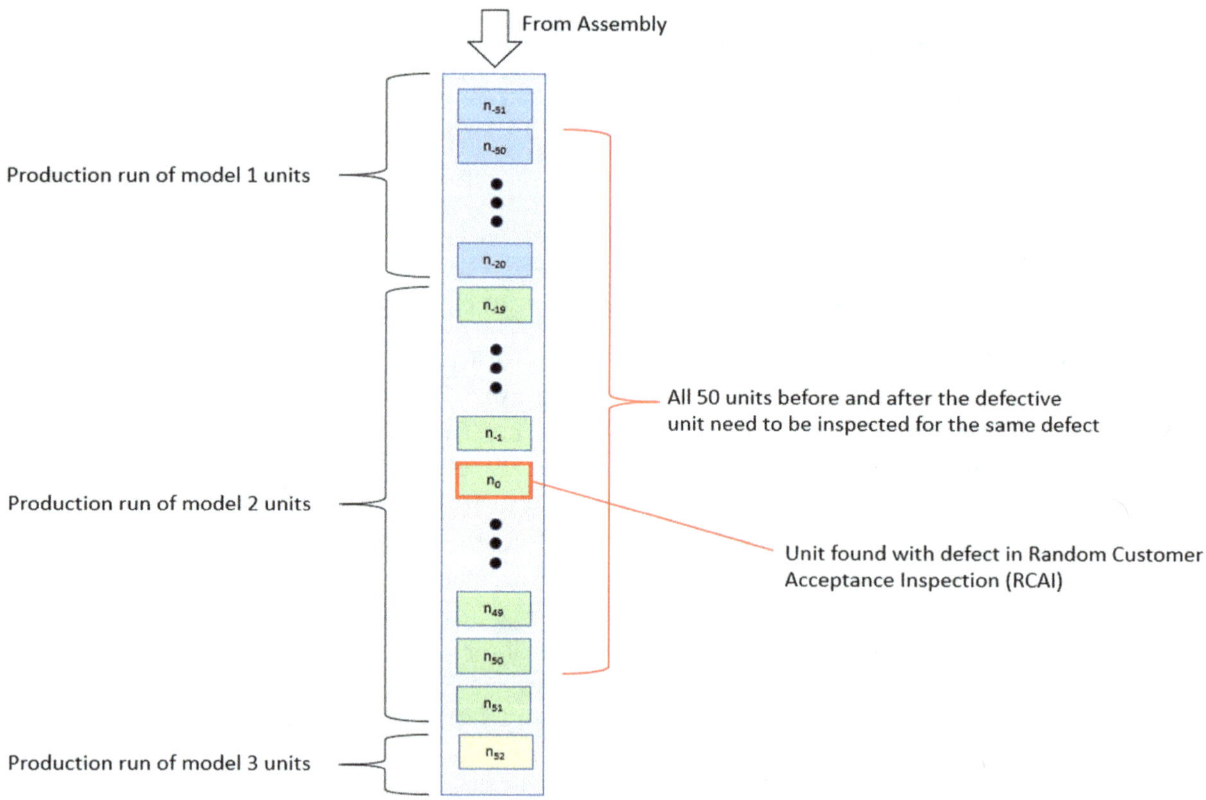

Figure 3. A depiction of the inspection and containment process for suspected units after one is found defective in RCAI.

Since the serial numbers are assigned sequentially to units on each line during assembly with zero loss of units before inspection, the unit's serial numbers can be used to measure their relative position on the production with respect to other units in this study.

During RCAI, units found with defects are recorded on a separate dataset which includes additional quality-related information such as unit model, nature of defect, inspector identifiers, and all other units before and after that were marked for inspection regarding the same defect. The data points recorded in the RCAI dataset have fields that contain unrestricted string inputs. As such, they tend to require manual dissemination to structure into definable features that can be used for classification. Thus, each RCAI defect is categorized into a number of headings.

The top three categories accounting for RCAI defects are described as follows:

3. New Product Introduction (NPI) Defect. These are defects found on models that have been introduced into production recently and thus usually contain a large variety of defects caused by design oversights or frequent model revisions. This category stops being applied to defects once occurrence of such defects reaches acceptable thresholds. Thus, there is little need to train classification models to detect these since the categorization of their occurrence depends on their rate of occurrence and would result in a trivial discovery.

4. Aesthetic Issues. These RCAI defects pertain to superficial issues such as scratches, dents, or discolorations. Multiple factory studies in these issues had found process related deviations and other random human error to be the cause of such defects. However, since the dataset chosen does not contain manufacturing process parameters for each unit, it would seem rather difficult to train an accurate prediction model to classify such occurrences. Nonetheless, there is scope for such a study once the required data becomes available.

5. Wrong or Missing Parts. These refer to defects in which parts were missing from the final product or if wrong parts were assembled instead. Early discussions regarding these occurrences had suggested causes such as model changeovers, worker fatigue in late shift, complexity of certain product models, among others. Since there are abundant data regarding most of the suggested causes, it could be possible to train a classification model to predict these occurrences and test the validity of suggested causes. As such, RCAI defects exclusively categorized as Wrong/Missing Parts would be used as class variables in the classification model discussed in this paper.

Collection of assembly and fabrication process parameters via vision systems and tool metrics has been implemented on newer low-throughput lines but was still in process of being installed on the main assembly line studied in this paper. Despite the availability of many more features on the new line, the reason for using the high-throughput line instead is because of the fact that the newer lines do not operate at fixed speeds and also because they only account for about 5% of the total factory's production as compared to the main line. Combined with the fact that each unit on the new lines is inspected and reworked on-the-spot with no guarantee of documentation, it would have made for a difficult task to build a classification model using the smaller and unreliable dataset available.

4. Results

These are the results of multiple classification models trained from the training dataset containing 60,508 instances and their resultant predictions on an independent test dataset containing 15,128 instances. The results pertain to various models trained from combinations of feature selection, feature construction, classification and hyperparameter tuning methods applied as per the workflow described in Figure 1.

All of the resulting models were trained from a dataset containing datapoints synthesized using SMOTE to address the massive class imbalance. Additionally, the same seed was set to ensure consistency regarding the cross-validation folds in order to better compare results.

All models were trained with the use of the caret, classification, and regression training R package by Max Kuhn, 2020 along with associated packages used for different classification models and visualization. On average, training the random forest models took 16.45 min and training the XGBoost models took 14.47 min.

Evaluation of the models was completed via Cohen's Kappa (κ), derived from confusion matrices. Graphical representations of performance are also depicted in the form of ROC curves. Predicting the response of the independent test dataset took on average 0.32 s for random forest models and 0.05 s for XGBoost models.

4.1. Model A: Initial Classification with no Feature Construction or Selection

Model A was learned via random forest and XGBoost algorithm from the initially obtained clean dataset, with no feature selection or additional features constructed. Figure 4 and Table 3 show the results of its evaluation.

Despite the generally high accuracy derived from the confusion matrix, it does not indicate a good prediction model since the majority class accounts for 93.23% of the dataset. Thus, a random classifier would likely achieve similar or better accuracy.

The XGBoost classifier performs much better than the random forest classifier when it comes to predicting the minority class. However, it still fails to correctly predict 58.89% of the suspicious units from the test dataset.

Table 3. Confusion matrices for the test results of XGBoost and random forest classifiers without any constructed features.

		Random Forest				**XGBoost**	
		Reference				**Reference**	
		No	Yes			No	Yes
Prediction	No	13,916	855	**Prediction**	No	13,632	603
	Yes	188	169		Yes	472	421
Accuracy: 0.9311				Accuracy: 0.9289			
Cohen's Kappa: 0.2174				Cohen's Kappa: 0.4015			

Primarily, basic features like model or platform type were used in the initial classification to make predictions on the likelihood of defects. This would be similar to what a quality inspector would conclude based on knowledge of previous inspection data pertaining to problematic models or product platforms. However, a model like this would not be sufficient to make any decisions to change inspection strategies.

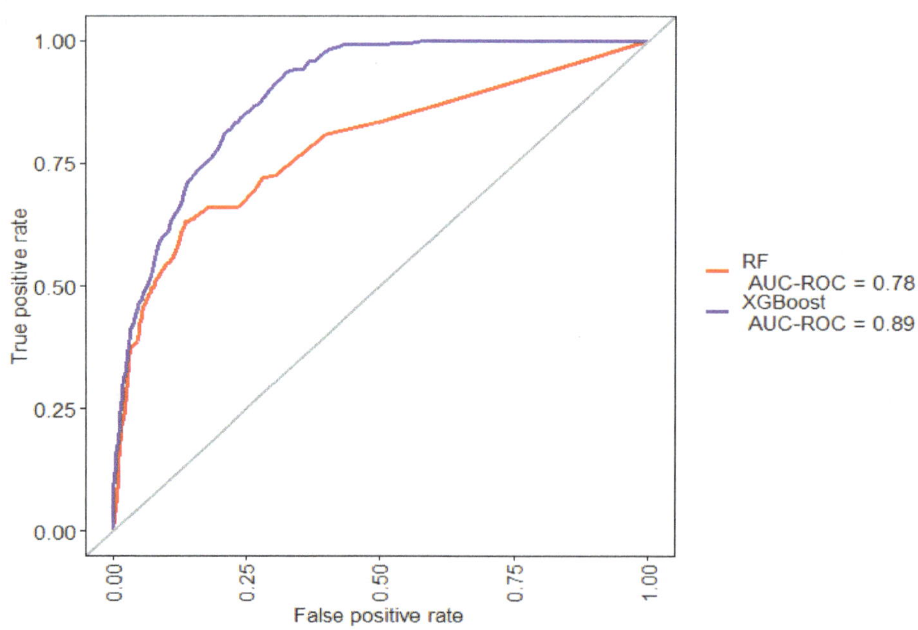

Figure 4. ROC curves for the results of XGBoost and random forest classifiers without any constructed features.

4.2. Model B: Classification with Model Changeover Feature

In order to explore some of the possible root causes discussed regarding RCAI defects, Model B was trained on a dataset in which irrelevant features like Seq.Number were removed. Additionally, a new feature, model_change is constructed in a manner similar to that described in this paper's method section. The new feature denotes if a unit immediately follows another unit of different model in the production line. Figure 5 and Table 4 show the results of the changes.

Table 4. Confusion matrices for the test results of XGBoost and Random Forest Classifiers using model_change feature.

	Random Forest				XGBoost		
		Reference				Reference	
		No	Yes			No	Yes
Prediction	No	13,907	833	Prediction	No	13,657	627
	Yes	197	191		Yes	447	397
Accuracy: 0.9319				Accuracy: 0.9290			
Cohen's Kappa: 0.2424				Cohen's Kappa: 0.3876			

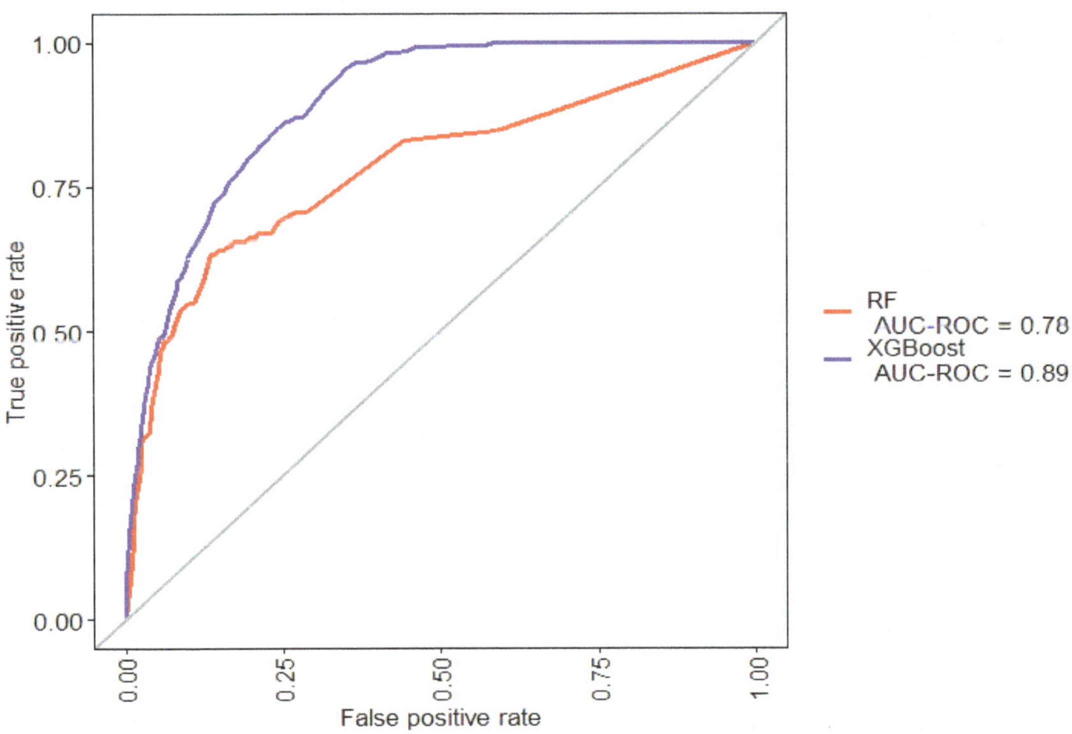

Figure 5. ROC curves for the results of XGBoost and Random Forest Classifiers using model_change feature.

There is a slight improvement in the performance of the random forest model and slight worsening of the XGBoost model, but this is mostly negligible in both cases and could be attributed to randomness.

Likely, the problem with the constructed feature is that only the first unit after each model changeover is marked whereas multiple suspicious units are identified after each RCAI failure. This is supported by the fact that even the minority class constitutes more data points than the constructed feature. Therefore, the constructed feature would be a heavy minority and would not result in significant improvements in the classification model.

4.3. Model C: Classification with Proximity to Model Changeover Feature

To correct the problem identified in Model B, another feature has been derived from the model_change feature for Model C, using the method described in this paper. This feature, batch_seq, represents the number of units between the product pertaining to the data point and the last unit which had model_change = 1. The results of training classification models with the newly constructed feature are shown in Figure 6 and Table 5.

Table 5. Confusion matrices for the test results of XGBoost and Random Forest Classifiers using batch_seq feature.

Random Forest				XGBoost			
		Reference				Reference	
		No	Yes			No	Yes
Prediction	No	13,882	408	**Prediction**	No	13,937	17
	Yes	222	616		Yes	167	1007
Accuracy: 0.9584				Accuracy: 0.9878			
Cohen's Kappa: 0.6397				Cohen's Kappa: 0.9098			

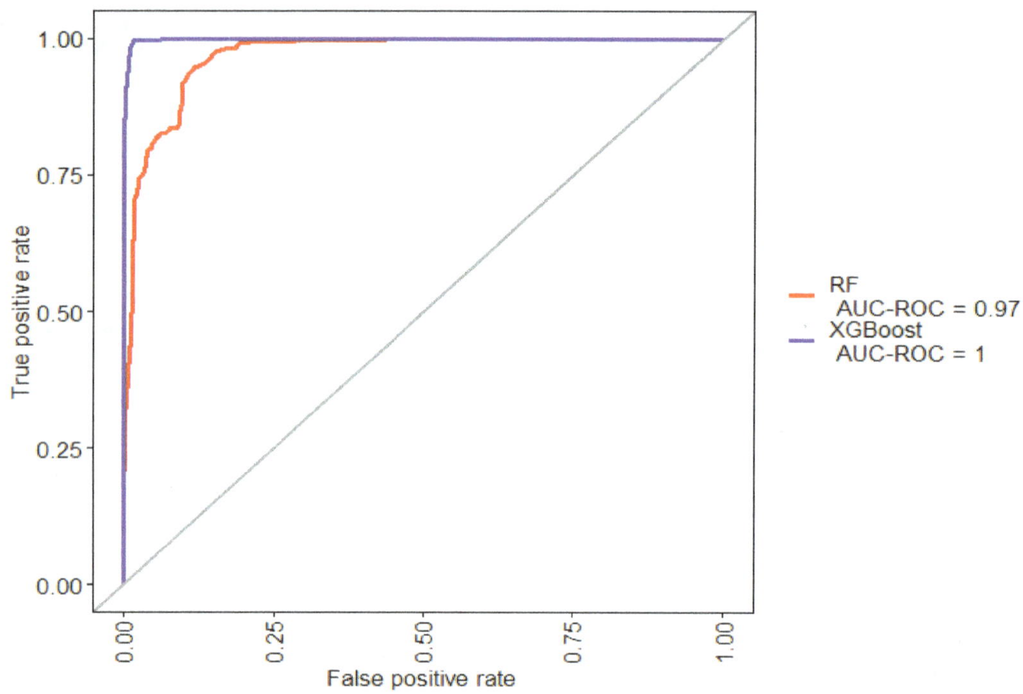

Figure 6. ROC curves for the results of XGBoost and Random Forest Classifiers using batch_seq feature.

The results show a significant improvement in the performance of both models in predicting the independent test dataset. The XGBoost model could predict suspected defective units with 98.34% accuracy. This likely confirms the correlation between model changeovers and the occurrence of RCAI defects that were categorized as missing or wrong parts.

The results also show the differences between boosting methods such as XGBoost and bagging methods like random forest. While bagging methods tend to suffer from increased bias, boosting methods tend to have less bias but suffer from overfitting. Table 5 shows the fact that despite the drastically improved performance as compared to the previous model, the random forest classifier still only classified 60.16% of the minority class correctly but had a bias towards classifying the majority class instead.

The XGBoost algorithm performed much better than the random forest classifier since it correctly predicted 98.34% of the minority class in the independent test dataset and had a comparable performance when evaluated within the cross-validation loop in the training dataset.

4.4. Model D: Classification with Normalized Proximity to Model Changeover Feature

Despite the success of Model C trained from the earlier constructed feature, the results could possibly be improved further by processing the constructed feature even more.

A limitation of the batch_seq feature constructed earlier is the fact that its value depends on the number of units between the product and previous model change. As such, in the factory with production runs of variable lengths, an adjusted measure of the unit's position in the run could prove more helpful than an absolute measure like batch_seq.

Thus, in Model D, the feature batch_seqperc is derived from batch_seq to replace it such that:

batch_seqperc = batch_seq/total units in production run

This feature accounts for production runs of varying size. Figure 7 and Table 6 show the results of training the classification models on the newly constructed batch_seqperc feature.

Table 6. Confusion matrices for the test results of XGBoost and Random Forest Classifiers using batch_seqperc feature.

	Random Forest				XGBoost		
		Reference				Reference	
		No	Yes			No	Yes
Prediction	No	13,900	558	Prediction	No	13,970	36
	Yes	204	466		Yes	134	988
	Accuracy: 0.9496				Accuracy: 0.9888		
	Cohen's Kappa: 0.5247				Cohen's Kappa: 0.9147		

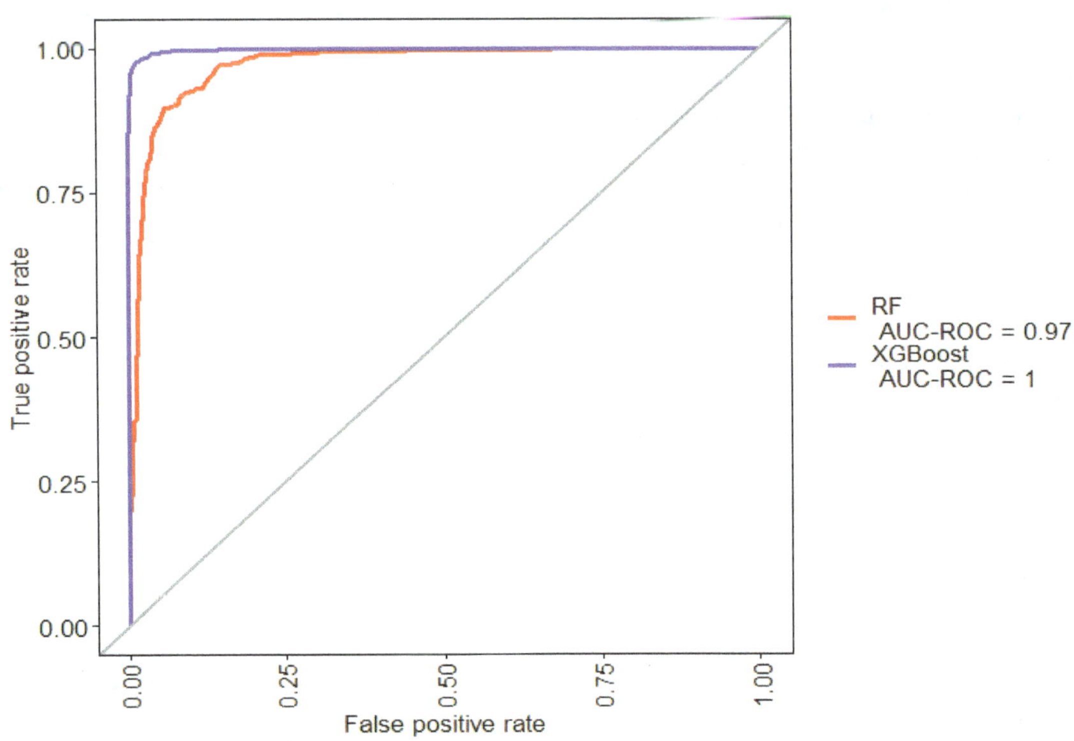

Figure 7. ROC curves for the results of XGBoost and Random Forest Classifiers using batch_seqperc feature.

The results show slight improvement in the Cohen's Kappa of the XGBoost model and a worsening of the random forest model. Both of the models become worse at predicting the minority class after using the batch_seqperc feature as opposed to the batch_seq feature.

This likely suggests that the class value of suspicious units can be better explained by an absolute measure of the unit's position in the production run rather than an adjusted measure. Possibly, the occurrence of human error when assembling wrong parts in units is highest in the units immediately after model change regardless of the total number of units in the same-model production run.

5. Conclusions

Despite the common use of machine learning and data mining techniques at various stages of a supply chain, this paper's results indicate that it is viable to use similar methods to predict compliance quality from data pertaining to multiple processes. The case study pertaining to a manufacturing industry provided a dataset that could be used to predict product containing wrong or missing parts with significant accuracy and Cohen's Kappa, using the proposed method. Using production and quality data from the case studied, machine learning techniques were used to predict the compliance quality of a manufactured unit in the context of end inspection. With the increasing availability of process data due to Industry 4.0 implementations in modern manufacturing plants, future applications of the proposed method are likely to be successful. Thus, product compliance quality can be predicted as well as process-specific quality, with sufficiently large and feature-rich datasets available.

This paper builds upon similar research conducted regarding the application of machine learning in various manufacturing tasks, by expanding its scope to predict the overall compliance quality of product in final inspection of a production line. It also improves upon similar research in multistage manufacturing quality prediction by addressing class imbalance and the need for feature enrichment via feature engineering. The classification model trained from data pertaining to the case study can be used in its relevant factory to preemptively contain product units that would be defective due to wrong or missing parts in assembly, thus significantly saving on costs for product recalls and transportation due to escaped units. The prediction models could be trained in approximately 15 min using a personal computer whereas predictions took less than 1 s for 15,128 datapoints. The time required for training could vary significantly with the nature of the training dataset, algorithm and hyperparameters chosen but the time required to predict responses using the trained model was negligibly low for the case in this paper. Thus, the model could be trained in advance using the available data and prediction could be done in real-time. As such, this could provide an additional layer of quality control with relatively low investment in equipment or time.

The aim of this paper was to explore the use of machine learning to predict manufacturing compliance quality as per the outcome of quality inspections. It was possible to train a prediction model with defective unit prediction accuracy of 98.34%, overall accuracy of 98.78% and Cohen's Kappa of 0.91, using a combination of feature construction and ensemble classification algorithms. The results obtained indicate that there is a significant improvement in the prediction model's performance when it uses a dataset with features constructed to signify the position of each unit within the production line's run of a product model. It also appears that using an absolute measure of the unit's sequential order produced slightly better prediction of the minority class as compared to a normalized variable to represent the unit's position in the batch.

The performance of the XGBoost model had been consistently better than that learned via random forest and significantly better at predicting the minority class. However, the case study in this paper indicates that regardless of the algorithm chosen, the trained model seemed to perform significantly better when specific features were constructed using prior domain knowledge regarding the nature of the dataset.

This study is subject to a few limitations, which suggests future research direction in the following: Firstly, for the case studied, improvement can be possible by constructing features to signify a change in model color or platform, similar to the model change. Other applications of the proposed method might require more features in order to obtain a prediction model with the required values of accuracy and Cohen's Kappa. Secondly, analysis of the models to understand what caused quality defects can be conducted. This was not conducted due to the design of this study and availability of the data. Future studies using datasets with more features corresponding to manufacturing process data can be conducted to understand the causes for quality non-compliances as well as predictions. Thirdly, due to dataset limitations, only quality defects pertaining to products with wrong or missing parts could be feasibly predicted with the available features. With sufficient process data, machine learning methods researched in process-specific applications could be incorporated into end quality inspection as well. Finally, this paper did not explore the use of popularly used prediction algorithms such as neural networks due to the early success of the gradient boosting method and the fact that neural networks tend to perform better with larger and more feature-rich datasets, as opposed to that used in this paper's case study. However, this suggests the possibility of future research to evaluate the performance of neural networks alongside algorithms used in this paper.

Author Contributions: Methodology, S.S.; writing—original draft preparation, S.S.; writing—review and editing, S.S. and G.H.; supervision, G.H. Both authors have read and agreed to the published version of the manuscript.

References

1. Campanella, J. Principles of Quality Costs: Principles, implementation, and use. In *Proceedings of the ASQ World Conference on Quality and Improvement Proceedings*; American Society for Quality: Milwaukee, WI, USA, 1999; p. 507.
2. Moldovan, D.; Cioara, T.; Anghel, I.; Salomie, I. Machine learning for sensor-based manufacturing processes. In Proceedings of the 2017 13th IEEE International Conference on Intelligent Computer Communication and Processing (ICCP), Cluj-Napoca, Romania, 7–9 September 2017; IEEE: New York, NY, USA, 2017; pp. 147–154.
3. Kerdprasop, K.; Kerdprasop, N. Feature selection and boosting techniques to improve fault detection accuracy in the semiconductor manufacturing process. In Proceedings of the World Congress on Engineering 2012, London, UK, 4–6 July 2012; International Association of Engineers (IAENG): Kowloon, China, 2010; Volume 2188, pp. 398–403.
4. Kao, H.-A.; Hsieh, Y.-S.; Chen, C.-H.; Lee, J. Quality prediction modeling for multistage manufacturing based on classification and association rule mining. In Proceedings of the 2nd International Conference on Precision Machinery and Manufacturing Technology (ICPMMT), Kenting, Taiwan, 19–21 May 2017; EDP Sciences: Les Ulis, France, 2017; Volume 123.
5. Arif, F.; Suryana, N.; Hussin, B. A data mining approach for developing quality prediction model in multi-stage manufacturing. *Int. J. Comput. Appl.* **2013**, *69*, 35–40. [CrossRef]
6. Köksal, G.; Batmaz, İ.; Testik, M.C. A review of data mining applications for quality improvement in manufacturing industry. *Expert Syst. Appl.* **2011**, *38*, 13448–13467. [CrossRef]
7. Saura, J.R. Using Data Sciences in Digital Marketing: Framework, methods, and performance metrics. *J. Innov. Knowl.* **2020**. [CrossRef]
8. Jain, A.K.; Murty, M.N.; Flynn, P.J. Data clustering: A review. *ACM Comput. Surv. CSUR* **1999**, *31*, 264–323. [CrossRef]
9. Piatetsky-Shapiro, G. Discovery, analysis, and presentation of strong rules. *Knowledge Discovery Databases* **1991**, *11*, 229–238.
10. Schuh, G.; Reinhart, G.; Prote, J.-P.; Sauermann, F.; Horsthofer, J.; Oppolzer, F.; Knoll, D. Data mining definitions and applications for the management of production complexity. *Procedia CIRP* **2019**, *81*, 874–879. [CrossRef]
11. Kotsiantis, S.B.; Zaharakis, I.; Pintelas, P. Supervised machine learning: A review of classification techniques. *Emerg. Artif. Intell. Appl. Comput. Eng.* **2007**, *160*, 3–24.

12. Scime, L.; Beuth, J. Anomaly detection and classification in a laser powder bed additive manufacturing process using a trained computer vision algorithm. *Addit. Manuf.* **2018**, *19*, 114–126. [CrossRef]
13. Moretti, C.; Steinhaeuser, K.; Thain, D.; Chawla, N.V. Scaling up classifiers to cloud computers. In Proceedings of the 2008 Eighth IEEE International Conference on Data Mining, Pisa, Italy, 15–19 December 2008; IEEE: New York, NY, USA, 2008; pp. 472–481.
14. Fernández, A.; García, S.; Herrera, F. Addressing the classification with imbalanced data: Open problems and new challenges on class distribution. In Proceedings of the 6th International conference on hybrid artificial intelligence systems, Wroclaw, Poland, 23–25 May 2011; Springer: Berlin/Heidelberg, Germany, 2011; pp. 1–10.
15. Kim, A.; Oh, K.; Jung, J.-Y.; Kim, B. Imbalanced classification of manufacturing quality conditions using cost-sensitive decision tree ensembles. *Int. J. Comput. Integr. Manuf.* **2018**, *31*, 701–717. [CrossRef]
16. Zhu, H.; Liu, G.; Zhou, M.; Xie, Y.; Abusorrah, A.; Kang, Q. Optimizing Weighted Extreme Learning Machines for Imbalanced Classification and Application to Credit Card Fraud Detection. *Neurocomputing* **2020**. [CrossRef]
17. Jia, F.; Lei, Y.; Lu, N.; Xing, S. Deep normalized convolutional neural network for imbalanced fault classification of machinery and its understanding via visualization. *Mech. Syst. Signal. Process.* **2018**, *110*, 349–367. [CrossRef]
18. Zhang, X.; Wei, X.; Li, F.; Hu, F.; Jia, W.; Wang, C. Fuzzy Support Vector Machine with Imbalanced Regulator and its Application in Stroke Classification. In Proceedings of the 2019 IEEE Fifth International Conference on Big Data Computing Service and Applications (BigDataService), Newark, CA, USA, 4–9 April 2019; IEEE: New York, NY, USA, 2019; pp. 290–295.
19. Mahanipour, A.; Nezamabadi-Pour, H. Improved PSO-based feature construction algorithm using Feature Selection Methods. In Proceedings of the 2017 2nd Conference on Swarm Intelligence and Evolutionary Computation (CSIEC), Kerman, Iran, 7–9 March 2017; IEEE: New York, NY, USA, 2017; pp. 1–5.
20. Chawla, N.V.; Bowyer, K.W.; Hall, L.O.; Kegelmeyer, W.P. SMOTE: Synthetic minority over-sampling technique. *J. Artif. Intell. Res.* **2002**, *16*, 321–357. [CrossRef]
21. Tao, X.; Li, Q.; Guo, W.; Ren, C.; Li, C.; Liu, R.; Zou, J. Self-adaptive cost weights-based support vector machine cost-sensitive ensemble for imbalanced data classification. *Inf. Sci.* **2019**, *487*, 31–56. [CrossRef]
22. Feng, W.; Dauphin, G.; Huang, W.; Quan, Y.; Liao, W. New margin-based subsampling iterative technique in modified random forests for classification. *Knowl. Based Syst.* **2019**, *182*, 104845. [CrossRef]
23. Faris, H.; Abukhurma, R.; Almanaseer, W.; Saadeh, M.; Mora, A.M.; Castillo, P.A.; Aljarah, I. Improving financial bankruptcy prediction in a highly imbalanced class distribution using oversampling and ensemble learning: A case from the Spanish market. *Prog. Artif. Intell.* **2020**, *9*, 31–53. [CrossRef]
24. Zhao, H.; Sinha, A.P.; Ge, W. Effects of feature construction on classification performance: An empirical study in bank failure prediction. *Expert Syst. Appl.* **2009**, *36*, 2633–2644. [CrossRef]
25. Muñoz, E.; Capón-García, E. Systematic approach of multi-label classification for production scheduling. *Comput. Chem. Eng.* **2019**, *122*, 238–246. [CrossRef]
26. Bodunov, O.; Schmidt, F.; Martin, A.; Brito, A.; Fetzer, C. Real-time Destination and ETA Prediction for Maritime Traffic. In Proceedings of the 12th ACM International Conference on Distributed and Event-based Systems, Hamilton, New Zealand, 25–29 June 2018; ACM: New York, NY, USA, 2018; pp. 198–201.
27. Esbensen, K.H.; Geladi, P. Principles of proper validation: Use and abuse of re-sampling for validation. *J. Chemom.* **2010**, *24*, 168–187. [CrossRef]
28. Kohavi, R. A study of cross-validation and bootstrap for accuracy estimation and model selection. In Proceedings of the 1995 IJCAI 14th International Joint Conference on Artificial Intelligence, Montreal, QC, Canada, 20–25 August 1995; IJCAI: Saxony, Germany, 1995; Volume 14, pp. 1137–1145.
29. Lan, H.; Pan, Y. A crowdsourcing quality prediction model based on random forests. In Proceedings of the 2019 IEEE/ACIS 18th International Conference on Computer and Information Science (ICIS), Beijing, China, 12–14 June 2019; IEEE Computer Society: Washington, DC, USA, 2019; pp. 315–319.
30. Chen, T.; Guestrin, C. Xgboost: A scalable tree boosting system. In Proceedings of the 22nd Acm Sigkdd International Conference on Knowledge Discovery and Data Mining, San Francisco, CA, USA, 13–17 August 2016; ACM: New York, NY, USA, 2016; pp. 785–794.
31. Xu, H.; Wang, H. Identifying diseases that cause psychological trauma and social avoidance by Xgboost. In Proceedings of the 2019 IEEE International Conference on Bioinformatics and Biomedicine (BIBM), San Diego, CA, USA, 18–21 November 2019; IEEE: New York, NY, USA, 2019; pp. 1809–1813.

32. Yang, L.; Li, Y.; Di, C. Application of XGBoost in Identification of Power Quality Disturbance Source of Steady-state Disturbance Events. In Proceedings of the 2019 IEEE 9th International Conference on Electronics Information and Emergency Communication (ICEIEC), Beijing, China, 12–14 July 2019; IEEE: New York, NY, USA, 2019; pp. 1–6.
33. Drucker, H.; Burges, C.J.; Kaufman, L.; Smola, A.J.; Vapnik, V. Support vector regression machines. In *Advances in Neural Information Processing Systems*; MIT Press: Cambridge, MA, USA, 1997; pp. 155–161.
34. Tan, Y.; Xia, Y.; Wang, J. Neural network realization of support vector methods for pattern classification. In Proceedings of the IEEE-INNS-ENNS International Joint Conference on Neural Networks. IJCNN 2000. Neural Computing: New Challenges and Perspectives for the New Millennium, Como, Italy, 27 July 2000; IEEE: New York, NY, USA, 2000; Volume 6, pp. 411–416.
35. Cheldelin, B.; Ishii, K. Mixed-model assembly quality: An approach to prevent human errors. In Proceedings of the ASME International Mechanical Engineering Congress and Exposition, Anaheim, CA, USA, 13–19 November 2004; ASME: New York, NY, USA, 2004; Volume 47055, pp. 109–119.
36. Neumann, W.P.; Kolus, A.; Wells, R.W. Human factors in production system design and quality performance—A systematic review. *IFAC Pap.* **2016**, *49*, 1721–1724. [CrossRef]
37. Hasan, M.; Ibrahimy, M.; Motakabber, S.; Shahid, S. Classification of Multichannel EEG Signal by Single Layer Perceptron Learning Algorithm. In Proceedings of the 2014 International Conference on Computer and Communication Engineering, Kuala Lumpur, Malaysia, 23–25 September 2014; IEEE: New York, NY, USA, 2014; pp. 255–257.
38. Chinchor, N.; Sundheim, B.M. MUC-5 Evaluation Metrics. In Proceedings of the Fifth Message Understanding Conference (MUC-5), Baltimore, MD, USA, 25–27 August 1993.
39. Chicco, D.; Jurman, G. The advantages of the Matthews correlation coefficient (MCC) over F1 score and accuracy in binary classification evaluation. *BMC Genom.* **2020**, *21*, 6. [CrossRef]
40. Vieira, S.M.; Kaymak, U.; Sousa, J.M. Cohen's kappa coefficient as a performance measure for feature selection. In Proceedings of the International Conference on Fuzzy Systems, Barcelona, Spain, 18–23 July 2010; IEEE: New York, NY, USA, 2010; pp. 1–8.
41. Chou, P.B.; Rao, A.R.; Sturzenbecker, M.C.; Wu, F.Y.; Brecher, V.H. Automatic defect classification for semiconductor manufacturing. *Mach. Vis. Appl.* **1997**, *9*, 201–214. [CrossRef]

3

Pricing Decision within an Inventory Model for Complementary and Substitutable Products

Ata Allah Taleizadeh [1], Masoumeh Sadat Babaei [2], Shib Sankar Sana [3] and Biswajit Sarkar [4,*]

1. School of Industrial Engineering, College of Engineering, University of Tehran, Tehran 456311155, Iran; taleizadeh@ut.ac.ir
2. Department of Industrial Engineering, Islamic Azad University, South Tehran Branch, Tehran 1584743311, Iran; st_m_babaei@azad.ac.ir
3. Principal, Kishore Bharati Bhagini Nivedita College, Ramkrishna Sarani, Behala, Kolkata 700060, India; shib_sankar@yahoo.com
4. Department of Industrial & Management Engineering, Hanyang University, Ansan, Gyeonggi-do 15588, Korea

* Correspondence: bsbiswajitsarkar@gmail.com.

Abstract: A combination of substitutable and complementary products is very important for any business industry to make all-round profit from different aspects. How deterioration affects complementary products or substitutable products is discussed in this study. This study investigates the pricing and inventory decisions for complementary and substitutable items which are deteriorating in nature. Four models are analyzed where the demand of one product is dependent upon the selling price and the price of another product. This paper tries to compute the optimum prices and order quantities to optimize the total profit, which is the main aim. Theoretically, this model is solved by a classical optimization method. Numerical examples demonstrate the applicability of this model. Results conclude that the total profit is dependent on the degree of substitutability and complementarity. A sensitivity analysis of optimal solutions is given to test the stability of the proposed model.

Keywords: inventory; pricing; complementary products; substitutable products; deterioration

---

1. Introduction

Everyday life starts with the use of a toothbrush and toothpaste, which are complementary products. Then, the energy for the whole day is derived from the breakfast where at least one seasonal fruit is mandatory for good health. The seasonal fruit can be replaced by another fruit based on the availability and price and is substitutable product. Whenever the increasing price of one product indicates the increasing sales rate of another product, then those products substitute each other [1]. Increasing the price of one item can help to increase the sale of another substitutable item. For example, Coca-Cola and Pepsi are two substitutable products as both can satisfy the purpose. One customer can buy Pepsi instead of Coca-Cola in any circumstances. Complementary products are used in combination with another product [2]. Single use of any part of complementary products has limited usage. The overall utility is increased whenever both products are used together. The market demand of complementary products depends upon both products [3]. Car and fuel, camera and memory card, toothpaste and brush—these are a few examples of complementary products. A car cannot move without proper fuel, which implies that the car is useless without fuel, and fuel is no longer important if the car is not there. To earn in all aspects, the manufacturer introduces substitutable as well as complementary products in the business. From the perspective of the inventory, complementary products need more storage and inventory, as both products are important. In reality, substitutable products give the manufacturer tough competition, as it is important to choose alternative products very wisely such that it can satisfy the demand as well as the quality of the original product. The pricing factor is the next important thing for substitutable products. The price of the alternative should be less than or equal to the original product. The situation may differ in case of the reverse situation when the alternative's price is greater than the original price. The situation gets more complicated when the product deteriorates in nature.

There are several researches who focus on substitutable and complementary products, while none of them have investigated both inventory and pricing decisions of the products simultaneously. Moreover, this investigation is done for deteriorating products. In this paper, four models are discussed for pricing and inventory decisions of two types of items, which may be complementary or substitutable, and deteriorating or non-deteriorating. The rest of the present paper is organized as follows: Section 2 presents a literature review, whereas problem definition, notation, and assumptions are defined in Section 3, and the mathematical modeling is presented in Section 4. Numerical examples, a sensitivity analysis, and managerial insights are provided in Section 5. Section 6 concludes the achievements of this proposed model. The last section provides the References of the study.

2. Literature Review

Pricing policy is an important part in the field of supply chain management [4–7]. Several research articles have been studied on inventory and pricing decisions for single or multi-products inventory models [8–10]. However, few of them studied pricing decisions for complementary or substitutable products. In such cases, the price, quality, and durability of the products play a crucial role to attract customers [5,9]. On the other hand, retailers face the cross-selling phenomenon for complementary products. McGillivray and Silver [11] investigated the full substitutability and its effect on total cost within an inventory model. Zhang et al. [12] illustrated an inventory system when the demand of a minor and a major item is correlated with cross-selling. Zhang [13] discussed an inventory problem for multiple products with the inventory replenishment. They proposed a search method which gives a global optimum solution. Yue et al. [14] investigated a duopoly market dealing with complementary products.

An inventory model was discussed by Liu and Yuan [15] for two types of products where the consumption rates of those products are correlated, and the replenishment rates are coordinated. Wei et al. [9] analyzed a non-coordination supply chain management consisting of two manufactures with one retailer for two complementary type products with the help of five different game theories: manufacturer-leader Bertrand, manufacturer-leader Stackelberg, retailer-leader Bertrand, retailer-leader Stackelberg, and Nash game (NG) models. Guchhait et al. [16] studied an advertisement-dependent demand scenario within supply chain management. They used a Stackelberg game policy to solve the model. Shavandi et al. [17] extended the pricing and lot sizing model of Abad [18] for multiple products. They proposed several pricing strategies of multi-products for deteriorating items which may be complementary, substitutable, or independent. The aim of the research was to maximize the profit optimization, along with optimum prices and production quantities.

An inventory model was studied by Balkhi and Benkherouf [19] for deteriorating products. The market demand was stock dependent and time dependent. A procedure was proposed by them to achieve the optimum replenishment schedule. Anjos et al. [20] considered perishable products and presented a continuous pricing strategy where the optimal pricing strategy can be explicitly characterized and easily implemented. Dye [21] extended an inventory model for perishable items with a time-dependent backlog by assuming a price-sensitive demand and variable deterioration with time. In that article, an improved algorithm was proposed by them to find out the optimal selling price of a product and replenishment schedule. Panda et al. [22] considered a single-item economic order quantity (EOQ) model with discounted sale as products were perishable in nature, and the demand was stock dependent. Pang [23] studied optimal dynamic pricing policies and inventory control policies for deteriorating products within a periodic type inventory model. Sarkar and Saren [24] investigated the effect of the trade-credit policy for exponentially deteriorating products. Sarkar et al. [25] developed a two-echelon supply chain model for deteriorating items, where single-setup-multi-delivery (SSMD) transportation policy was adopted.

Sett et al. [26] and Sarkar et al. [27] studied fixed lifetime products and their replenishment policy. The deterioration rate was variable, and the backlog was time dependent. Sarkar et al. [28] applied preservation technology for seasonal deteriorating products while Ullah et al. [29] used the preservation for waste generation. Recycling of the deteriorated products was investigated by Iqbal and Sarkar [30]. The market demand can be affected by the deterioration and a backlog scenario may appear in that case. The service level can help to get rid of this situation [31].

Gürler and Yilmaz [32] studied a supply chain management with two substitutable products in a newsboy problem. Demands for both products were independent, and the aim of the research was to maximize the retailer's and manufacturer's profits jointly. Kim and Bell [33] proposed that the production system of a firm sold their products to multiple market segments for a specific period. A price-driven scenario was examined by them, and the effect on the pricing strategy and production capacity of the firm were discussed.

Stavrulaki [34] investigated an inventory model for substitutable products. They explained the combined effect of the demand stimulation for stochastic demand. Karakul and Chan [35] studied the analytical implications of product substitutability and procurement decisions in a newsboy problem with two products. Thereafter, they established the unimodality of the expected profit of their previous work [36] with respect to the procurement and price of quantities. Birge et al. [37] developed an inventory model for substitutable products where market demand follows a uniform distribution. Netessine et al. [38] determined the optimum procurement for multi-substitutable items under exogenous prices of an integrated inventory-pricing problem.

Parlar and Goyal [39] developed two integrated production–inventory models of two substitutable products under different prices and assumptions [40]. Tang and Yin [41] discussed an improved procedure to find the optimum selling price within an inventory framework where demand is price sensitive for substitutable products. Xia [42] suggested a two-echelon supply chain model including competitive multiple suppliers and multiple buyers, where each supplier offers one type of substitutable product to the buyers. Netessine et al. [43] discussed about centralized inventory policy where market demand is substitutable. Table 1 gives the contribution from different authors associated with the area of research.

Table 1. Author's contributions table.

Author(s)	Inventory	Pricing	Deterioration	Complementary Products	Substitutable Products
Sarkar and Lee [2]		√		√	
Sana [4]	√	√			
Smith et al. [7]	√	√			
Wei at al. [9]		√		√	
Mcgillivray and Silver [11]	√				√
Yue et al. [14]		√		√	
Balkhi and Benkherouf [19]	√		√		
Dye [21]	√	√	√		
Karakul and Chan [35]	√	√			√
Tang and Yin [41]	√	√			√
This study	√	√	√	√	√

3. Problem Description, Notation, and Assumptions

3.1. Problem Definition

Consider a market where the retailer intends to determine the order quantities of different products. The products may be complementary or substitutable, for which two different scenarios exist.

In case of complementary products, the retailer expects a higher consumption rate of a product when the consumption rate of its complementary product increases and vice versa. Moreover, the demand of every complementary product depends upon the price of the other such that the high price of one of them decreases the demand of the complementary product and vice versa. Obviously, the degree of complementarity between two products, which is between 0 and 1, has the main effect on the changes between their respective demands and prices. In this situation, the retailer determines the optimal order quantity and selling price of each product in such a way that the retailer can maximize the total profit.

In the second scenario, each product can be used instead of another one in the case of substitutable products. In this situation, the retailer faces the situation that the decreasing demand of a substitutable product increases the other product's demand. The demand of each substitutable product depends upon its own price and the price of another one such that the higher price of one of them increases the demand of its substitutable product and vice versa because the customer intends to use the substitutable product, which has a lower price compared to the other. Generally, the degree of substitutability between two products, which is between 0 and 1, has the main effects on the changes between their demands and prices. Therefore, the retailer should determine the optimum order quantity such that the total profit should maximize.

3.2. Notation

The notation related to this study is given in the Nomenclature section at the end of the study.

3.3. Assumptions

Assumptions are used to formulate the model are as follows:

1. A retailer maximizes the profit based on an economic order quantity (EOQ) model. The demand is dependent upon selling prices and lot sizes of two types of products. The demand of two complementary products is dependent upon each product's selling price and the degree of complementarity.
2. After a certain time, products start to deteriorate with a constant deterioration rate. As deterioration is considered in this model, deterioration cost per time unit for the product m is η'_m (\$ per unit). The deterioration rate of product 1 and product 2 is D_r.
3. The degree of complementarity between two products is given by θ, where $0 \leq \varphi \leq 1$, and the degree of substitutability between two products is given by d, where $0 \leq \omega \leq 1$.
4. The lead time is zero, and there are no shortages for both types of products.
5. The time horizon is infinite.

4. Mathematical Model

The scenarios for both non-deteriorating (Figure 1) and deteriorating (Figure 2) products are discussed.

In all cases, the aim is to find the optimal values of the order quantity, the cycle length, and selling prices for profit maximization.

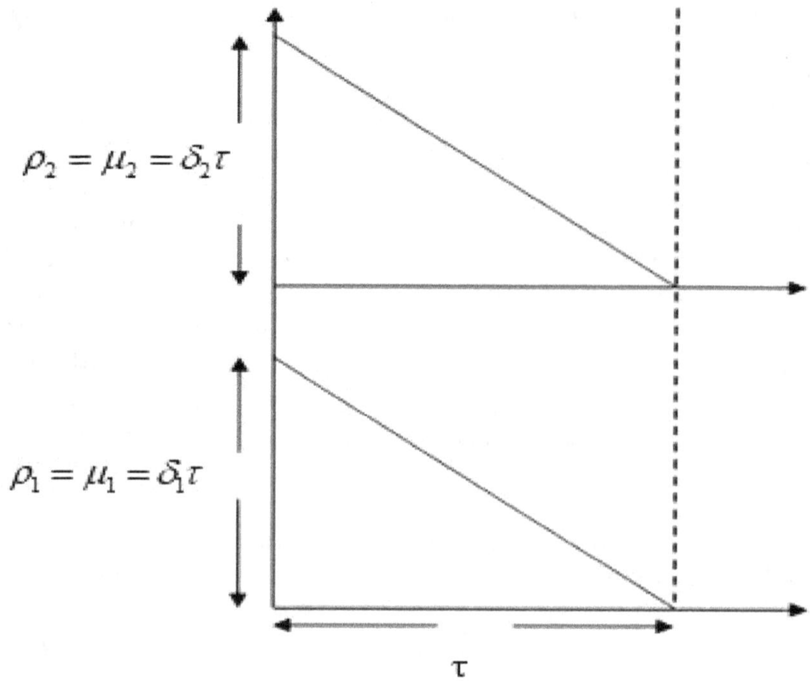

Figure 1. Inventory level of two products (complementary or substitutable) without shortage under EOQ policy.

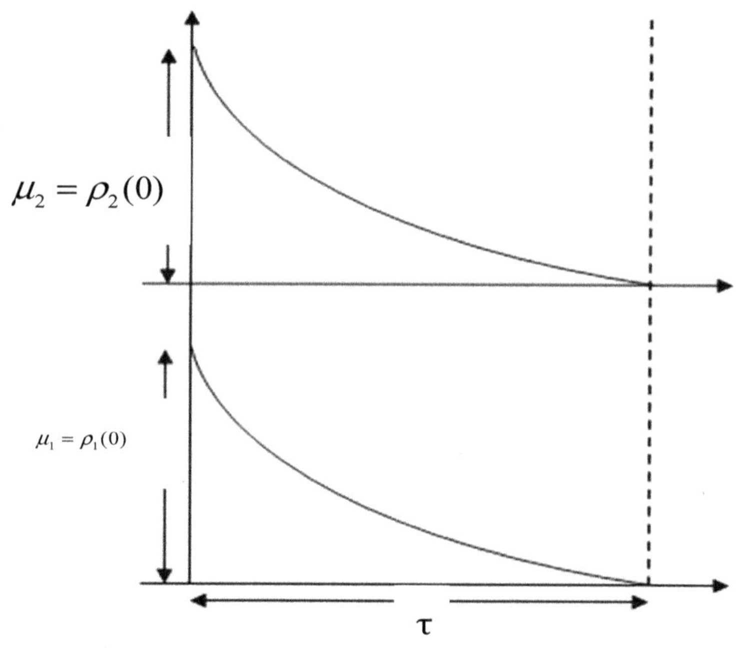

Figure 2. Inventory level of two deteriorating products (complementary or substitutable) without shortage under EOQ policy.

4.1. Non-Deteriorating Complementary Products

This section discusses the complementary products when products are of a non-deteriorating type. The demand [35,44] of two complementary products, product 1 and product 2, are

$$\delta_1(\gamma_1, \gamma_2) = \alpha - \beta\gamma_1 - \beta\varphi\gamma_2 \tag{1}$$

$$\delta_2(\gamma_1, \gamma_2) = \alpha - \beta\gamma_2 - \beta\varphi\gamma_1 \tag{2}$$

respectively. Here, all parameters (α, β, φ) are non-negative. The retailer is trying to obtain the optimal values of decision variables. Therefore, the average profit is

$$\text{Maximize}$$
$$\pi_C(\tau, \gamma_1, \gamma_2) = (\gamma_1 - \eta_1)\delta_1(\gamma_1, \gamma_2) + (\gamma_2 - \eta_2)\delta_2(\gamma_1, \gamma_2) - \left(\frac{G_1+G_2}{\tau} + \frac{[\lambda_1\delta_1(\gamma_1,\gamma_2)+\lambda_2\delta_2(\gamma_1,\gamma_2)]}{2}\tau\right) \tag{3}$$

Expression (1) gives the revenue from product 1. Expression (2) explains the revenue from product 2, and Expression (3) defines the summation of ordering costs and holding costs. The purpose is to maximize the profit, using the following theorem:

Theorem 1. *$\pi_C(\tau, \gamma_1, \gamma_2)$ is strictly concave if*

$$G_1 + G_2 > \frac{\tau\beta}{2}\left[\tau(\lambda_1(\gamma_1 + \gamma_2\varphi) + \lambda_2(\gamma_2 + \gamma_1\varphi)) - 4\varphi\gamma_1\gamma_2 - 2(\gamma_1^2 + \gamma_2^2)\right].$$

Proof. See Appendix A.

Substituting the values of $\delta_1(\gamma_1, \gamma_2)$ and $\delta_2(\gamma_1, \gamma_2)$ within the first order derivative of the Equation (3) with respect to the decision variable τ, one can obtain

$$\frac{\partial \pi_C(\tau, \gamma_1, \gamma_2)}{\partial \tau} = \frac{G_1 + G_2}{\tau^2} + \frac{\lambda_1(\beta\gamma_1 - \alpha + \beta\varphi\gamma_2) + \lambda_2(\beta\gamma_2 - \alpha + \beta\varphi\gamma_1)}{2} \tag{4}$$

The first order derivatives of Equation (3) with respect to the variables γ_1 and γ_2 are given by

$$\frac{\partial \pi_C(\tau, \gamma_1, \gamma_2)}{\partial \gamma_1} = \frac{\tau\beta(\lambda_1 + \varphi\lambda_2)}{2} + \alpha + \beta\eta_1 - 2\beta\gamma_1 + \beta\varphi\eta_2 - 2\beta\varphi\gamma_2 \tag{5}$$

$$\frac{\partial \pi_C(\tau, \gamma_1, \gamma_2)}{\partial \gamma_2} = \frac{\tau\beta(\lambda_2 + \varphi\lambda_1)}{2} + \alpha + \beta\eta_2 - 2\beta\gamma_2 + \beta\varphi\eta_1 - 2\beta\varphi\gamma_1 \tag{6}$$

The required optimum values of γ_1 and γ_2 are as given by the following expressions as:

$$\gamma_1^* = \frac{2\alpha + \beta(1 + \varphi)(\lambda_1\tau + 2\eta_1)}{4\beta(1 + \varphi)} \tag{7}$$

$$\gamma_2^* = \frac{2\alpha + \beta(1 + \varphi)(\lambda_2\tau + 2\eta_2)}{4\beta(1 + \varphi)} \tag{8}$$

Using Equations (7) and (8) on the right-hand side of Equation (4) and equating them to zero, one can obtain

$$\alpha_1\tau^3 + \alpha_2\tau^2 + \alpha_4 = 0 \tag{9}$$

where

$$\alpha_1 = \beta({\lambda_1}^2 + {\lambda_2}^2 + 2\varphi\lambda_1\lambda_2) \tag{10}$$

$$\alpha_2 = 2[\beta(\lambda_1\eta_1 + \lambda_2\eta_2) + \beta\varphi(\lambda_1\eta_2 + \lambda_2\eta_1) - \alpha(\lambda_1 + \lambda_2)] \tag{11}$$

$$\alpha_4 = 8(G_1 + G_2) \tag{12}$$

Equation (9) is a cubic polynomial and the discriminant is

$$\Delta = -4{\alpha_2}^3\alpha_4 - 27{\alpha_1}^2{\alpha_4}^2 \tag{13}$$

According to Δ, the discriminant has three types of values. $\Delta > 0$ gives three real roots of distinct quality, $\Delta = 0$ gives multiple real roots, and $\Delta < 0$ gives one real root and two complex roots. The following improved solution procedure has been developed to solve the problem numerically.

Solution algorithm

Step 1. Using Equations (10)–(12), calculate the coefficients of polynomial shown in Equation (9).
Step 2. All positive real roots of Equation (9) are found by using MATLAB software. Go to the Step (3).
Step 3. Determine the values (γ_1, γ_2) from Step (2).
Step 4. For all combinations of $(\tau, \gamma_1, \gamma_2)$, obtain the profit and check the concavity clause (Theorem 1). Select the maximum value as an optimal value of the profit. Then the related optimum decision variables are τ^*, $\gamma_1{}^*$ and $\gamma_2{}^*$.
Step 5. Determine the optimal values of the order quantities using $\mu_1^* = \delta_1(\gamma_1{}^*, \gamma_2{}^*)\tau^*$ and $\mu_2^* = \delta_2(\gamma_1{}^*, \gamma_2{}^*)\tau^*$.

4.2. Non-Deteriorating Substitutable Products

This section discusses about non-deteriorating substitutable products. The demand of two substitutable products, product 1 and product 2, is defined as follows:

$$\delta_1(\gamma_1, \gamma_2) = \alpha - \beta\gamma_1 + \beta\omega\gamma_2 \tag{14}$$

$$\delta_2(\gamma_1, \gamma_2) = \alpha - \beta\gamma_2 + \beta\omega\gamma_1 \tag{15}$$

where all parameters are non-negative. The average profit is

$$\pi_S(\tau, \gamma_1, \gamma_2) = (\gamma_1 - \eta_1)\delta_1(\gamma_1, \gamma_2) + (\gamma_2 - \eta_2)\delta_2(\gamma_1, \gamma_2) - \left(\frac{G_1 + G_2}{\tau} + \frac{[\lambda_1\delta_1(\gamma_1, \gamma_2) + \lambda_2\delta_2(\gamma_1, \gamma_2)]}{2}\tau\right) \tag{16}$$

The following theorem is used to maximize the above profit.

Theorem 2. *$\pi_S(\tau, \gamma_1, \gamma_2)$ is strictly concave*

$$G_1 + G_2 > \frac{\tau\beta}{2}\left[\tau\lambda_1(\gamma_1 - \gamma_2\omega) + \tau\lambda_2(\gamma_2 - \gamma_1\omega) + 4\omega\gamma_1\gamma_2 - 2(\gamma_1^2 + \gamma_2^2)\right].$$

Proof. See Appendix B.

Now, Equation (16) gives the following expression with respect to the variables τ, γ_1, and γ_2.

$$\frac{\partial\pi_S(\tau, \gamma_1, \gamma_2)}{\partial\tau} = \frac{G_1 + G_2}{\tau^2} + \frac{\lambda_1(\beta\gamma_1 - \alpha - \beta\omega\gamma_2) + \lambda_2(\beta\gamma_2 - \alpha - \beta\omega\gamma_1)}{2} \tag{17}$$

$$\frac{\partial\pi_S(\tau, \gamma_1, \gamma_2)}{\partial\gamma_1} = \frac{\tau\beta(\lambda_1 - \omega\lambda_2)}{2} + \alpha - 2\beta\gamma_1 + \beta\eta_1 + 2\beta\omega\gamma_2 - \beta\omega\eta_2 \tag{18}$$

$$\frac{\partial\pi_S(\tau, \gamma_1, \gamma_2)}{\partial\gamma_2} = \frac{\tau\beta(\lambda_2 - \omega\lambda_1)}{2} + \alpha - 2\beta\gamma_2 + \beta\eta_2 + 2\beta\omega\gamma_1 - \beta\omega\eta_1 \tag{19}$$

Equations (18) and (19) give the optimum values of γ_1 and γ_2 as

$$\gamma_1^* = \frac{2\alpha + \beta(1+\omega)(\lambda_1\tau + 2\eta_1)}{4\beta(1+\omega)}, \textit{ where } 0 \le \omega < 1 \tag{20}$$

$$\gamma_2^* = \frac{2\alpha + \beta(1+\omega)(\lambda_2\tau + 2\eta_2)}{4\beta(1+\omega)}, \textit{ where } 0 \le \omega < 1 \tag{21}$$

Using Equations (20) and (21) on the right-hand side of Equation (17) and equating them to zero, one can obtain

$$\alpha_1\tau^3 + \alpha_2\tau^2 + \alpha_4 = 0 \quad (22)$$

where

$$\alpha_1 = \beta({\lambda_1}^2 - 2\omega\lambda_1\lambda_2 + {\lambda_2}^2) \quad (23)$$

$$\alpha_2 = 2[-\alpha(\lambda_1 + \lambda_2) + \beta(\lambda_1\eta_1 + \lambda_2\eta_2) - \beta\omega(\lambda_1\eta_2 + \lambda_2\eta_1)] \quad (24)$$

$$\alpha_4 = 8(G_1 + G_2) \quad (25)$$

According to the discussion about the discriminant of cubic polynomial above, the following solution procedure is suggested for the substitutable products case.

Solution algorithm

Step 1. Using Equations (23)–(25), calculate the coefficients of the polynomial in Equation (22).
Step 2. After finding all possible roots of Equation (22) with the help of MATLAB software, go to the Step (3).
Step 3. From Step 2, determine the values of (γ_1, γ_2).
Step 4. For all values of $(\tau, \gamma_1, \gamma_2)$, determine the total profit and check the concavity clause (Theorem 2). Then select the maximum value as the optimum one. The related decision variables associated with the maximum profit are $\tau^*, {\gamma_1}^*$, and ${\gamma_2}^*$.
Step 5. Determine optimal values of order quantities using $\mu_1^* = \delta_1({\gamma_1}^*, {\gamma_2}^*)\tau^*$ and $\mu_2^* = \delta_2({\gamma_1}^*, {\gamma_2}^*)\tau^*$.

4.3. Deteriorating Complementary Products

In this case, it is assumed that the retailer is doing the business with two deteriorating complementary products. For both products, the deterioration rates are constant. The differential equation $\frac{d\rho(t)}{dt} = -D_r\rho(t) - \delta(\gamma)$ shows that the inventory level changes over time, where $\rho(\tau) = 0$. Solving this differential equation yields $\rho(t) = \frac{\delta(\gamma)}{D_r}(e^{D_r(\tau-t)} - 1)$, and according to Figure 2, the order quantity of each product is equal to $\mu = \rho(0) = \frac{\delta(\gamma)}{D_r}\left(e^{D_r\tau} - 1\right)$. The holding cost is given by

$$\lambda\int_0^\tau \rho(t) = \lambda\int_0^\tau \frac{\delta(\gamma)}{D_r}(e^{D_r(\tau-t)} - 1)dt = \frac{\lambda\delta(\gamma)}{{D_r}^2}(e^{D_r\tau} - D_r\tau - 1) \quad (26)$$

Moreover, the deterioration cost is

$$\eta'\left(\underbrace{\rho(0) - \delta(\gamma)\tau}_{\text{Deterioration Quantity}}\right) = \eta'\left(\frac{\delta(\gamma)}{D_r}(e^{D_r\tau} - 1) - \delta(\gamma)\tau\right) \quad (27)$$

The total cost after utilizing the approximation of the Taylor series expansion, $e^{D_r\tau} = 1 + D_r\tau + \frac{(D_r\tau)^2}{2}$, is

$$TC = \frac{1}{\tau}\left[G + \overbrace{\frac{\lambda\delta(\gamma)}{{D_r}^2}(e^{D_r\tau} - D_r\tau - 1)}^{\text{Holding Cost}} + \underbrace{\frac{\eta'\delta(\gamma)}{D_r}(e^{D_r\tau} - 1) - \eta'\delta(\gamma)\tau}_{\text{Deterioration Cost}}\right]$$

$$TC = \frac{1}{\tau}\left[G + \frac{\lambda\delta(\gamma)}{2}\tau^2 + \frac{\eta'\delta(\gamma)D_r\tau^2}{2}\right] \overset{e^{D_r\tau}=1+D_r\tau+\frac{(D_r\tau)^2}{2}}{\longrightarrow} = \frac{G}{\tau} + \frac{(\lambda+\eta' D_r)\delta(\gamma)}{2}\tau^2 \quad (28)$$

The demand of two deteriorating complementary products, product 1 and product 2, is as follows (as defined in Section 4.1):

$$\delta_1(\gamma_1,\gamma_2) = \alpha - \beta\gamma_1 + \beta\varphi\gamma_2 \quad (29)$$

$$\delta_2(\gamma_1,\gamma_2) = \alpha - \beta\gamma_2 + \beta\varphi\gamma_1 \quad (30)$$

Finally, the average profit is

$$\begin{aligned}\pi_C(\tau,\gamma_1,\gamma_2) = \ & (\gamma_1 - \eta_1)\delta_1(\gamma_1,\gamma_2) + (\gamma_2 - \eta_2)\delta_2(\gamma_1,\gamma_2) \\ & -[\tfrac{G_1+G_2}{\tau} + \tfrac{\tau(\lambda_1+\eta_1' D_r)\delta_1(\gamma_1,\gamma_2)}{2} + \tfrac{\tau(\lambda_2+\eta_2' D_r)\delta_2(\gamma_1,\gamma_2)}{2}]\end{aligned} \quad (31)$$

To optimize the profit, the following theorem is used.

Theorem 3. *$\pi_C(\tau,\gamma_1,\gamma_2)$ is strictly concave if*

$$G_1 + G_2 > \tfrac{\tau\beta}{2}[-4\varphi\gamma_1\gamma_2 - 2(\gamma_1^2+\gamma_2^2) + \tau\gamma_1(\lambda_1+\varphi\lambda_2) + \tau\gamma_2(\lambda_2+\varphi\lambda_1) + \tau\eta_1' D_r(\gamma_1+\varphi\gamma_2) + \tau\eta_2' D_r(\gamma_2+\varphi\gamma_1)]$$

Proof. See Appendix C.

Here, the first order derivative of Equation (31) with respect to T yields

$$\frac{\partial\pi_c(\tau,\gamma_1,\gamma_2)}{\partial\tau} = \frac{G_1+G_2}{\tau^2} + \frac{(\lambda_1+\eta_1' D_r)(\beta\gamma_1 - \alpha + \beta\varphi\gamma_2) + (\lambda_2+\eta_2' D_r)(\beta\gamma_2 - \alpha + \beta\varphi\gamma_1)}{2} \quad (32)$$

The first order derivatives of the Equation (31) for variables p_1 and p_2 are given by

$$\frac{\partial\pi_C(\tau,\gamma_1,\gamma_2)}{\partial\gamma_1} = \frac{\tau\beta((\lambda_1+\eta_1' D_r) + \varphi(\lambda_2+\eta_2' D_r))}{2} + \alpha - 2\beta(\gamma_1+\varphi\gamma_2) + \beta(\eta_1+\varphi\eta_2) \quad (33)$$

$$\frac{\partial\pi_C(\tau,\gamma_1,\gamma_2)}{\partial\gamma_2} = \frac{\tau\beta((\lambda_2+\eta_2' D_r) + \varphi(\lambda_1+\eta_1' D_r))}{2} + \alpha - 2\beta(\gamma_2+\varphi\gamma_1) + \beta(\eta_2+\varphi\eta_1) \quad (34)$$

Optimum values of the decision variables p_1 and p_2 are

$$\gamma_1^* = \frac{2\alpha + \beta(1+\varphi)(2\eta_1 + \tau\lambda_1 + \tau\eta_1' D_r)}{4\beta(1+\varphi)} \quad (35)$$

$$\gamma_2^* = \frac{2\alpha + \beta(1+\varphi)(2\eta_2 + \tau\lambda_2 + \tau\eta_2' D_r)}{4\beta(1+\varphi)} \quad (36)$$

Using Equations (35) and (36) in Equation (32) and equating them to zero for optimality, one can obtain

$$\alpha_1\tau^3 + \alpha_2\tau^2 + \alpha_4 = 0 \quad (37)$$

where

$$\alpha_1 = \beta\,[(\gamma_1^2+\gamma_2^2) + D_r^2(\eta_1'^2+\eta_2'^2) + 2\eta_1' D_r(\gamma_1+\varphi\lambda_2) + 2\eta_2' D_r(\lambda_2+\varphi\lambda_1) + 2\varphi(\lambda_1\lambda_2 + \eta_1'\eta_2' {D_r}^2)] \quad (38)$$

$$\alpha_2 = -2\,[(\gamma_1+\eta_1' D_r)(\alpha - \beta(\eta_1+\varphi\eta_2)) + (\gamma_2+\eta_2' D_r)(\alpha - \beta(\eta_2+\varphi\eta_1))] \quad (39)$$

$$\alpha_4 = 8(G_1+G_2) \quad (40)$$

According to the discussion about the discriminant of cubic polynomial provided above (Section 4.1.), the following improved solution procedure is suggested for finding optimum solutions.

Solution algorithm

Step 1. Using Equations (38)–(40), calculate coefficients of polynomial shown in the Equation (37).
Step 2. Equation (37) gives all roots by using MATLAB software and then go to the Step (3).
Step 3. Determine all values of (γ_1, γ_2) for period from the Step (2).
Step 4. For all possible combinations of $(\tau, \gamma_1, \gamma_2)$, determine the total profit and check the concavity clause (Theorem 3). Select the maximum value as the optimal value of the profit. Then optimal values of the decision variables are $\tau^*, \gamma_1{}^*$, and $\gamma_2{}^*$.

4.4. Deteriorating Substitutable Products

This case discusses deteriorating substitutable products. The demand of two deteriorating substitutable products, product 1 and product 2, is

$$\delta_1(\gamma_1, \gamma_2) = \alpha - \beta\gamma_1 + \beta\omega\gamma_2 \tag{41}$$

$$\delta_2(\gamma_1, \gamma_2) = \alpha - \beta\gamma_2 + \beta\omega\gamma_1 \tag{42}$$

where all parameters are non-negative. The average profit function is calculated as

$$\begin{aligned}\pi_S(\tau, \gamma_1, \gamma_2) = \; & (\gamma_1 - \eta_1)\delta_1(\gamma_1, \gamma_2) + (\gamma_2 - \eta_2)\delta_2(\gamma_1, \gamma_2) \\ & -[\tfrac{G_1+G_2}{\tau} + \tfrac{\tau(\lambda_1+\eta_1' D_r)\delta_1(\gamma_1,\gamma_2)}{2} + \tfrac{\tau(\lambda_2+\eta_2' D_r)\delta_2(\gamma_1,\gamma_2)}{2}]\end{aligned} \tag{43}$$

With the help of the following theorem, this profit can be optimized.

Theorem 4. *$\pi_S(T, p_1, p_2)$ is strictly concave if*

$$G_1 + G_2 > \tfrac{\tau\beta}{2}[4\omega\gamma_1\gamma_2 - 2(\gamma_1^2 + \gamma_2^2) + \tau\gamma_1(\lambda_1 - \omega\lambda_2) + \tau\gamma_2(\lambda_2 - \omega\lambda_1) + \tau\eta_1' D_r(\gamma_1 - \omega\gamma_2) + \tau\eta_2' D_r(\gamma_2 - \omega\gamma_1)]$$

holds.

Proof. See Appendix D.

From the necessary condition, Equation (43) gives the following expressions

$$\frac{\partial\pi_s(\tau, \gamma_1, \gamma_2)}{\partial\tau} = \frac{G_1 + G_2}{\tau^2} + \frac{\big(\lambda_1 + \eta_1' D_r\big)(\beta\gamma_1 - \alpha - \beta\omega\gamma_2) + \big(\lambda_2 + \eta_2' D_r\big)(\beta\gamma_2 - \alpha - \beta\omega\gamma_1)}{2} \tag{44}$$

$$\frac{\partial\pi_s(\tau, \gamma_1, \gamma_2)}{\partial\gamma_1} = \frac{\tau\beta((\lambda_1 + \eta_1' D_r) - \omega(\lambda_2 + \eta_2' D_r))}{2} + \alpha - 2\beta(\gamma_1 - \omega\gamma_2) + \beta(\eta_1 - \omega\eta_2) \tag{45}$$

$$\frac{\partial\pi_s(\tau, \gamma_1, \gamma_2)}{\partial\gamma_2} = \frac{\tau\beta((\lambda_2 + \eta_2' D_r) - \omega(\lambda_1 + \eta_1' D_r))}{2} + \alpha - 2\beta(\gamma_2 - \omega\gamma_1) + \beta(\eta_2 - \omega\eta_1) \tag{46}$$

Setting Equations (45) and (46) equal to zero, p_1 and p_2 are given as follows:

$$\gamma_1^* = \frac{2\alpha + \beta(1-\omega)(2\eta_1 + \tau\lambda_1 + \tau\eta_1' D_r)}{4\beta(1-\omega)}, \textit{ where } 0 \le d < 1 \tag{47}$$

$$\gamma_2^* = \frac{2\alpha + \beta(1-\omega)(2\eta_2 + \tau\lambda_2 + \tau\eta_2' D_r)}{4\beta(1-\omega)}, \textit{ where } 0 \le d < 1 \tag{48}$$

Using Equations (47) and (48) on the right-hand side of the Equation (44) and equating them to zero, one can obtain

$$\alpha_1\tau^3 + \alpha_2\tau^2 + \alpha_4 = 0 \quad (49)$$

where

$$\alpha_1 = \beta\,[(\gamma_1^2+\gamma_2^2) + D_r^2(\eta_1'^2+\eta_2'^2) + 2\eta_1' D_r(\gamma_1-\omega\lambda_2) + 2\eta_2' D_r(\lambda_2-\omega\lambda_1) - 2\omega(\lambda_1\lambda_2+\eta_1'\eta_2' D_r{}^2)] \quad (50)$$

$$\alpha_2 = -2\,[(\gamma_1+\eta_1' D_r)(\alpha-\beta(\eta_1-\omega\eta_2)) + (\gamma_2+\eta_2' D_r)(\alpha-\beta(\eta_2-\omega\eta_1))] \quad (51)$$

$$\alpha_4 = 8(G_1+G_2) \quad (52)$$

According to the discussion about discriminant of cubic polynomial provided above (Section 4.1), the following solution procedure can help to find the numerical solution for substitutable products.

Solution algorithm

Step 1. Calculate the coefficients of polynomial from the Equation (49) by using Equations (50) to (52).
Step 2. Equation (49) gives the roots of the equation using MATLAB software. Move to the Step 3.
Step 3. Find (γ_1, γ_2) using Step 2.
Step 4. For $(\tau, \gamma_1, \gamma_2)$, obtain the profit and check the concavity clause (Theorem4). Choose the maximum value as the optimal value of the profit. Then related decision variables for the maximum profit are given by $\tau^*, \gamma_1{}^*$, and $\gamma_2{}^*$.
Step 5. Determine optimal values of order quantities using $\mu_1^* = \delta_1(\gamma_1{}^*, \gamma_2{}^*)\tau^*$ and $\mu_2^* = \delta_2(\gamma_1{}^*, \gamma_2{}^*)\tau^*$.

5. Numerical Examples and Sensitivity Analysis

Four numerical examples are given to justify the mathematical model. All examples explain effects of changes of the degree of complementarity or substitutability.

5.1. Example 1: Non-Deteriorating Complementary Products

The values of all parameters are $G_1 = 120$, $G_2 = 100$, $\lambda_1 = 6$,, $\lambda_2 = 3$, $\alpha = 100$, $\beta = 0.4$, $\eta_1 = 20$, and $\eta_2 = 10$. By using Steps (1) to (5) of the proposed solution algorithm in Section 4.1, optimal values of τ, γ_1, γ_2, μ_1, and μ_2 are obtained. These values with some variation of φ are depicted in the Table 2. '*' demonstrates optimum results. Table 2 reveals the increasing value of φ gives decreasing values of $\gamma_1^*, \gamma_2^*, \mu_1^*, \mu_2^*$, and $\pi_C(\tau^*, \gamma_1^*, \gamma_2^*)$, but gives an increasing value of τ^*. More precisely, these changes on γ_1^* and γ_2^* are shown in Figure 3. Bold font indicates the optimum results.

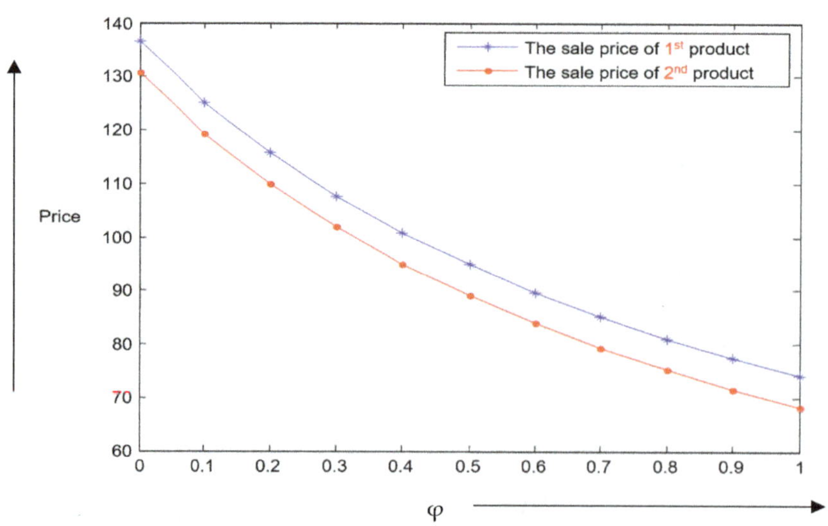

Figure 3. Changes of the complementarity degree upon selling prices.

Table 2. Optimum results for some variations of φ.

φ	Concavity	τ	γ_1	γ_2	μ_1	μ_2	Π_C
	True	93.3221	274.9832	199.9916	−932.5924	1866.8	1247.6
0	True	**1.0290 ***	**136.5438 ***	**130.7719 ***	**46.7087 ***	**49.0849 ***	**10621 ***
	False	−1.0180	-	-	-	-	-
	True	85.9137	252.5069	183.0716	−715.2833	1432.3	944.4092
0.1	True	**1.0327 ***	**125.1854 ***	**119.4109 ***	**46.6255 ***	**48.7723 ***	**8548.6 ***
	False	−1.0229	-	-	-	-	-
	True	79.5269	233.4570	168.8118	−547.7619	1097.4	715.6247
0.2	True	**1.0362 ***	**115.7210 ***	**109.9438 ***	**46.4588 ***	**48.1415 ***	**7752.7 ***
	False	−1.0229	-	-	-	-	-
	True	73.9641	217.1000	156.6269	-416.8036	835.5901	539.8298
0.3	True	**1.0398 ***	**107.7135 ***	**101.9337 ***	**46.4588 ***	**48.1415 ***	**7752.7 ***
	False	−1.0253	-	-	-	-	-
	True	69.0754	202.8988	146.0923	−313.2069	628.5371	402.6596
0.4	True	**1.0433 ***	**100.8607 ***	**95.0682 ***	**46.3753 ***	**47.5031 ***	**6481.3 ***
	False	−1.0278	-	-	-	-	-
	True	64.7452	190.4512	136.8923	−230.4240	463.1133	294.2216
0.5	True	**1.0470 ***	**94.9038 ***	**89.1186 ***	**46.2917 ***	**47.5031 ***	**6481.3 ***
	False	−1.0303	-	-	-	-	-
	True	60.8831	179.4496	128.7873	−163.7025	329.8139	207.5357
0.6	True	**1.0606 ***	**89.7010 ***	**83.9130 ***	**46.2079 ***	**47.1809 ***	**5965.9 ***
	False	−1.0328	-	-	-	-	-
	True	57.4169	169.6547	121.5921	−109.5329	221.6201	137.5717
0.7	True	**1.0644 ***	**85.1110 ***	**79.3202 ***	**46.0401 ***	**46.5305 ***	**5108.5 ***
	False	−1.0353	-	-	-	-	-
	True	54.2887	160.8775	115.1610	−65.2832	133.2681	80.6367
0.8	True	**1.0581 ***	**81.0316 ***	**75.2380 ***	**46.0401 ***	**46.5305 ***	**5108.5 ***
	False	−1.0379	-	-	-	-	-
	True	51.4514	152.9666	109.3780	−28.9524	60.7553	33.9747
0.9	True	**1.0619 ***	**77.3824 ***	**71.5859 ***	**45.9560 ***	**46.2022 ***	**4748.4 ***
	False	−1.0404	-	-	-	-	-
	True	48.8661	145.7992	104.1496	1.0005	1.0005	−4.5010
1.0	True	**1.0658 ***	**74.0987 ***	**68.2993 ***	**45.8717 ***	**45.8717 ***	**4424.9 ***
	False	−1.0430	-	-	-	-	-

"-" means infeasible result; "*" means optimum solution.

5.2. *Example 2: Non-Deteriorating Substitutable Products*

Values of parameters are $G_1 = 150$, $G_2 = 155$, $\lambda_1 = 4.5$, $\lambda_2 = 4$, $\alpha = 100$, $\beta = 0.3$, $\eta_1 = 15$, and $\eta_2 = 13$. Steps (1) to (5) of the solution algorithm proposed in the Section 4.2 help to obtain optimum values of $\tau, \gamma_1, \gamma_2, \mu_1$, and μ_2 for different values of d. Table 3 reveals that when ω increases, τ^*, γ_1^*, and γ_2^* decrease, but μ_1^*, μ_2^*, and $\pi_S(\tau^*, \gamma_1^*, \gamma_2^*)$ increase. The changes on γ_1^* and γ_2^* are shown in Figure 4. Figure 5 reveals that the profit decreases whenever the degree of complementarity increases and vice-versa. Therefore, the retailer who intends to work on complementary products should select two products with a smaller degree of complementary to make more profit. But if he/she intends to work on substitutable products, he/she should select two products with a larger degree of substitutability.

Table 3. Optimum results for some variations of ω.

ω	Concavity	τ	γ_1	γ_2	μ_1	μ_2	Π_C
	True	149.7187	342.6002	322.8854	−416.2286	469.2757	56.4733
0	True	**1.2292** *	**175.5495**	**174.3959** *	**58.1837** *	**58.6091** *	**14,799** *
	False	−1.2170	-	-	-	-	-
	True		380.5094	358.6400	−566.5791	638.3149	77.4294
0.1	True		**194.0644** *	**192.9112** *	**58.3172** *	**58.7838** *	**16,647** *
	False	−1.2170	-	-	-	-	-
	True	188.4626	427.8537	403.2959	−783.6958	882.4670	107.5095
0.2	True	**1.2228** *	**217.2090** *	**216.0561** *	**58.4503** *	**58.9578** *	**18,957** *
	False	−1.2149	-	-	-	-	-
	True	216.0547	488.6568	460.6500	−1110.2	1249.7	152.5353
0.3	True		**246.9673** *	**245.8149** *	**58.5830** *	**59.1312** *	**21,929** *
	False	−1.2107	-	-	-	-	-
	True	252.7386	569.6088	537.0164	−1627.8	1831.9	223.6789
0.4	True	**1.2165** *	**286.6464** *	**285.4943** *	**58.7153** *	**59.3039** *	**25,894** *
	False	−1.2107	-	-	-	-	-
	True	303.8947	682.7149	643.7281	−2508.7	2822.8	344.5513
0.5	True	**1.2134** *	**342.1984** *	**341.0467** *	**58.8472** *	**59.4761** *	**31,445** *
	False	−1.2086	-	-	-	-	-
	True	380.1918	851.8825	803.3585	−4167.0	4688.3	572.1289
0.6	True	**1.2103** *	**425.5283** *	**424.3770** *	**58.9787** *	**59.6475** *	**39,774** *
	False	−1.2065	-	-	-	-	-
	True	506.2033	1132.5	1068.3	−7808.6	8784.9	1073.6
0.7	True	**1.2073** *	**564.4137** *	**563.2628** *	**59.1098** *	**59.8184** *	**53,659** *
	False	−1.2044	-	-	-	-	-
	True	753.9936	1689.1	1593.8	−18,250	20,531	2525.4
0.8	True	**1.2043** *	**842.1881** *	**841.0376** *	**59.2405** *	**59.9887** *	**81,433** *
	False	−1.2013	-	-	-	-	-
	True	1465.2	3322.5	3138.4	−72,373	81,420	10,300
0.9	True	**1.2013** *	**1675.5** *	**1674.4** *	**59.3708** *	**60.1583** *	**164,760** *
	False	-	-	-	-	-	-
	Undefined	22659	∞	∞	∞	∞	Undefined
1.0	Undefined	1.1983	$-\infty$	∞	∞	$-\infty$	Undefined
	Undefined	−1.1982	$-\infty$	∞	$-\infty$	∞	Undefined

"-" means infeasible result; "*" means optimum solution.

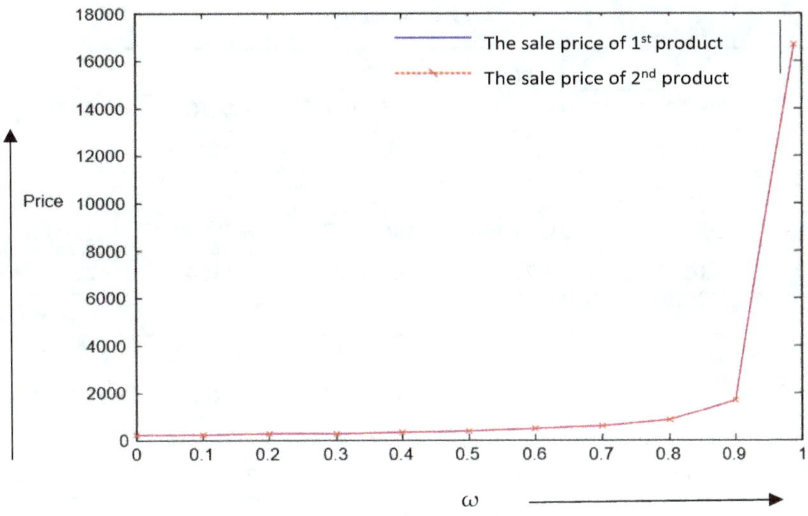

Figure 4. Changes in the degree of substitutability upon selling prices.

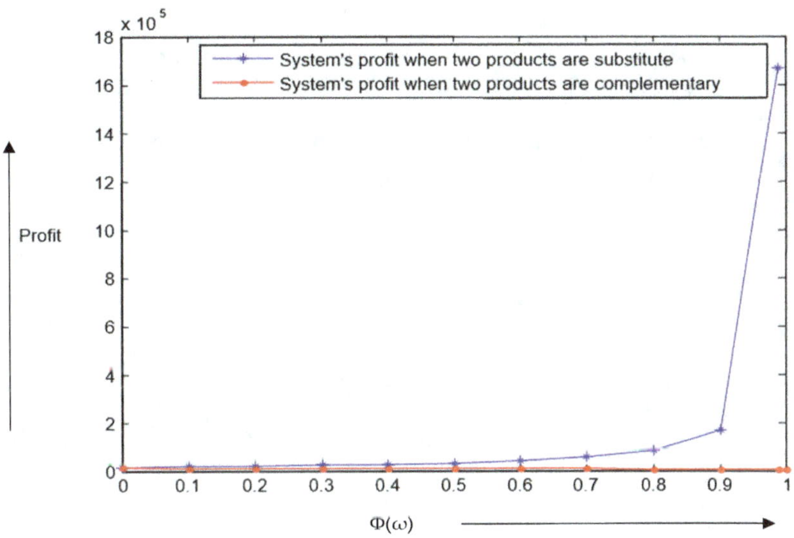

Figure 5. Effects of the changes of complementarity and substitutability degrees on profits.

5.3. Example 3: Deteriorating Complementary Products

The values of parameters for this example are $G_1 = 120$, $G_2 = 100$, $\lambda_1 = 6$, $\lambda_2 = 3$, $\alpha = 100$, $\beta = 0.4$, $\eta_1 = 20$, $\eta_2 = 10$, $\eta_1' = 10$, $\eta_2' = 5$ and $D_r = 0.01$. Using Steps (1) to (5) of the proposed solution algorithm in Section 4.3, the optimal values of $\tau, \gamma_1, \gamma_2, \mu_1$, and μ_2 are obtained. Table 4 shows that $\gamma_1^*, \gamma_2^*, \mu_1^*, \mu_2^*$, and $\pi_C(\tau^*, \gamma_1^*, \gamma_2^*)$ decrease when φ increases, but at that time, τ^* increases (similar to Section 5.1). The changes on γ_1^* and γ_2^* are shown in Figure 6 precisely.

Table 4. Optimum results for some variations of φ (deterioration case).

φ	Concavity	τ	γ_1	γ_2	μ_1	μ_2	Π_C
	Truc	91.7921	274.9829	199.9914	−1503.0	3008.7	1247.6
0	True	**1.0208 ***	**136.5567 ***	**130.7784 ***	**46.5584 ***	**48.9299 ***	**10,618 ***
	False	−1.0096	-	-	-	-	-
	True	84.5050	252.5065	183.0715	−1105.7	2214.1	944.3665
0.1	True	**1.0242 ***	**125.1983 ***	**119.4174 ***	**46.4762 ***	**48.6187 ***	**9486.7 ***
	False	−1.0120	-	-	-	-	-

Table 4. *Cont.*

φ	Concavity	τ	γ_1	γ_2	μ_1	μ_2	Π_C
	True	78.2230	233.4567	168.8117	−817.1057	1637.0	715.5786
0.2	True	**1.0277 ***	**115.7340 ***	**109.9503 ***	**46.3938 ***	**48.3057 ***	**8545.1 ***
	False	−1.0144	-	-	-	-	-
	True	72.7513	217.0996	156.6267	−602.9090	1208.7	539.7803
0.3	True	**1.0313 ***	**107.7265 ***	**101.9402 ***	**46.3114 ***	**47.9908 ***	**7749.2 ***
	False	−1.0168	-	-	-	-	-
	True	67.9427	202.8984	146.0921	−441.0527	885.1454	402.6065
0.4	True	**1.0348 ***	**100.8638 ***	**95.0748 ***	**46.2288 ***	**47.6741 ***	**7067.7 ***
	False	−1.0193	-	-	-	-	-
	True	63.6836	190.4508	136.8921	−316.9006	636.9689	294.1649
0.5	True	**1.0384 ***	**94.9169 ***	**89.1251 ***	**46.1462 ***	**47.3553 ***	**6477.9 ***
	False	−1.0218	-	-	-	-	-
	True	59.8847	179.4492	128.7871	−220.4678	444.2343	207.4755
0.6	True	**1.0421 ***	**89.7142 ***	**83.9196 ***	**46.0634 ***	**47.0346 ***	**5962.4 ***
	False	−1.0243	-	-	-	-	-
	True	56.4753	169.6543	121.5918	−144.7758	292.9846	137.5079
0.7	True	**1.0458 ***	**85.1242 ***	**79.3268 ***	**45.9805 ***	**46.7119 ***	**5508.2 ***
	False	−1.0268	-	-	-	-	-
	True	53.3984	160.8770	115.1607	−84.8435	173.2576	80.5692
0.8	True	**1.0495 ***	**81.0449 ***	**75.2447 ***	**45.8975 ***	**46.3871 ***	**5105.1 ***
	False	−1.0293	-	-	-	-	-
	True	50.6076	152.9660	109.3778	−37.0492	77.8092	33.9034
0.9	True	**1.0533 ***	**77.3957 ***	**71.5926 ***	**45.8144 ***	**46.0602 ***	**4745.0 ***
	False	−1.0318	-	-	-	-	-
	True	48.0647	145.7986	104.1493	1.2845	1.2845	−4.5761
1.0	True	**1.0571 ***	**74.1121 ***	**68.3060 ***	**45.7312 ***	**45.7312 ***	**4421.4 ***
	False	−1.0344	-	-	-	-	-

"-" means infeasible result; "*" means optimum solution.

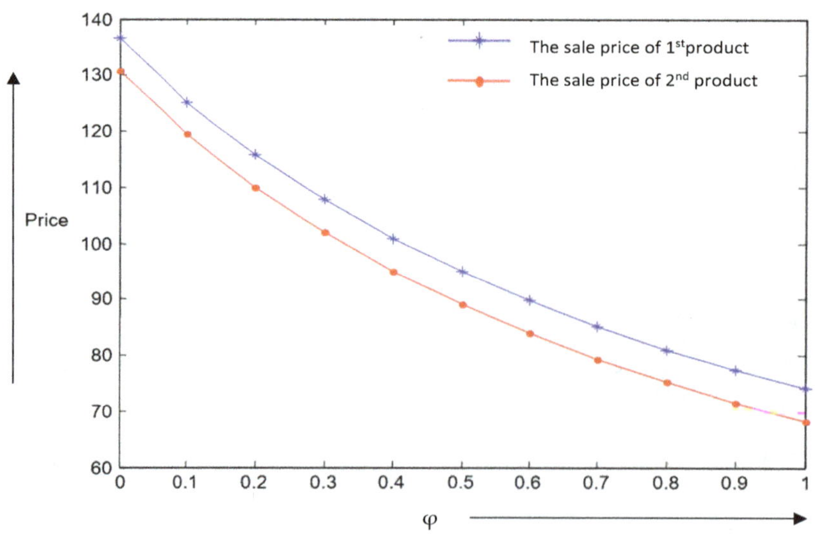

Figure 6. Effects of changes of complementarity degree on selling prices of deteriorating products.

5.4. *Example 4: Deteriorating Substitutable Products*

Fixed parametric values are $G_1 = 150$, $G_2 = 155$, $\lambda_1 = 4.5$, $\lambda_2 = 4$, $\alpha = 100$, $\beta = 0.3$, $\eta_1 = 15$, $\eta_2 = 13$, $\eta_1' = 7$, $\eta_2' = 6$, and $D_r = 0.01$. For different values of ω, one can use Steps (1) to (5) of proposed solution algorithm in Section 4.4. From the Table 5, it can be concluded that τ^*, γ_1^*, and γ_2^* decrease with increased values of ω, but μ_1^*, μ_2^*, and $\pi_S(\tau^*, \gamma_1^*, \gamma_2^*)$ increase (similar Section 5.2). The changes on γ_1^* and γ_2^* are presented in the Figure 7. Figure 8 describes that the profit decreases when the complementarity degree increases, but profit increases while the degree of substitutability increases. As a result, the retailer who intends to work on complementary products should select two products with a smaller degree of complementary in order to make more profit. If he/she intends to work on substitutable products, he/she should select two products with a larger degree of substitutability.

Table 5. Optimum results for some variations of ω (deterioration case).

ω	Concavity	τ	γ_1	γ_2	μ_1	μ_2	Π_S
	True	147.4584	342.6379	322.8369	−940.472	1060.9	56.9563
0	True	**1.2199 ***	**175.5604 ***	**174.4049 ***	**58.0953 ***	**58.5208 ***	**14,795 ***
	False	−1.2099	-	-	-	-	-
	True	164.4331	380.5500	358.5848	−1423.5	1604.6	78.1010
0.1	True	**1.2167 ***	**194.0753 ***	**192.9202 ***	**58.2279 ***	**58.6945 ***	**16,643 ***
	False	−1.2078	-	-	-	-	-
	True	185.6144	427.8978	403.2320	−2254.3	2539.8	108.4521
0.2	True	**1.2136 ***	**217.2198 ***	**216.0651 ***	**58.3600 ***	**58.8676 ***	**18,954 ***
	False	−1.2057	-	-	-	-	-
	True	212.7872	488.7046	460.5742	−3816.6	4298.5	153.8840
0.3	True	**1.2104 ***	**246.9782 ***	**245.8238 ***	**56.4918 ***	**59.0400 ***	**21,926 ***
	False	−1.2036	-	-	-	-	-
	True	248.9126	569.6604	536.9240	−7146.8	8047.2	225.6693
0.4	True	**1.2073 ***	**286.6572 ***	**285.5032 ***	**58.6231 ***	**59.2118 ***	**25,890 ***
	False	−1.2015	-	-	-	-	-
	True	299.2879	682.7698	643.6106	−15,703	17,678	347.6315
0.5	True	**1.2043 ***	**342.2092 ***	**341.0556 ***	**58.7540 ***	**59.3829 ***	**31441 ***
	False	−1.1994	-	-	-	-	-
	True	374.4167	851.9377	803.1996	−45,423	51134	577.2581
0.6	True	**1.2012 ***	**425.5390 ***	**424.3859 ***	**58.8845 ***	**59.5534 ***	**39,770 ***
	False	−1.1974	-	-	-	-	-
0.7	True	498.4871	1132.6	1068.0	−224,880	253,130	1083.2

"-" means infeasible result; "*" means optimum solution.

Managerial insights

Numerical studies justify the theoretical implications of this research in each case separately. The degree of complementarity of a product has an inverse impact on the selling price, i.e., the increasing complementarity gives less profit. This implies that if one product is more dependent or complement to the other product, the selling of the one product is dependent upon the other product. In the case that one of the complementary products is not available at the right time, the chances to sale of the other product are smaller. To handle this situation, industry managers need to take more care of the inventory and safety stock of the complementary products.

For the case of substitutable products, the increasing rate of the degree of substitutability has a direct impact on the profit. This implies that if customers can get any similar substitutable product, the customer is satisfied with the service. The absence of the original product does not have any impact on the profit. This study suggests to industry managers that they should introduce similar types of alternative products to the shops or their centers such that they can survive facing losses from original products. It helps to keep the brand image of the industry as well as the retailer. The scenario is the same for the case of deteriorated substitutable and complementary products. Therefore, the industry needs to create their strategies carefully for complementary products rather than substitutable products.

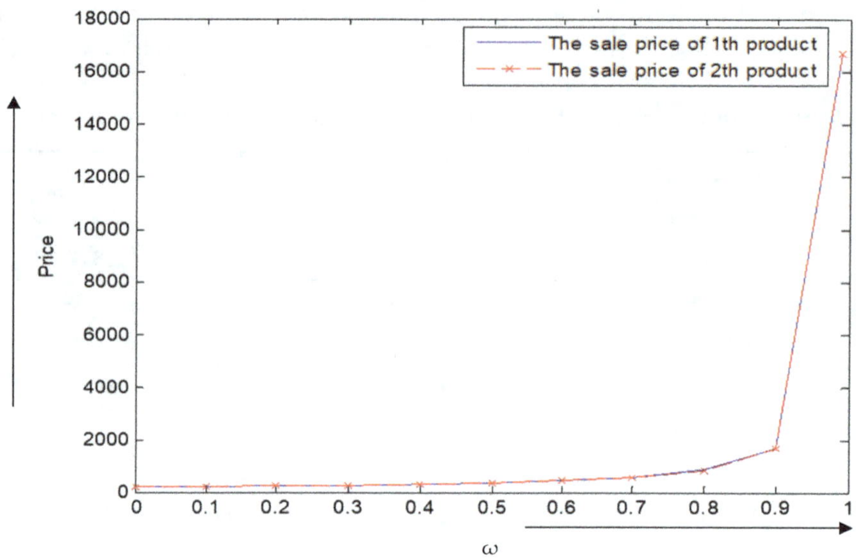

Figure 7. Changes in the degree of substitutability upon selling prices of deteriorating products.

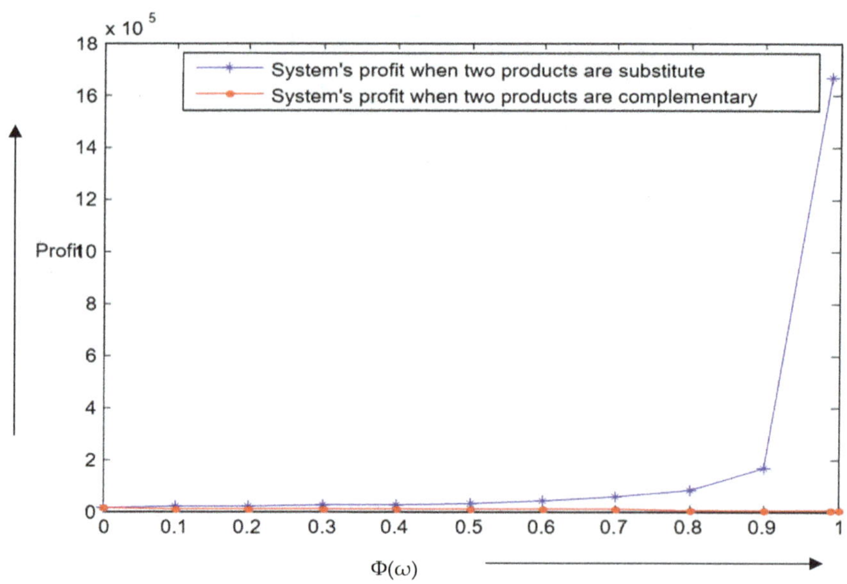

Figure 8. Changes of the complementarily and substitutability degrees upon profits of deteriorating products.

6. Conclusions

Four pricing models for two types of products which may be complementary or substitutable and non-deteriorating or deteriorating were studied analytically as well as numerically. For all cases, it was observed that selling prices and lot sizes of both products as well as the average profit were decreased, and the period length was increased while the degree of complementarity was increased and vice versa. The selling prices and order quantities of both products as well as average profit were increased, and the cycle length was decreased with increased values of the degree of substitutability. This study concluded that sell of complementary products should be with some smaller degree of complementarity and substitutable products with bigger degree of substitutability in order to make more profit. The present model can help a manager to fix optimal selling prices, ordered quantities, and cycle length for different scenarios so that the joint profit is maximized. In the existing literature, some research articles are focused on substitutable and complementary products but none of them have investigated both inventory and pricing decisions of products simultaneously. Therefore, the present article is quite new compared to existing literature. Lead time is considered as zero in this study, but in reality it is not possible. This study can be extended with random lead time and backorder. The business can be spread through multiple number of retailers. Extra service facilities can be included within the study [45–48] for deteriorating products and non-zero random lead time. The quality improvement [49,50] and preservation technology [28,29] for deteriorating products can improve the study in a realistic way.

Author Contributions: Methodology, data curation, writing—original draft preparation, M.S.B.; formal analysis, conceptualization, visualization S.S.S.; investigation, S.S.S. and B.S.; resources, validation, supervision, project administration, A.A.T.; writing—review and editing, funding acquisition, B.S.

Nomenclature

Index	Description
m	1, 2
Decision variables	
γ_m	product m's selling price ($/unit)
μ_m	order quantity of product m (units)
τ	time interval between two replenishments
Parameters	
$\delta_1(\gamma_1, \gamma_2)$	market demand for product 1 (units)
$\delta_2(\gamma_1, \gamma_2)$	demand for product 2 (units)
ρ_m	inventory level at time τ of product m
$\propto$	basic potential demand (units)
β	self-price sensitivity
λ_m	holding cost of product m ($/unit/unit time)
η_m	purchasing cost of product m ($/unit)
η'_m	deterioration cost of product m ($/unit)
D_r	deterioration rate of product 1 and product 2
φ	degree of complementarity of two products, $0 \leq \varphi \leq 1$
ω	degree of substitutability of two products, $0 \leq \omega \leq 1$

Appendix A

Proof of Theorem 1. $\pi_C(\tau, \gamma_1, \gamma_2)$ is strictly concave, if $X.H.X^T = [\ \tau \quad \gamma_1 \quad \gamma_2\ ] \times H \times [\ \tau \quad \gamma_1 \quad \gamma_2\ ]^T < 0$ where

$$H = \begin{bmatrix} \frac{\partial^2 \pi_C}{\partial \tau^2} & \frac{\partial^2 \pi_C}{\partial \tau \partial \gamma_1} & \frac{\partial^2 \pi_C}{\partial \tau \partial \gamma_1} \\ \frac{\partial^2 \pi_C}{\partial \gamma_1 \partial \tau} & \frac{\partial^2 \pi_C}{\partial \gamma_1^2} & \frac{\partial^2 \pi_C}{\partial \gamma_1 \partial \gamma_2} \\ \frac{\partial^2 \pi_C}{\partial \gamma_2 \partial \tau} & \frac{\partial^2 \pi_C}{\partial \gamma_2 \partial \gamma_1} & \frac{\partial^2 \pi_C}{\partial \gamma_2^2} \end{bmatrix} = \begin{bmatrix} -\frac{2(G_1+G_2)}{\tau^3} & \frac{\beta(\lambda_1+\lambda_2\varphi)}{2} & \frac{\beta(\lambda_1\varphi+\lambda_2)}{2} \\ \frac{\beta(\lambda_1+\lambda_2\varphi)}{2} & -2\beta & -2\beta\varphi \\ \frac{\beta(\lambda_1\varphi+\lambda_2)}{2} & -2\beta\varphi & -2\beta \end{bmatrix} \tag{A1}$$

Now,

$$X \times H \times X^T = -\frac{2(G_1+G_2)}{\tau} + \tau\beta\lambda_1(\gamma_1+\gamma_2\varphi) + \tau\beta\lambda_2(\gamma_2+\gamma_1\varphi) - 4\beta\varphi\gamma_1\gamma_2 - 2\beta(\gamma_1^2+\gamma_2^2) \quad \text{(A2)}$$

It has to show that the expression in Equation (A2) is negative. Therefore, the profit function is strictly concave if

$$G_1+G_2 > \frac{\tau\beta}{2}\left[\tau(\lambda_1(\gamma_1+\gamma_2\varphi)+\lambda_2(\gamma_2+\gamma_1\varphi)) - 4\varphi\gamma_1\gamma_2 - 2(\gamma_1^2+\gamma_2^2)\right]$$

holds; hence the proof.

Appendix B

Proof of Theorem 2. $\pi_S(\tau,\gamma_1,\gamma_2)$ is a concave whenever $X.H.X^T = \begin{bmatrix} \tau & \gamma_1 & \gamma_2 \end{bmatrix} \times H \times \begin{bmatrix} \tau & \gamma_1 & \gamma_2 \end{bmatrix}^T < 0$ where

$$H = \begin{bmatrix} \frac{\partial^2\pi_S}{\partial\tau^2} & \frac{\partial^2\pi_S}{\partial\tau\partial\gamma_1} & \frac{\partial^2\pi_S}{\partial\tau\partial\gamma_1} \\ \frac{\partial^2\pi_S}{\partial\gamma_1\partial\tau} & \frac{\partial^2\pi_S}{\partial\gamma_1^2} & \frac{\partial^2\pi_S}{\partial\gamma_1\partial\gamma_2} \\ \frac{\partial^2\pi_S}{\partial\gamma_2\partial\tau} & \frac{\partial^2\pi_S}{\partial\gamma_2\partial\gamma_1} & \frac{\partial^2\pi_S}{\partial\gamma_2^2} \end{bmatrix} = \begin{bmatrix} -\frac{2(G_1+G_2)}{\tau^3} & \frac{\beta(\lambda_1-\lambda_2\omega)}{2} & \frac{\beta(-\lambda_1\omega+\lambda_2)}{2} \\ \frac{\beta(\lambda_1-\lambda_2\omega)}{2} & -2\beta & 2\beta\omega \\ \frac{\beta(-\lambda_1\omega+\lambda_2)}{2} & 2\beta\omega & -2\beta \end{bmatrix} \quad \text{(A3)}$$

Thus, one can have

$$X \times H \times X^T = -\frac{2(G_1+G_2)}{\tau} + \tau\beta\lambda_1(\gamma_1-\gamma_2\omega) + \tau\beta\lambda_2(\gamma_2-\gamma_1\omega) + 4\beta\omega\gamma_1\gamma_2 - 2\beta(\gamma_1^2+\gamma_2^2) \quad \text{(A4)}$$

Then, it has to show that $X \times H \times X^T < 0$.
Hence, the profit function $\pi_S(\tau,\gamma_1,\gamma_2)$ is concave if

$$G_1+G_2 > \frac{\tau\beta}{2}\left[\tau(\lambda_1(\gamma_1-\gamma_2\omega)+\lambda_2(\gamma_2-\gamma_1\omega)) + 4\omega\gamma_1\gamma_2 - 2(\gamma_1^2+\gamma_2^2)\right]$$

holds. The proof is completed here.

Appendix C

Proof of Theorem 3. $\pi_C(\tau,\gamma_1,\gamma_2)$ is strictly concave if $X.H.X^T = \begin{bmatrix} \tau & \gamma_1 & \gamma_2 \end{bmatrix} \times H \times \begin{bmatrix} \tau & \gamma_1 & \gamma_2 \end{bmatrix}^T < 0$ where

$$H = \begin{bmatrix} \frac{\partial^2\pi_C}{\partial\tau^2} & \frac{\partial^2\pi_C}{\partial\tau\partial\gamma_1} & \frac{\partial^2\pi_C}{\partial\tau\partial\gamma_1} \\ \frac{\partial^2\pi_C}{\partial\gamma_1\partial\tau} & \frac{\partial^2\pi_C}{\partial\gamma_1^2} & \frac{\partial^2\pi_C}{\partial\gamma_1\partial\gamma_2} \\ \frac{\partial^2\pi_C}{\partial\gamma_2\partial\tau} & \frac{\partial^2\pi_C}{\partial\gamma_2\partial\gamma_1} & \frac{\partial^2\pi_C}{\partial\gamma_2^2} \end{bmatrix}$$

$$= \begin{bmatrix} -\frac{2(G_1+G_2)}{\tau^3} & \frac{\beta(\gamma_1+\eta_1'D_r)}{2} + \frac{\beta\varphi(\gamma_2+\eta_2'D_r)}{2} & \frac{\beta(\gamma_2+\eta_2'D_r)}{2} + \frac{\beta\varphi(\gamma_1+\eta_1'D_r)}{2} \\ \frac{\beta(\gamma_1+\eta_1'D_r)}{2} + \frac{\beta\varphi(\gamma_2+\eta_2'D_r)}{2} & -2\beta & -2\beta\varphi \\ \frac{\beta(\gamma_2+\eta_2'D_r)}{2} + \frac{\beta\varphi(\gamma_1+\eta_1'D_r)}{2} & -2\beta\varphi & -2\beta \end{bmatrix} \quad \text{(A5)}$$

Now,

$$\begin{aligned} X \times H \times X^T = & -\tfrac{2(G_1+G_2)}{\tau} - 4\beta\varphi\gamma_1\gamma_2 - 2\beta(\gamma_1^2+\gamma_2^2) + \tau\beta\gamma_1(h_1+\varphi h_2) + \tau\beta\gamma_2(h_2+\varphi h_1) \\ & +\tau\beta\eta_1'D_r(\gamma_1+\varphi\gamma_2) + \tau\beta\eta_2'D_r(\gamma_2+\varphi\gamma_1) \end{aligned} \quad \text{(A6)}$$

Now, it is needed to show that $X \times H \times X^T < 0$.
Therefore, the profit function $\pi_C(\tau,\gamma_1,\gamma_2)$ is concave if

$$G_1+G_2 > \frac{\tau\beta}{2}[-4\varphi\gamma_1\gamma_2 - 2(\gamma_1^2+\gamma_2^2) + \tau\gamma_1(\lambda_1+\varphi\lambda_2) + \tau\gamma_2(\lambda_2+\varphi\lambda_1) + \tau\eta_1'D_r(\gamma_1+\varphi\gamma_2) + \tau\eta_2'D_r(\gamma_2+\varphi\gamma_1)]$$

holds. The proof is completed here.

Appendix D

Proof of Theorem 4. $\pi_S(\tau,\gamma_1,\gamma_2)$ is a strictly concave function if $X.H.X^T = [\ \tau \quad \gamma_1 \quad \gamma_2\ ] \times H \times [\ \tau \quad \gamma_1 \quad \gamma_2\ ]^T < 0$ holds. where

$$H = \begin{bmatrix} \frac{\partial^2 \pi_C}{\partial \tau^2} & \frac{\partial^2 \pi_C}{\partial \tau \partial \gamma_1} & \frac{\partial^2 \pi_C}{\partial \tau \partial \gamma_1} \\ \frac{\partial^2 \pi_C}{\partial \gamma_1 \partial \tau} & \frac{\partial^2 \pi_C}{\partial \gamma_1^2} & \frac{\partial^2 \pi_C}{\partial \gamma_1 \partial \gamma_2} \\ \frac{\partial^2 \pi_C}{\partial \gamma_2 \partial \tau} & \frac{\partial^2 \pi_C}{\partial \gamma_2 \partial \gamma_1} & \frac{\partial^2 \pi_C}{\partial \gamma_2^2} \end{bmatrix} = \begin{bmatrix} -\frac{2(G_1+G_2)}{\tau^3} & \frac{\beta(\lambda_1+\eta_1' D_r)}{2} - \frac{\beta\omega(\lambda_2+\eta_2' D_r)}{2} & \frac{\beta(\lambda_2+\eta_2' D_r)}{2} - \frac{\beta\omega(\lambda_1+\eta_1' D_r)}{2} \\ \frac{\beta(\lambda_1+\eta_1' D_r)}{2} - \frac{\beta\omega(\lambda_2+\eta_2' D_r)}{2} & -2\beta & 2\beta\omega \\ \frac{\beta(\lambda_2+\eta_2' D_r)}{2} - \frac{\beta\omega(\lambda_1+\eta_1' D_r)}{2} & 2\beta\omega & -2\beta \end{bmatrix} \tag{A7}$$

Thus,

$$\begin{aligned} X \times H \times X^T = & -\frac{2(G_1+G_2)}{\tau} + 4\beta\omega\gamma_1\gamma_2 - 2\beta(\gamma_1^2+\gamma_2^2) + \tau\beta\gamma_1(\lambda_1-\omega\lambda_2) + \tau\beta\gamma_2(\lambda_2-\omega\lambda_1) \\ & +\tau\beta\eta_1' D_r(\gamma_1-\omega\gamma_2) + \tau\beta\eta_2' D_r(\gamma_2-\omega\gamma_1) \end{aligned} \tag{A8}$$

Then, it has to show that $X \times H \times X^T < 0$.
Then, the profit function $\pi_S(\tau,\gamma_1,\gamma_2)$ is concave if

$$G_1+G_2 > \frac{\tau\beta}{2}[4\omega\gamma_1\gamma_2 - 2(\gamma_1^2+\gamma_2^2) + \tau\gamma_1(\lambda_1-\omega\lambda_2) + \tau\gamma_2(\lambda_2-\omega\lambda_1) + \tau\eta_1' D_r(\gamma_1-\omega\gamma_2) + \tau\eta_2' D_r(\gamma_2-\omega\gamma_1)]$$

holds. The proof is completed here.

References

1. Shocker, A.D.; Bayus, B.L.; Kim, N. Product complments and substitutes in the real world: The relevence of "other products". *J. Market.* **2004**, *68*, 28–40. [CrossRef]
2. Sarkar, M.; Lee, Y.H. Optimum pricing strategy for complementary products with stochastic reservation price in a supply chain model. *J. Ind. Manag. Optim.* **2017**, *13*, 1553–1586.
3. Avgeropoulos, S.; Sammut-Bonnici, T.; McGee, J. Complementary products. *Encycl. Manag.* **2015**, *12*. [CrossRef]
4. Sana, S.S. Price-sensitive demand for perishable items—An EOQ model. *Appl. Math. Comput.* **2011**, *217*, 6248–6259.
5. Sarkar, B.; Saren, S.; Wee, H.M. An inventory model with variable demand, component cost and selling price for deteriorating items. *Econ. Model.* **2013**, *30*, 306–310. [CrossRef]
6. Salvietti, L.; Smith, N.R.; Cárdenas-Barrón, L.E. A stochastic profit-maximising economic lot scheduling problem with price optimisation. *Eur. J. Oper. Res.* **2014**, *8*, 193–221. [CrossRef]
7. Smith, N.R.; Robles, J.L.; Cárdenas-Barrón, L.E. Optimal pricing and production master planning in a multiperiod horizon considering capacity and inventory constraints. *Math. Probl. Eng.* **2009**, *932676*, 15. [CrossRef]
8. Soon, W. A review of multi-product pricing models. *Appl. Math. Comput.* **2011**, *217*, 8149–8165. [CrossRef]
9. Wei, J.; Zhao, J.; Li, Y. Pricing decisions for complementary products with firms' different market powers. *Eur. J. Oper. Res.* **2013**, *224*, 507–519. [CrossRef]
10. Dey, B.; Sarkar, B.; Sarkar, M.; Pareek, S. An integrated inventory model involving discrete setup cost reduction, variable safety factor, selling-price dependent demand, and investment. *RAIRO Oper. Res.* **2019**, *53*, 39–57. [CrossRef]
11. Mcgillivray, R.; Silver, E. Some concepts for inventory control under substitutable demand. *Inf. Syst. Oper. Res.* **1978**, *16*, 47–63. [CrossRef]
12. Zhang, R.Q.; Kaku, I.; Xiao, Y.Y. Deterministic EOQ with partial backordering and correlated demand caused by cross-selling. *Eur. J. Oper. Res.* **2011**, *210*, 537–551. [CrossRef]
13. Zhang, R.Q. An extension of partial backordering EOQ with correlated demand caused by cross-selling considering multiple minor items. *Eur. J. Oper. Res.* **2012**, *220*, 876–881. [CrossRef]

14. Yue, X.; Mukhopadhyay, S.K.; Zhu, X. A Bertrand model of pricing of complementary goods under information asymmetry. *J. Bus. Res.* **2006**, *59*, 1182–1192. [CrossRef]
15. Liu, L.; Yuan, X.M. Coordinated replenishments in inventory systems with correlated demands. *Eur. J. Oper. Res.* **2000**, *123*, 490–503. [CrossRef]
16. Guchhait, R.; Sarkar, M.; Sarkar, B.; Pareek, S. Single-vendor multi-buyer game theoretic model under multi-factor dependent demand. *Int. J. Inventory Res.* **2017**, *4*, 303–332. [CrossRef]
17. Shavandi, H.; Mahlooji, H.; Nosratian, N.E. A constrained multi-product pricing and inventory control problem. *Appl. Soft Comput.* **2012**, *12*, 2454–2461. [CrossRef]
18. Abad, P.L. Optimal pricing and lot-sizing under conditions of perishability, finite production and partial backordering and lost sale. *Eur. J. Oper. Res.* **2003**, *144*, 677–685. [CrossRef]
19. Balkhi, Z.T.; Benkherouf, L. On an inventory model for deteriorating items with stock dependent and time-varying demand rates. *Comput. Oper. Res.* **2004**, *31*, 223–240. [CrossRef]
20. Anjos, M.F.; Cheng, R.C.; Currie, C.S. Optimal pricing policies for perishable products. *Eur. J. Oper. Res.* **2005**, *166*, 246–254. [CrossRef]
21. Dye, C.Y. Joint pricing and ordering policy for a deteriorating inventory with partial backlogging. *Omega* **2007**, *35*, 184–189. [CrossRef]
22. Panda, S.; Saha, S.; Basu, M. An EOQ model for perishable products with discounted selling price and stock dependent demand. *Cent. Eur. J. Oper. Res.* **2009**, *17*, 31. [CrossRef]
23. Pang, Z. Optimal dynamic pricing and inventory control with stock deterioration and partial backordering. *Oper. Res. Lett.* **2011**, *39*, 375–379. [CrossRef]
24. Sarkar, B.; Majumder, A.; Sarkar, M.; Dey, B.K.; Roy, G. Two-echelon supply chain model with manufacturing quality improvement and setup cost reduction. *J. Ind. Manag. Optim.* **2017**, *13*, 1085–1104. [CrossRef]
25. Sarkar, B.; Saren, S. Partial trade-credit policy of retailer with exponentially deteriorating items. *Int. J. Appl. Comput. Math.* **2015**, *1*, 343–368. [CrossRef]
26. Sett, B.K.; Sarkar, B.; Sarkar, B.; Yun, W.Y. Optimal replenishment policy with variable deterioration for fixed lifetime products. *Sci. Iran.* **2016**, *23*, 2318–2329.
27. Sarkar, B.; Sett, B.K.; Roy, G.; Goswami, A. Flexible setup cost and deterioration of products in a supply chain model. *Int. J. Appl. Comput. Math.* **2016**, *2*, 25–40. [CrossRef]
28. Sarkar, B.; Mandal, B.; Sarkar, S. Preservation of deteriorating seasonal products with stock-dependent consumption rate and shortages. *J. Ind. Manag. Optim.* **2017**, *13*, 187–206. [CrossRef]
29. Ullah, M.; Sarkar, B.; Asghar, I. Effects of preservation technology investment on waste generation in a two-echelon supply chain model. *Mathematics* **2019**, *7*, 20. [CrossRef]
30. Iqbal, M.W.; Sarkar, B. Recycling of lifetime dependent deteriorated products through different supply chains. *RAIRO-Operat. Res.* **2019**, *53*, 129–156. [CrossRef]
31. Dey, B.K.; Sarkar, B.; Pareek, S. A two-echelon supply chain management with setup time and cost reduction, quality improvement and variable production rate. *Mathematics* **2019**, *7*, 25. [CrossRef]
32. Gürler, Ü.; Yılmaz, A. Inventory and coordination issues with two substitutable products. *Appl. Math. Model.* **2010**, *34*, 539–551. [CrossRef]
33. Kim, S.W.; Bell, P.C. Optimal pricing and production decisions in the presence of symmetrical and asymmetrical substitution. *Omega* **2011**, *39*, 528–538. [CrossRef]
34. Stavrulaki, E. Inventory decisions for substitutable products with stock-dependent demand. *Int. J. Prod. Econ.* **2011**, *129*, 65–78. [CrossRef]
35. Karakul, M.; Chan, L.M.A. Analytical and managerial implications of integrating product substitutability in the joint pricing and procurement problem. *Eur. J. Oper. Res.* **2008**, *190*, 179–204. [CrossRef]
36. Karakul, M.; Chan, L.M.A. Joint pricing and procurement of substitutable products with random demands–a technical note. *Eur. J. Oper. Res.* **2010**, *201*, 324–328. [CrossRef]
37. Birge, J.R.; Drogosz, J.; Duenyas, I. Setting single-period optimal capacity levels and prices for substitutable products. *Int. J. Flex. Manuf. Syst.* **1998**, *10*, 407–430. [CrossRef]
38. Netessine, S.; Dobson, G.; Shumsky, R.A. Flexible service capacity: Optimal investment and the impact of demand correlation. *Oper. Res.* **2002**, *50*, 375–388. [CrossRef]
39. Parlar, M.; Goyal, S. Optimal ordering decisions for two substitutable products with stochastic demands. *Opsearch* **1984**, *21*, 1–15.

40. Pasternack, B.A.; Drezner, Z. Optimal inventory policies for substitutable commodities with stochastic demand. *Nav. Res. Logist.* **1991**, *38*, 221–240. [CrossRef]
41. Tang, C.S.; Yin, R. Joint ordering and pricing strategies for managing substitutable products. *Prod. Oper. Manag.* **2007**, *16*, 138–153. [CrossRef]
42. Xia, Y. Competitive strategies and market segmentation for suppliers with substitutable products. *Eur. J. Oper. Res.* **2011**, *210*, 194–203. [CrossRef]
43. Netessine, S.; Rudi, N. Centralized and competitive inventory models with demand substitution. *Oper. Res.* **2003**, *51*, 329–335. [CrossRef]
44. Gupta, S.; Loulou, R.J.M.S. Process innovation, product differentiation, and channel structure: Strategic incentives in a duopoly. *Market. Sci.* **1998**, *17*, 301–316. [CrossRef]
45. Sarkar, M.; Guchhait, R.; Sarkar, B. Modelling for service solution of a closed-loop supply chain with the presence of third party logistics. In *Advances in Production Management Systems. Production Management for Data-Driven, Intelligent, Collaborative, and Sustainable Manufacturing*; APMS 2018; IFIP Advances in Information and Communication Technology; Springer: Cham, Switzerland, 2018; Volume 535, pp. 320–327.
46. Khanna, A.; Kishore, A.; Sarkar, B.; Jaggi, C. Supply chain with customer-based two-level credit policies under an imperfect quality environment. *Mathematics* **2018**, *6*, 35. [CrossRef]
47. Guchhait, R.; Pareek, S.; Sarkar, B. Application of distribution-free approach in integrated and dual-channel supply chain under buyback contract. In *Handbook of Research on Promoting Business Process Improvement Through Inventory Control Techniques*; IGI Global: Hershey, PA, USA, 2018; pp. 388–426.
48. Guchhait, R.; Dey, B.K.; Bhuniya, S.; Mandal, B.; Bachar, R.; Chauduri, K.S.; Wee, H.M.; Sarkar, B. Investment for process quality improvement and setup cost reduction in an imperfect production process with warranty policy and shortages. *RAIRO-Oper. Res.* **2019**. [CrossRef]
49. Shin, D.; Guchhait, R.; Sarkar, B.; Mittal, M. Controllable lead time, service level constraint, and transportation discounts in a continuous review inventory model. *RAIRO-Oper. Res.* **2016**, *50*, 921–934. [CrossRef]
50. Majumder, A.; Guchhait, R.; Sarkar, B. Manufacturing quality improvement and setup cost reduction in a vendor-buyer supply chain model. *Eur. J. Ind. Eng.* **2017**, *11*, 588–612. [CrossRef]

4

Interactive Fuzzy Multi Criteria Decision Making Approach for Supplier Selection and Order Allocation in a Resilient Supply Chain

Sonia Irshad Mari [1], Muhammad Saad Memon [1,*], Muhammad Babar Ramzan [2,*], Sheheryar Mohsin Qureshi [3] and Muhammad Waqas Iqbal [4]

1 Department of Industrial Engineering and Management, Mehran University of Engineering and Technology, Jamshoro, Sindh 76062, Pakistan; sonia.irshad@faculty.muet.edu.pk
2 Department of Garment Manufacturing, National Textile University, Faisalabad 37610, Pakistan
3 Department of Industrial and Manufacturing Engineering, NED University of Engineering and Technology, Karachi 75270, Pakistan; sheheryar@neduet.edu.pk
4 Department of Industrial Engineering, Hongik University, Seoul 04066, Korea; waqastextilion@gmail.com
* Correspondence: saad.memon@faculty.muet.edu.pk (M.S.M.); babar_ramzan@yahoo.com (M.B.R.);

Abstract: Modern supply chains are vulnerable to high impact, low probability disruption risks. A supply chain usually operates in such a network of entities where the resilience of one supplier is critical to overall supply chain resilience. Therefore, resilient planning is a key strategic requirement in supplier selection decisions for a competitive supply chain. The aim of this research is to develop quantitative resilient criteria for supplier selection and order allocation in a fuzzy environment. To serve the purpose, a possibilistic fuzzy multi-objective approach was proposed and an interactive fuzzy optimization solution methodology was developed. Using the proposed approach, organizations can tradeoff between cost and resilience in supply networks. The approach is illustrated using a supply chain case from a garments manufacturing company.

Keywords: resilient supply chain; supplier selection; fuzzy optimization; disruption risks

---

1. Introduction

Outsourcing is a competitive strategy in the global supply chain. Evaluating and selecting the best set of suppliers is a challenging decision in outsourcing and it plays a significant role in supply chain performance [1,2]. Traditionally the supplier selection and order allocation decision is made based on cost and quality criteria. However, modern supply chains are more prone to unexpected High Impact Low Probability (HILP) and Low Impact High Probability (LIHP) disruption events [3]. HILP disruption events are commonly known as random disruptions risks such as man-made and natural disasters, whereas LIHP disruptions are targeted disruptions such as day-to-day operational risks. Tang and Tomlin [4] proposed six disruption sources in the supply chain and among them, supplier performance is most frequent. The role of the supplier selection decision in these supply chain risk has only been partially explored in the literature [5]. Multi-sourcing strategies and are now common to many supply chains in order to minimize the supplier's disruption risks [6]. For example, during a fire in a plant of Philips Electronics in 2010 disrupt two of its major customers: Ericsson and Nokia. Ericsson lost about a month of production and suffered $200 million while Nokia recovered due to its multi-sourcing ability [7]. Toyota Motor Corp. lost billions of dollars in 2010 during product recall due to its part sourcing from one supplier for many car models. These examples show that multi-sourcing strategies work well. On the contrary, multi-sourcing strategies failed during some

HILP disruption events. For example, Japan earthquake disrupted many semiconductor supply chains. Chinese Firm ZTE Corp. faced shortages of batteries and LCD screens due to all of its suppliers in the affected region. Ford Motor Co. and General Motor Co. faced shortages of auto parts and stop production due to the shutdown of two Hitachi Ltd.'s plants. These historical events suggest that the supplier selection criteria should be extended to new resilience capabilities [8]. Thus it crucial to provide a reliable level of resilience to the supply side to protect such shortages especially during HILP events [9].

Several studies have been conducted to consider resilience in the supply chain [8,10,11]. The concept of resilience in specific to the supplier selection problem has also been discussed by several authors [12–25]. Most of these studies focused on multiple sourcing and operational performance of suppliers. However, to the best of authors knowledge, this is the first study which focuses on supply network by considers supply density, resilience score of supplier's locations, and transit time in addition to other operational criteria. This paper aims to develop a supplier selection and order allocation model to build a resilient supply chain in response to HILP disruptions. To do so, a possibilistic multi-objective fuzzy optimization-based model with a new resilience objective is proposed which consists of supply density, resilience score, and transit time. Furthermore, the proposed model is solved using Tiwari, et al. [26] weighted additive approach and Werners [27] fuzzy and operator methods. Fuzzy based multi-objective approaches are widely used in supplier selection problem to deal with uncertain information [28,29]. This research answers the following questions: (i) Which supplier is selected based on the importance given to each objective? (ii) How much to purchase from each selected supplier?

The remainder of the paper is organized as follows. Section 2 provides related literature. Section 3 includes problem description and mathematical model for supplier selection and order allocation with a new resilience objective. Section 4 comprises of the proposed possibilistic fuzzy based solution methodology. Section 5 presents a numerical example to show the application of proposed supplier selection and order allocation model. Sections 6 and 7 discussed the results of the proposed mathematical model and solution methodology. Finally, Section 8 presents some conclusions and future directions drawn from the study.

2. Literature Review

The word resilience first coined by Holling [30] in the context of ecology. According to Holling [30], the resilience is the ability of a system to absorb changes in state variables, driving variable and parameters, and still persist. Due to the increase in complexity and uncertainty in the business environment, several studies shown interest in the concept of resilience in a managerial perspective. Hamel and Valikangas [31] defined resilience as a capacity for continuous reconstruction. Sheffi [32] defined resilience in terms of enterprise resilience as the ability of an organization to successfully confront the unforeseen. Sutcliffe and Vogus [33] stated that resilience is the (1) ability to absorb strain and improve the functionality of organization despite the presence of difficulty or (2) ability to bounce back after disturbances. More recently, Woods [34] defined resilience in simple terms as system's ability to bounce back after disruptions and to bounce forward through learning from those disruption events and increase the system's adaptive capacity for handling uncertain events.

The concept of supply chain resilience gained prominent importance during recent years in supply chain risk management research [35]. Supply chain resilience is a relatively new concept to mitigate risks that can be defined as the ability to reduce the probability of a disruption, to reduce the impact of disruption, and to reduce the recovery time to normal performance [36]. Supply chain resilience has been defined by several authors in simpler and broader terms. Most of these studies described supply chain resilience as the ability to withstand disruptions and converge to the original state or to a new desirable state. Despite the increasing number of publication in supply chain resilience area, most of the researches provided qualitative insights and there is a limited number of quantitative modelling techniques available [37]. These qualitative models used different performance measures for designing resilient supply chains. Priya Datta, et al. [38] developed the framework to improve operational resilience. Falasca, Zobel and Cook [36] proposed three determinants (density, complexity, and node criticality) of supply chain resilience for supply chain design. Azevedo, et al. [39] proposed GResilient index to assess supply chain resilience using the Delphi technique. Miller-Hooks, et al. [40] proposed a transportation network resilience model using stochastic programming. They presented the expected fraction of demand fulfilment after disruption as resilience metric.

Supply chain usually functions in the system of parties where the resilience of one party (e.g. a supplier) is critical for overall supply chain resilience. As discuss earlier, many major disruptions break down supply networks and it takes a long time to recover. Whereas, the probability of disruption may be reduced by developing a resilient network and it takes considerably less recovery time [41]. Suppliers constitute the most important role in the performance of the supply chain, therefore, the resilience of the supply network is expected to contribute and increase overall supply chain resilience [42]. Despite its importance, there is very limited research conducted which consider resilience of supply network. The literature on resilient supplier selection is summarized in Table 1. Haldar, Ray, Banerjee and Ghosh [20] developed a resilient supplier selection model using AHP-QFD method to rate the suppliers. They used five resilience criteria (density, complexity, node criticality, responsiveness and re-engineering) for the supplier selection process. More recently, Haldar, Ray, Banerjee and Ghosh [21] proposed a fuzzy group decision-making approach for resilient supplier selection using triangular and trapezoidal fuzzy numbers. They considered investment, responsiveness, and capacity of holding inventory as resilient criteria. Sawik [22] proposed resilience strategies for supplier selection and order allocation problem under disruption risks. Recently, Sawik [23] proposed a stochastic mixed integer model for supplier selection and customer order scheduling using disruption risks for single and dual sourcing. Yilmaz-Börekçi, İşeri Say and Rofcanin [24] proposed a scale for measuring supplier resilience within supply networks. Mari, et al. [43] proposed a resilient and sustainable supply chain network model. They used expected disruption cost as a resilient metric. Torabi, Baghersad and Mansouri [9] developed a bi-objective stochastic model to trade-off the resilience level of the supply network and system cost. Memon, Lee and Mari [28] proposed supplier selection model using grey and uncertainty theories. Rajesh and Ravi [25] proposed a resilient supplier selection method using grey relational analysis. They consider supplier's performance, responsiveness, risk reduction, technical capabilities and sustainability as selection criteria. Sahu, Datta and Mahapatra [14] proposed fuzzy-VIKOR based supplier selection framework by combining the general selection strategy and resilience strategy. They considered investment capacity, responsiveness, and inventory capacity as resilience criteria. Hosseini and Al Khaled [13] developed a resilience score of suppliers based on eight criteria namely backup supplier contracting, surplus inventory, location separation, robustness, reliability, reorganization, rerouting, and restoration. They proposed a hybrid ensemble and AHP approach to select a suitable supplier. Pramanik, et al. [44] developed a resilient supplier selection model using AHP-TOPSIS and Sen, et al. [45] proposed g-resilient supplier selection model using dominance-based fuzzy decision making. Wang, Zhang, Chong and Wang [18] developed a resilient supplier selection model for construction supply chain by integrating building information modelling and geographical information system. Parkouhi and Ghadikolaei [12] proposed FANP and grey-VIKOR based resilient supplier selection model. They considered benefits, opportunities, costs, and risks as resilience criteria. López and Ishizaka [15] proposed a coupled method of FCM and AHP for supplier selection for a resilient supply chain. Recently, Jabbarzadeh, Fahimnia and Sabouhi [17] proposed stochastic bi-objective optimization model considering fuzzy c-means to developed sustainable supply chain network that performs resiliently. Sabouhi, Pishvaee and Jabalameli [16] presented integrated stochastic and fuzzy DEA based model for supplier selection considering resilience. They considered multi-sourcing, supplier fortification, and emergency inventory as resilience criteria. More recently, Parkouhi, et al. [46] proposed a resilient supplier selection framework using grey-DEMATEL approach.

Table 1. The classification of literature in resilient supplier selection.

Authors	Resilience Criteria	Types of Risks		Solution Methodology				
		Network Risks	Operational Risks	Possibilistic Programming	Stochastic Programming	Fuzzy Programming	Grey Programming	Other
Wang, Herty and Zhao [6]	Multi-supplier, production capacity, product quality, production cost	-	✓	-	-	-	-	Fluid-dynamic models
Torabi, Baghersad and Mansouri [9]	Multiple sourcing, fortifying supplier, pre-positioned inventories, backup supplier, and supplier's business continuity	-	✓	✓	✓	-	-	-
Parkouhi and Ghadikolaei [12]	Benefits, Opportunities, costs, and risks	-	✓	-	-	✓	✓	-
Hosseini and Al Khaled [13]	Absorptive capacity, adaptive capacity, and restorative capacity	✓	✓	-	-	-	-	Predictive analytics models
Sahu, Datta and Mahapatra [14]	Investment capacity, Responsiveness, and Inventory capacity	-	✓	-	-	✓	-	-
López and Ishizaka [15]	Flexibility, Visibility, Anticipation, Recovery, Security, Adaptability, Financial strength, Market position, and Collaboration	-	✓	-	-	✓	-	-
Sabouhi, Pishvaee and Jabalameli [16]	multi-sourcing, supplier fortification, and emergency inventory	-	✓	-	✓	✓	-	-
Jabbarzadeh, Fahimnia and Sabouhi [17]	Extra production capacities, Multiple sourcing, and Backup suppliers	-	✓	-	✓	✓	-	-
Hosseini and Barker [19]	Absorptive capacity, adaptive capacity, and restorative capacity	✓	✓	-	-	-	-	Bayesian network
Haldar, Ray, Banerjee and Ghosh [21]	Investment capacity, Responsiveness, and Emergency inventory holding capacity	-	✓	-	-	✓	-	-
Rajesh and Ravi [25]	Responsiveness, risk reduction, and Technical support	-	✓	-	-	-	✓	-
Parkouhi, Ghadikolaei and Lajimi [46]	Safety, Visibility, Environmental Controls, Trust, Flexibility, Support Services, Future Manufacturing Capabilities, and others	-	✓	-	-	-	✓	-
This paper	Supply density, Transit time, Resilience score of supplier's locations	✓	✓	✓	-	✓	-	-

Most of the above studies considered operational risks for resilient supplier selection problem. This is the first time that a possibilistic fuzzy multi-objective model is proposed for resilient supplier selection and order allocation problem. Furthermore, for the first time in the literature supply density, resilience index score, and transit time are considered as supply selection criteria for resilient supply network.

Falasca, Zobel and Cook [36] proposed three characteristics (node criticality, supply chain complexity, and supply chain density) for building a resilient supply chain. Among them, supply chain density is the most important resilient criteria when designing supply networks under HILP disruptions. This is due to fact that denser supply networks are vulnerable to HILP disruption risks [36]. For example, 1999s Taiwan earthquake ended up having a significant effect on the entire global PC supply chain, because of the high concentration of computer component manufacturers in Hsinchu, Taiwan [47]. This example shows that the selection of a large number of suppliers from each region is a vulnerable multi-sourcing strategy, hence, this paper proposed supply density-based approach to tackle this problem. Furthermore, every country or territory has different resilient capabilities FMGlobal [48] and it affects the performance of the supply chain. Therefore, this study also proposed resilience index score-based criteria to supplier selection. The resilience index score is proposed by FMGlobal [48] is a data-driven tool to rank the countries to supply chain disruption risks. Nine key drivers of supply chain risks are considered and grouped into three categories namely: economic, risk quality, and supply chain factors. These nine drivers include local supplier quality, quality of fire risk management, GDP per capita, oil intensity, quality of hazard risk management, exposure to natural hazards, corruption control, infrastructure, and political risks [49]. Transit time is the last resilient criteria for supplier selection considered in this study. Transit time is an important indicator of supply chain flexibility [50]. Transit time reduction is one of the wildly used criteria to mitigate supply risks [51].

3. Problem Formulation

In this study, a garment manufacturer is assumed which want to select a suitable set of suppliers for the required material. All the model parameters are considered as fuzzy parameters. Five objectives considered in this study, namely: a cost which includes purchase and transportation costs, the rejection rate of suppliers, transit time from suppliers, supply density, and supplier resilience score based on their locations.

3.1. Mathematical Model Notations

• **Indices**	
s *existing suppliers*	$s = 1, 2, \ldots S$
l *number of objective functions*	$l = 1, 2, \ldots L$
• **Parameters**	
$\widetilde{d}$	Total demand of required material
$\widetilde{u}_s$	Purchase cost of required material from supplier s
$\widetilde{c}_s$	Transportation cost of material from supplier s
$\widetilde{\rho}_s$	Percentage of the rejected material delivered by the supplier s
$\widetilde{v}_s$	Capacity of supplier s
$\widetilde{k}_s$	Minimum acceptable purchase quantity from supplier s
mx_s	Maximum number of suppliers selected for required material
$\widetilde{t}_s$	Transit time from supplier s
ds^{ab}	Distance between selected supplier $a \in s$ and selected supplier $b \in s$ $(a \neq b)$
rs_s	Resilience index score of supplier s
• **Decision Variables**	
q_s	Purchase quantity from supplier s
$\Phi_s = \begin{cases} 1 \\ 0 \end{cases}$	1, If supplier s is selected, otherwise 0.
$\begin{matrix} \omega_{ab} \\ \omega'_{ab} \end{matrix} = \begin{cases} 1, \text{ If supplier } a \text{ and supplier } b \text{ are selected.} \\ 0, \text{otherwise} \end{cases}$	$\forall (a, b) \in s$ and $a \neq b$

3.2. Model Objectives

The Equation (1) represents the objective function for the cost. It is the sum of procurement cost and transportation cost from selected suppliers. The objective function of the total rate of rejection is estimated in Equation (2). The objective function for transit time from all selected supplier is calculated as shown in Equation (3). The Equation (4) shows the supply density for all selected suppliers. Total resilience index score objective is estimated in Equation (5). Where rs_s represents the resilience index score of supplier location obtained from FMGlobal [48].

$$\text{Minimize } f_{cost} = \sum_{s} (\widetilde{u}_s + \widetilde{c}_s)q_s \tag{1}$$

$$\text{Minimizes } f_{rej} = \sum_{s} \widetilde{\rho}_s q_s \tag{2}$$

$$\text{Minimizes } f_{time} = \sum_{s} \widetilde{t}_s \Phi_s \tag{3}$$

$$\text{Maximize } f_{den} = \frac{1}{\widetilde{d}} \left(\sum_{a \in s} \sum_{\substack{b \in s \\ a \neq b}} ds^{ab} \Phi_s \right) \tag{4}$$

$$\text{Maximize } f_{res} = \sum_{s} \frac{rs_s q_s}{\widetilde{d}} \tag{5}$$

3.3. Model Constraints

Constraint (6) ensures that total procured material should satisfy its demand.

$$\sum_{s} q_s = \widetilde{d} \tag{6}$$

Constraint (7) is capacity restrictions on the supplier. Also, it controls the flow between the supplier and the buyer through a binary variable. Constraint (8) ensures that purchase quantity from the selected supplier will be more than its acceptable order quantity limit. Constraint (9) restricts the maximum allowable supplier to be selected for the required material.

$$q_s \leq \widetilde{v}_s \Phi_s \tag{7}$$

$$q_s \geq \widetilde{k}_s \Phi_s \tag{8}$$

$$\sum_{s} \Phi_s \leq mx_s \tag{9}$$

Constraints (10) and (11) determine the intra-stage flow between suppliers and buyer. If buyer received material from both supplier a and supplier b then $\omega_{ab} = \Phi_{a\epsilon s} = \Phi_{b\epsilon s} = 1$ and $\omega'_{ab} = 0$. On the contrary, if buyer not received material from both supplier a and supplier b then $\omega'_{ab} = 1$, $\omega_{ab} = 0$ and $\Phi_{a\epsilon s} \neq \Phi_{b\epsilon s}$.

$$2\omega_{ab} + \omega'_{ab} = \Phi_{a\in s} + \Phi_{b\in s} \quad \forall (a,b) \in s, \text{ and } a \neq b \tag{10}$$

$$\omega_{ab} + \omega'_{ab} \leq 1 \quad \forall (a,b) \in s, \text{ and } a \neq b \tag{11}$$

4. Fuzzy Based Solution Methodology

Fuzzy based programming methods are highly used for multi-objective optimization because of their capability in measuring and adjusting the decision maker's satisfaction level of each objective function explicitly. In addition, the fuzzy theory is helpful to tackle the uncertain parameters related to supply chain optimization problem [52]. The main advantage of interactive fuzzy based approaches is that decision maker can efficiently achieve his/her preferences by controlling the search direction. The proposed solution methodology consists of the following steps.

Step 1: Convert uncertain mathematical model to equivalent auxiliary crisp

The proposed mathematical model is converted to an equivalent auxiliary crisp model. In this study, Jiménez, Arenas, Bilbao and Rodrı [51] approach is used which is based on an expected interval (EI) and expected value (EV) of fuzzy numbers. According to Jiménez, Arenas, Bilbao and Rodrı [51], the EI and EV of triangular fuzzy number (TFN) can be defined as in equation (12) and (13) respectively. Where ϑ^{pes} is the pessimistic value, ϑ^{mos} is the most likely value, and ϑ^{opt} is the optimimum value of triangular fuzzy number (ϑ). This research considered TFN because it is frequently used for a practical purpose [53].

$$EI(\widetilde{\vartheta}) = [E_1^{\vartheta}, E_2^{\vartheta}] = \left[\int_0^1 f_{\vartheta}^{-1}(x)dx, \int_0^1 g_{\vartheta}^{-1}(x)dx, \right] = \left[\frac{1}{2}(\vartheta^{pes} + \vartheta^{mos}), \frac{1}{2}(\vartheta^{mos} + \vartheta^{opt}) \right] \tag{12}$$

$$EV(\widetilde{\vartheta}) = \frac{E_1^{\vartheta} + E_1^{\vartheta}}{2} = \frac{\vartheta^{pes} + 2\vartheta^{mos} + \vartheta^{opt}}{4} \tag{13}$$

Using the above Equations (12) and (13), the equivalent auxiliary crisp model can be formulated as follows.

$$\text{Minimize } f_{cost} = \sum_s \left(\frac{u_s^{pes} + 2u_s^{mos} + u_s^{opt} + c_s^{pes} + 2c_s^{mos} + c_s^{opt}}{4} \right) q_s \tag{14}$$

$$\text{Minimizes } f_{rej} = \sum_s \left(\frac{\rho_s^{pes} + 2\rho_s^{mos} + \rho_s^{opt}}{4} \right) q_s \tag{15}$$

$$\text{Minimizes } f_{time} = \sum_s \left(\frac{t_s^{pes} + 2t_s^{mos} + t_s^{opt}}{4} \right) \Phi_s \tag{16}$$

$$\text{Maximize } f_{den} = \frac{1}{\frac{d^{pes}+2d^{mos}+d^{opt}}{4}} \left(\sum_{a \in s} \sum_{\substack{b \in s \\ a \neq b}} ds^{ab} \Phi_s \right) \tag{17}$$

$$\text{Maximize } f_{res} = \sum_s \frac{rs_s q_s}{\frac{d^{pes}+2d^{mos}+d^{opt}}{4}} \tag{18}$$

Subject to

$$\sum_s q_s \geq \left[\frac{\alpha}{2} \left(\frac{d^{mos} + d^{opt}}{2} \right) + (1 - \frac{\alpha}{2}) \left(\frac{d^{pes} + d^{mos}}{2} \right) \right] \tag{19}$$

$$\sum_s q_s \leq \left[\frac{\alpha}{2} \left(\frac{d^{pes} + d^{mos}}{2} \right) + (1 - \frac{\alpha}{2}) \left(\frac{d^{mos} + d^{opt}}{2} \right) \right] \tag{20}$$

$$q_s \leq \Phi_s \left[\alpha \left(\frac{v_s^{pes} + v_s^{mos}}{2} \right) + (1 - \alpha) \left(\frac{v_s^{mos} + v_s^{opt}}{2} \right) \right] \tag{21}$$

$$q_s \geq \Phi_s \left[\alpha \left(\frac{k^{mos} + k^{opt}}{2} \right) + (1 - \alpha) \left(\frac{k^{pes} + k^{mos}}{2} \right) \right] \tag{22}$$

$$\sum_s \Phi_s \leq m x_s \tag{23}$$

$$2\omega_{ab} + \omega'_{ab} = \Phi_{a\in s} + \Phi_{b\in s} \tag{24}$$

$$\omega_{ab} + \omega'_{ab} \leq 1 \tag{25}$$

Step 2: Determine α- extreme solutions

To estimate the upper (α-UB) and lower (α-LB) bounds to each objective, the crisp model developed in Step 1 is solved for each objective along with its constraint.

Step 3: Determine fuzzy membership function

Develop the fuzzy membership function for each objective using lower (α-LB) and upper (α-UB) bound values. The linear memberships for fuzzy goals are given as follows. It is assumed that membership functions are linear based on preferences and satisfaction level.

$$\mu_{cost}(x) = \begin{cases} 1, & f_{cost}(x) \leq f_{cost}^{\alpha-LB} \\ \frac{f_{cost}^{\alpha-UB} - f_{cost}(x)}{f_{cost}^{\alpha-UB} - f_{cost}^{\alpha-LB}}, & f_{cost}^{\alpha-LB} < f_{cost}(x) \leq f_{cost}^{\alpha-UB} \\ 0, & f_{cost}(x) \geq f_{cost}^{\alpha-UB} \end{cases} \tag{26}$$

$$\mu_{rej}(x) = \begin{cases} 1, & f_{rej}(x) \leq f_{rej}^{\alpha-LB} \\ \frac{f_{rej}^{\alpha-UB} - f_{rej}(x)}{f_{rej}^{\alpha-UB} - f_{rej}^{\alpha-LB}}, & f_{rej}^{\alpha-LB} \leq f_{rej}(x) \leq f_{rej}^{\alpha-UB} \\ 0, & f_{rej}(x) \geq f_{rej}^{\alpha-UB} \end{cases} \tag{27}$$

$$\mu_{time}(x) = \begin{cases} 1, & f_{time}(x) \leq f_{time}^{\alpha-LB} \\ \frac{f_{time}^{\alpha-UB} - f_{time}(x)}{f_{time}^{\alpha-UB} - f_{time}^{\alpha-LB}}, & f_{time}^{\alpha-LB} \leq f_{time}(x) \leq f_{time}^{\alpha-UB} \\ 0, & f_{time}(x) \geq f_{time}^{\alpha-UB} \end{cases} \tag{28}$$

$$\mu_{den}(x) = \begin{cases} 1, & f_{den}(x) \geq f_{den}^{\alpha-UB} \\ \frac{f_{den}(x) - f_{den}^{\alpha-LB}}{f_{den}^{\alpha-UB} - f_{den}^{\alpha-LB}}, & f_{den}^{\alpha-UB} \leq f_{den}(x) \leq f_{den}^{\alpha-LB} \\ 0, & f_{den}(x) \leq f_{den}^{\alpha-LB} \end{cases} \tag{29}$$

$$\mu_{res}(x) = \begin{cases} 1, & f_{res}(x) \geq f_{res}^{\alpha-UB} \\ \frac{f_{res}(x) - f_{res}^{\alpha-LB}}{f_{res}^{\alpha-UB} - f_{res}^{\alpha-LB}}, & f_{res}^{\alpha-UB} \leq f_{res}(x) \leq f_{res}^{\alpha-LB} \\ 0, & f_{res}(x) \leq f_{res}^{\alpha-LB} \end{cases} \tag{30}$$

where $f_l^{\alpha-LB}$ is a minimum value of $f_l(x)$ and $f_l^{\alpha-UB}$ is a maximum value of $f_l(x)$ with predefined α. These values the lth objective depends on its nature. $f_l^{\alpha-LB}$ are set as the aspiration level of cost, rejection rate, and transit time objective. Whereas $f_l^{\alpha-UB}$ are set as the aspiration level of supply density and resilience index.

Step 4: Convert the multi-objective model into a single objective

The proposed model is converted into the single objective in this stage. In this paper, two most popular fuzzy based approaches i.e., Tiwari, Dharmar and Rao [26] *weighted additive approach* and Werners [27] *fuzzy and* operator are implemented.

The weighted additive approach allows the buyer to assign different weights to objectives in the simple additive fuzzy achievement function. Mathematical formulation of the weighted additive method is as follows.

$$\left.\begin{array}{ll} \text{maximize} & \sum_{l} w_l \mu_{zl}(x) \\ \text{subject to} & \mu_{zl}(x) \in [0,1], \quad \forall l \\ & x \geq 0 \end{array}\right\} \tag{31}$$

where w_l represents the weight of ith objective. Selection of weights are subjected choice of decision makers and some good techniques can be used to determine the weights such as FAHP and structural equation modelling.

Werners' fuzzy and operator method are widely used interactive method. The advantage of this method is that it is positively related to the compensation rate due to its strong monotonicity. Additionally, it is easy to handle and has generated reasonable consistent results in applications [54]. By adopting the Werner's' method following a single objective model can be formed.

$$\left.\begin{array}{lll} \text{maximize} & \gamma\zeta_0 + (1-\gamma)\sum_{l}\zeta_1 & \\ \text{subject to} & \mu_1(x) \geq \zeta_0 + \zeta_1, & \forall l \\ & \zeta_0, \zeta_1, \gamma \in [0,1] & \end{array}\right\} \tag{32}$$

where, ζ_1 is the difference between satisfaction level of objectives their minimum satisfaction level ζ_0, That is, $\zeta_1 = \mu_1 - \zeta_0$. γ denotes the coefficient of compensation.

Step 5: Determine the solution method parameter

Determine the values of the relative importance of objectives (w_l) and coefficient of compensation (γ) to solve the mathematical model using both weighted additive approach and Werners' fuzzy and operator methods.

Step 6: Solve the model

In the last step, solve the model by using parameters of the mathematical model and solution methods. This process continues by varying the solution method parameters (i.e., γ and w_l) until decision makers are satisfied with the final solution. If decision makers want to modify the value of α, then restart the process from step 2.

5. An Illustration

The effectiveness of the proposed resilient supplier selection model and solution methodology is demonstrated in this section. The data relates to a realistic situation of a garment manufacturing sector as shown in Figure 1. The adopted situation can easily be extended to any other industry. Initially, the data are estimated as most likely values, these most likely values of the fuzzy parameter (fz^{mos}) are estimated using available information or hypothetically set based on realistic assumption. The pessimistic and optimistic values are estimated using $fz^{pes} = (1-\eta_1)fz^{mos}$ and, where two random numbers η_1 and η_2 are assumed between 0.2 and 0.8 to estimate. Table 2 shows the data set of unit purchase cost, transportation cost, rejection rate, capacity, and resilience index score of potential suppliers. The unit purchase costs from each supplier are hypothetical set based on labour cost, land value, resource availability at supplier locations. Transportation cost and transit time (see Table 3) from each supplier to manufacturer are estimated from sea rates (https://www.searates.com/). The resilience index score of each supplier is estimated from FM, Global resilience index data-driven tool (https://www.fmglobal.com/) based on the location of suppliers. The distance (as the crow flies) between suppliers (see Table 4) are calculated using the Google maps (https://www.google.com/maps). The demand for raw materials is assumed as most likely 8000 units, minimum acceptable order quantity is assumed as 1000 units, and maximum three suppliers can be selected for required material.

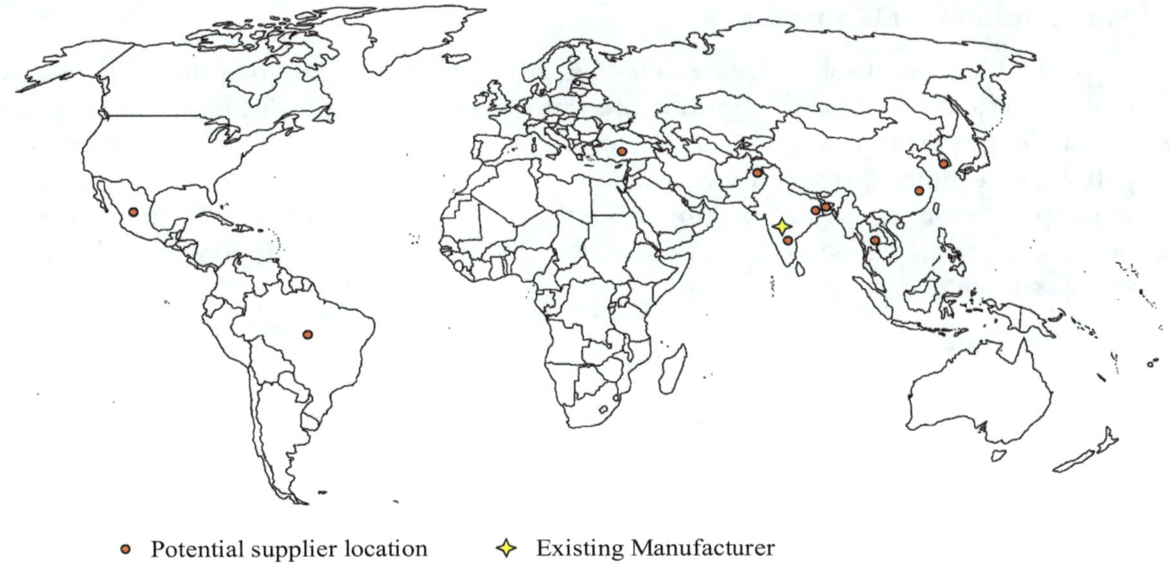

Figure 1. Supply network of the problem under consideration.

Table 2. Model input data.

Potential Supplier Location	Purchase Cost of Material ($/unit)	Transportation Cost of Material ($/unit)	Percentage of the Rejected Material	Capacity of Suppliers (1000 units)	Resilience Score
Korea	(6,8,10)	(0.08,0.13,0.21)	(0.01,0.01,0.02)	(3.08,5,8.08)	42.1
China	(1,2,4)	(0.11,0.19,0.30)	(0.04,0.06,0.10)	(3.7,6,9.7)	45.3
Thailand	(4,6,8)	(0.05,0.09,0.14)	(0.02,0.03,0.05)	(3.08,5,8.08)	39
Bangladesh	(1,3,5)	(0.11,0.17,0.28)	(0.02,0.03,0.05)	(3.7,6,9.7)	29
India-(Calcutta)	(3,5,7)	(0.10,0.16,0.26)	(0.01,0.02,0.03)	(2.46,4,6.46)	27.1
India-(Hyderabad)	(3,4,5)	(0.01,0.02,0.03)	(0.01,0.02,0.03)	(4.62,7.5,12.12)	27.1
Pakistan	(3,5,7)	(0.06,0.10,0.16)	(0.01,0.02,0.03)	(3.08,5,8.08)	22.2
Turkey	(8,10,12)	(0.08,0.12,0.20)	(0.01,0.01,0.02)	(3.08,5,8.08)	38.4
Brazil	(6,8,10)	(0.09,0.14,0.22)	(0.02,0.04,0.06)	(3.7,6,9.7)	47.8
Mexico	(7,9,11)	(0.11,0.17,0.28)	(0.02,0.03,0.05)	(4.31,7,11.31)	44.8

Table 3. Transit time from suppliers.

Supplier Location	Transit Time (days)
Korea	(7.39,12,19.39)
China	(9.24,15,24.24)
Thailand	(1.23,2,3.23)
Bangladesh	(1.85,3,4.85)
India-(Calcutta)	(1.23,2,3.23)
India-(Hyderabad)	(0.62,1,1.62)
Pakistan	(3.08,5,8.08)
Turkey	(8.01,13,21.01)
Brazil	(14.78,24,38.78)
Mexico	(19.71,32,51.71)

Table 4. The distance between suppliers (kilometers).

Suppliers' location.	Korea	China	Thailand	Bangladesh	India (Calcutta)	India (Hyderabad)	Pakistan	Turkey	Brazil	Mexico
Korea	N/A	1028.16	3723.83	3783.61	4037.39	5212.7	5768.78	7641.12	16,739.92	12,029.52
China	1028.16	N/A	2743.18	3059.41	3304.68	4478.2	5273.64	7587.05	17,589.43	13,051.91
Thailand	3723.83	2743.18	N/A	1526.77	1616.16	2388.26	3709.84	6926.22	17,348.59	15,721.92
Bangladesh	3783.61	3059.41	1526.77	N/A	250.06	1432.18	2374.24	5408.55	16,067.89	15,092.77
India (Calcutta)	4037.39	3304.68	1616.16	250.06	N/A	1180.87	2186.91	5327.76	15,902.02	15,299.76
India (Hyderabad)	5212.7	4478.2	2388.26	1432.18	1180.87	N/A	1461.88	4788.23	14,979.22	15,878.77
Pakistan	5768.78	5273.64	3709.84	2374.24	2186.91	1461.88	N/A	3334.57	13,710.04	14,838.79
Turkey	7641.12	7587.05	6926.22	5408.55	5327.76	4788.23	3334.57	N/A	10,723.38	11,983.21
Brazil	16,739.92	17,589.43	17,348.59	16,067.89	15,902.02	14,979.22	13,710.04	10,723.38	N/A	5692.39
Mexico	12,029.52	13,051.91	15,721.92	15,092.77	15,299.76	15878.77	14,838.79	11,983.21	5692.39	N/A

6. Model Solution and Result Analysis

The proposed fuzzy multi-objective model is solved in Lingo 14.0. Lingo optimization software has been widely used in supply chain optimization problems [55,56]. According to the steps of the proposed methodology, the payoff values are estimated by solving the single objective model. Table 5 shows payoff values estimated from the single objective model. The aspiration level of cost, rejection, transit time, supply density, and resilience score are estimated as $22,514.93, 0.025, 3.17 days, 0.286, and 47.89 respectively. Once the payoff values are estimated, the fuzzy membership functions of objectives are estimated as shown below.

$$\mu_{cost}(q_s) = \begin{cases} 1, f_{cost}(q_s) \leq 22514.93 \\ \frac{75796.12 - f_{cost}(q_s)}{75796.12 - 22514.93}, 22514.93 < f_{cost}(q_s) \leq 75796.12 \\ 0, f_{cost}(q_s) \geq 75796.12 \end{cases}$$

$$\mu_{rej}(q_s) = \begin{cases} 1, f_{rej}(q_s) \leq 0.025 \\ \frac{0.10 - f_{rej}(q_s)}{0.10 - 0.025}, 0.025 \leq f_{rej}(q_s) \leq 0.10 \\ 0, f_{rej}(q_s) \geq 0.10 \end{cases}$$

$$\mu_{time}(q_s) = \begin{cases} 1, f_{time}(q_s) \leq 3.17 \\ \frac{61.11 - f_{time}(q_s)}{61.11 - 3.17}, 3.17 \leq f_{time}(q_s) \leq 61.11 \\ 0, f_{time}(q_s) \geq 61.11 \end{cases}$$

$$\mu_{den}(q_s) = \begin{cases} 1, f_{den}(q_s) \geq 0.28 \\ \frac{f_{den}(q_s) - 0.017}{0.28 - 0.017}, 0.017 \leq f_{den}(q_s) \leq 0.28 \\ 0, f_{den}(q_s) \leq 0.017 \end{cases}$$

$$\mu_{res}(q_s) = \begin{cases} 1, f_{res}(q_s) \geq 47.89 \\ \frac{f_{res}(q_s) - 29.79}{47.89 - 29.79}, 29.79 \leq f_{res}(q_s) \leq 47.89 \\ 0, f_{res}(q_s) \leq 29.79 \end{cases}$$

Table 5. Payoff values.

Objective	Cost ($)	Rejection (%)	Transit Time (days)	Supply Density	Resilience Score
Minimize Cost	22,514.93	0.097	19.04	0.022	38.2
Minimize Rejection	75,796.12	0.025	26.45	0.056	39.22
Minimize Transit time	38,143.47	0.052	3.17	0.017	29.79
Maximize Supply density	70,159.31	0.1	61.11	0.286	43.41
Maximize Resilience score	50,560.14	0.1	41	0.12	47.89

The proposed mathematical model is solved using both methods, that is, a weighted additive approach and Werners' 'fuzzy and' operator methods. A solution of illustrated case example using both methods is shown in Table 6. The result shows that both methods produce a comprehensive optimal solution. However, the weighted additive approach considers the importance of objectives based on the weight given to each objective. For example, when more importance is given to resilience score (i.e., $w_5 = 0.3$), it results in 78.6% achievement of aspiration level. On the other hand, when all objectives and equal importance (Werners' method) than resilience score is given, the objective results in 34% achievement of aspiration level. Comparative analysis of both methods is shown in Figure 2.

Table 6. Solution of a case example.

Model Objectives	Weighted Additive Approach [a]		Werners' 'Fuzzy and' Operator [b]	
	$\mu_l(x)$	$f_l(x)$	$\mu_l(x)$	$f_l(x)$
Cost	0.346	78,238.03	0.195	65,374.99
Rejection	0.051	0.12	0.366	0.072
Transit time	0.457	42.0	0.383	38.89
Supply density	0.975	0.27	0.972	0.272
Resilience score	0.786	75.5	0.803	44.33

$\alpha = 0.9$, [a] $w_1 = 0.2, w_2 = 0.1, w_3 = 0.2, w_4 = 0.2, w_5 = 0.3$, [b] $\gamma = 0.7$.

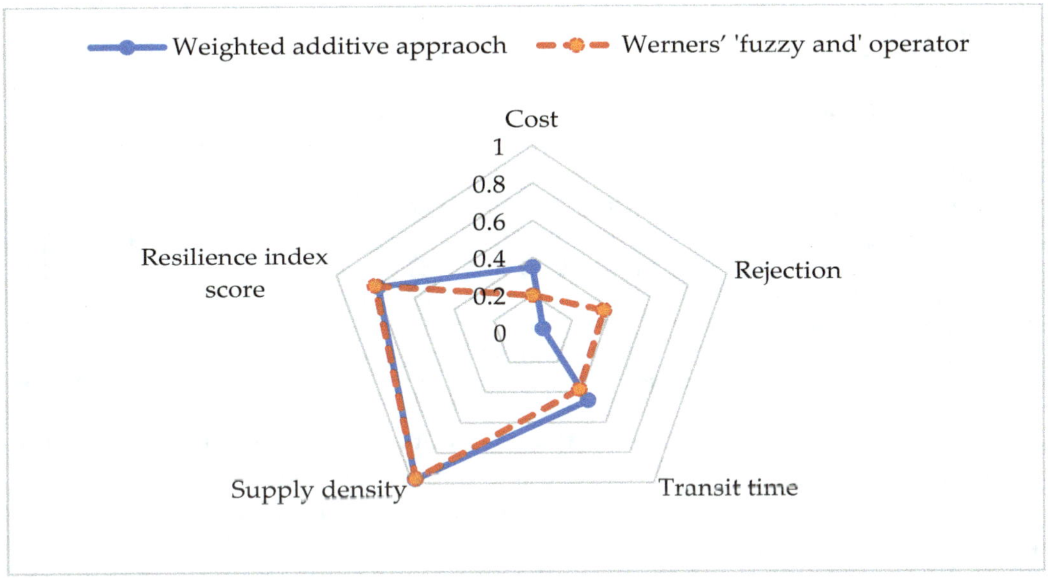

Figure 2. Achievement levels of objectives.

7. Sensitivity Analysis

Sensitivity analysis of both methods is carried out by varying methodology parameters (i.e., α, γ and w_l) as discussed in the final step of the proposed solution methodology. Tables 7 and 8 show the analysis result of the weighted additive approach and Werners' 'fuzzy and' operator method, respectively. Two cases are solved with the weighted additive approach: (1) more importance is given to resilience objective and (2) more importance given to cost objective.

The result shows that there is a tradeoff in economic objective and resilience objective. Hence, it can be said that the increase is supply chain resilience will tend to increase the total cost of the supply network. This is possible because economic supply networks are more dense networks in order to minimize transportation cost, hence any HILP disruption event such as earthquake or tsunami may disrupt more than one supplier. Therefore, it is suggested that companies should avoid denser supply networks to minimize risks from high impact low probability disruption event. Analysis result shows that 92% achievement of cost aspiration level will bring achievement of about only 6.1% supply density and 49.8% resilience index score aspiration levels respectively. On the other hand, 18.4% achievement of cost aspiration level will bring achievement of about 97% supply density and 88.1% resilience index score respectively.

Table 7. Sensitivity analysis of the weighted additive approach.

Alpha	Objective Weights	μ_{cost}	μ_{rej}	μ_{time}	μ_{den}	μ_{res}	f_{cost}	f_{rej}	f_{time}	f_{den}	f_{res}
0.9	$w_1 = 0.2, w_2 = 0.1, w_3 = 0.2, w_4 = 0.2, w_5 = 0.3$	0.184	0.200	0.365	0.97	0.881	65,941.28	0.085	39.9	0.27	45.7
	$w_1 = 0.3, w_2 = 0.2, w_3 = 0.2, w_4 = 0.1, w_5 = 0.2$	0.920	0.200	0.762	0.061	0.498	26,730.90	0.085	16.9	0.03	38.8
0.6	$w_1 = 0.2, w_2 = 0.1, w_3 = 0.2, w_4 = 0.2, w_5 = 0.3$	0.111	0.100	0.566	0.910	0.935	66,577.34	0.092	28.3	0.25	47.4
	$w_1 = 0.3, w_2 = 0.2, w_3 = 0.2, w_4 = 0.1, w_5 = 0.2$	0.851	0.200	0.762	0.061	0.634	27,727.65	0.085	16.9	0.03	42.6

Table 8. Sensitivity analysis of Werners' 'fuzzy and' operator.

α	γ	μ_{cost}	μ_{rej}	μ_{time}	μ_{den}	μ_{res}	f_{cost}	f_{rej}	f_{time}	f_{den}	f_{res}
	0.0–0.8	0.107	0.366	0.383	0.972	0.793	66,774.85	02.072	38.8	0.27	46.4
0.6	0.9	0.366	0.366	0.383	0.972	0.451	53,167.88	0.072	38.89	0.27	38.3
	1.0	0.407	0.566	0.543	0.407	0.406	51,054.51	0.057	29.62	0.12	37.3
	0.0–0.8	0.195	0.366	0.383	0.972	0.803	65,374.99	0.072	38.89	0.27	44.33
0.9	0.9	0.366	0.366	0.383	0.972	0.541	56,259.06	0.072	38.89	0.27	39.59
	1.0	0.392	0.566	0.543	0.407	0.392	54,899.34	0.057	29.6	0.12	36.88

Assume that decision makers required balance results and choose the best outcome at $\alpha = 0.9$, and $\gamma = 1.0$ (highlighted row in Table 8). Hence, the final decision of the illustrated example is shown in Figure 3. The optimal quantity of material purchase from suppliers are q (Korea) = 3751 units, q (Bangladesh) = 3655 units, and q (Turkey) = 1258 units.

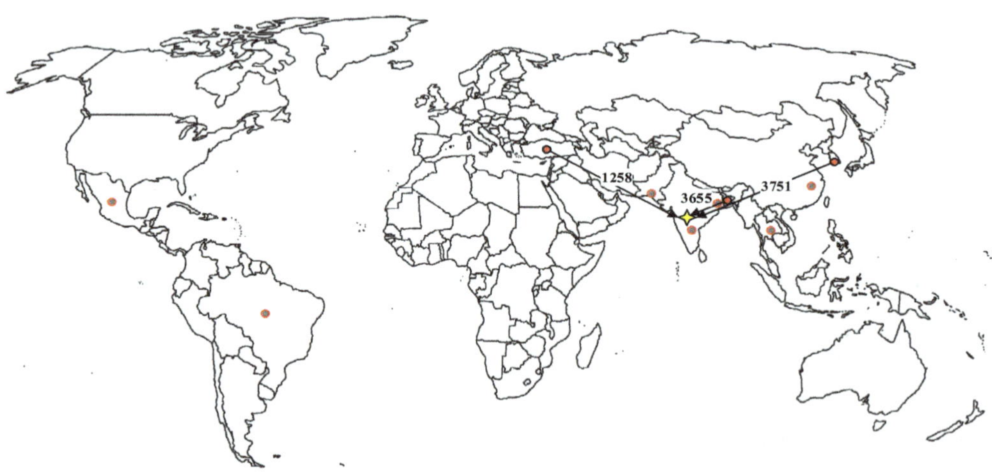

Figure 3. Final decision based on decision makers preferences.

8. Conclusion and Future Suggestions

This paper introduced a novel possibilistic fuzzy environment for resilient supplier selection and order allocation considering new resilience criteria (supply density, transit time and resilience index score) for the selection of suppliers. For this purpose, an interactive fuzzy multi-objective programming approach is introduced to reduce uncertainties inherent in the supplier selection decision.

A six-step solution methodology is designed to solve the proposed uncertain multi-objective model and a numerical case example is provided to show the applicability of the proposed model in a real situation.

This study significantly helps the practitioners who are trying to consider resilience in their supply network. The research results show the importance of supplier's location in order to minimize disruption risks. Furthermore, the proposed model will help the managers to effectively evaluate and select a suitable set of suppliers while considering cost and resilience simultaneously. Moreover, the proposed possibilistic fuzzy based solution methodology will be helpful for practitioners and academicians to tackle cognitive and stochastic uncertainties related to supplier evaluation and selection problem. This study also helps the academicians to analyze the importance of resilient supply networks under disruption risks. Additionally, the proposed multi-objective possibilistic fuzzy-based approach can be useful in another area of supply chain optimization.

Although this research gained important insights from the implementation of proposed resilience criteria and solution methodology, there are some limitations which may be considered in future research. This paper assumed triangular fuzzy numbers for model parameters, it will be interesting to compare the research results with other fuzzy numbers such as trapezoidal in future studies. Furthermore, this research only considered a disruption of the location where suppliers are located. However, it will be important to consider the disruption of transport links between suppliers and buyers. Another interesting possible direction for future research is to extend the proposed model to multi-commodity and multi-period planning horizon. Lastly, the analysis result indicates that economic networks are denser network and they may be vulnerable to disruption risks, it will be valuable to further investigate the relationship between the economic network (denser network) and disruption risks.

Author Contributions: Conceptualization, S.I.M.; Methodology, M.S.M.; Software, M.B.R.; Supervision, S.M.Q.; Validation, S.M.Q.; Writing—original draft, S.I.M.; Writing—review & editing, M.W.I.

Acknowledgments: This research was supported by the Higher Education Commission of Pakistan through the Startup Research Grant Program (SRPG#1299 and SRGP#1335).

References

1. Schoenherr, T.; Modi, S.B.; Benton, W.C.; Carter, C.R.; Choi, T.Y.; Larson, P.D.; Leenders, M.R.; Mabert, V.A.; Narasimhan, R.; Wagner, S.M. Research Opportunities in Purchasing and Supply Management. *Int. J. Prod. Res.* **2012**, *50*, 4556–4579. [CrossRef]
2. Kim, J.S.; Jeon, E.; Noh, J.; Park, J.H. A Model and an Algorithm for a Large-Scale Sustainable Supplier Selection and Order Allocation Problem. *Mathematics* **2018**, *6*, 325. [CrossRef]
3. Mari, I.S.; Lee, Y.H.; Memon, M.S.; Park, Y.S.; Kim, M. Adaptivity of Complex Network Topologies for Designing Resilient Supply Chain Networks. *Int. J. Ind. Eng.* **2015**, *22*, 102–116.
4. Tang, C.; Tomlin, B. The Power of Flexibility for Mitigating Supply Chain Risks. *Int. J. Prod. Econ.* **2008**, *116*, 12–27. [CrossRef]
5. Wang, C.-N.; Nguyen, V.T.; Thai, H.T.N.; Tran, N.N.; Tran, T.L.A. Sustainable Supplier Selection Process in Edible Oil Production by a Hybrid Fuzzy Analytical Hierarchy Process and Green Data Envelopment Analysis for the Smes Food Processing Industry. *Mathematics* **2018**, *6*, 302. [CrossRef]
6. Wang, X.; Herty, M.; Zhao, L. Contingent Rerouting for Enhancing Supply Chain Resilience from Supplier Behavior Perspective. *Int. Trans. Oper. Res.* **2016**, *23*, 775–796. [CrossRef]
7. Chopra, S.; Sodhi, M. Reducing the Risk of Supply Chain Disruptions. *Mit Sloan Manag. Rev.* **2014**, *55*, 72–80.
8. Barroso, A.P.; Machado, V.H.; Barros, A.R.; Machado, V.C. Toward a Resilient Supply Chain with Supply Disturbances. Presented at the 2010 IEEE International Conference on Industrial Engineering and Engineering Management (IEEM), Macao, China, 7–10 December 2010.
9. Torabi, S.A.; Baghersad, M.; Mansouri, S.A. Resilient Supplier Selection and Order Allocation under Operational and Disruption Risks. *Transp. Res. Part E: Logist. Transp. Rev.* **2015**, *79*, 22–48. [CrossRef]

10. Carvalho, H.; Azevedo, S.G.; Cruz-Machado, V. Agile and Resilient Approaches to Supply Chain Management: Influence on Performance and Competitiveness. *Logist. Res.* **2012**, *4*, 49–62. [CrossRef]
11. Ribeiro, P.J.; Barbosa-Povoa, A. Supply Chain Resilience: Definitions and Quantitative Modelling Approaches–a Literature Review. *Comput. Ind. Eng.* **2018**, *115*, 109–122. [CrossRef]
12. Parkouhi, V.S.; Ghadikolaei, A.S. A Resilience Approach for Supplier Selection: Using Fuzzy Analytic Network Process and Grey Vikor Techniques. *J. Clean. Prod.* **2017**, *161*, 431–451. [CrossRef]
13. Hosseini, S.; Al Khaled, A. A Hybrid Ensemble and Ahp Approach for Resilient Supplier Selection. *J. Intell. Manuf.* **2016**, 1–22. [CrossRef]
14. Sahu, A.K.; Datta, S.; Mahapatra, S.S. Evaluation and Selection of Resilient Suppliers in Fuzzy Environment: Exploration of Fuzzy-Vikor. *Benchmarking Int. J.* **2016**, *23*, 651–673. [CrossRef]
15. López, C.; Ishizaka, A. A Hybrid Fcm-Ahp Approach to Predict Impacts of Offshore Outsourcing Location Decisions on Supply Chain Resilience. *J. Bus. Res.* **2017**. [CrossRef]
16. Sabouhi, F.; Pishvaee, M.S.; Jabalameli, M.S. Resilient Supply Chain Design under Operational and Disruption Risks Considering Quantity Discount: A Case Study of Pharmaceutical Supply Chain. *Comput. Ind. Eng.* **2018**, *126*, 657–672. [CrossRef]
17. Jabbarzadeh, A.; Fahimnia, B.; Sabouhi, F. Resilient and Sustainable Supply Chain Design: Sustainability Analysis under Disruption Risks. *Int. J. Prod. Res.* **2018**, 1–24. [CrossRef]
18. Wang, T.-K.; Zhang, Q.; Chong, H.; Wang, X. Integrated Supplier Selection Framework in a Resilient Construction Supply Chain: An Approach Via Analytic Hierarchy Process (Ahp) and Grey Relational Analysis (Gra). *Sustainability* **2017**, *9*, 289. [CrossRef]
19. Hosseini, S.; Barker, K. A Bayesian Network Model for Resilience-Based Supplier Selection. *Int. J. Prod. Econ.* **2016**, *180*, 68–87. [CrossRef]
20. Haldar, A.; Ray, A.; Banerjee, D.; Ghosh, S. A Hybrid Mcdm Model for Resilient Supplier Selection. *Int. J. Manag. Sci. Eng. Manag.* **2012**, *7*, 284–292. [CrossRef]
21. Haldar, A.; Ray, A.; Banerjee, D.; Ghosh, S. Resilient Supplier Selection under a Fuzzy Environment. *Int. J. Manag. Sci. Eng. Manag.* **2014**, *9*, 147–156. [CrossRef]
22. Sawik, T. Selection of Resilient Supply Portfolio under Disruption Risks. *Omega* **2013**, *41*, 259–269. [CrossRef]
23. Sawik, T. Joint Supplier Selection and Scheduling of Customer Orders under Disruption Risks: Single vs. Dual Sourcing. *Omega* **2014**, *43*, 83–95. [CrossRef]
24. Yilmaz-Börekçi, D.; Say, A.İ.; Rofcanin, Y. Measuring Supplier Resilience in Supply Networks. *J. Chang. Manag.* **2015**, *15*, 64–82. [CrossRef]
25. Rajesh, R.; Ravi, V. Supplier Selection in Resilient Supply Chains: A Grey Relational Analysis Approach. *J. Clean. Prod.* **2015**, *86*, 343–359. [CrossRef]
26. Tiwari, R.N.; Dharmar, S.; Rao, J.R. Fuzzy Goal Programming—An Additive Model. *Fuzzy Sets Syst.* **1987**, *24*, 27–34. [CrossRef]
27. Werners, B.M. Aggregation Models in Mathematical Programming. In *Mathematical Models for Decision Support*; Springer: Berlin/Heidelberg, Germany, 1988; pp. 295–305.
28. Memon, M.S.; Lee, Y.H.; Mari, S.I. Group Multi-Criteria Supplier Selection Using Combined Grey Systems Theory and Uncertainty Theory. *Expert Syst. Appl.* **2015**, *42*, 7951–7959. [CrossRef]
29. Memon, M.S.; Mari, S.I.; Shaikh, F.; Shaikh, S.A. A Grey-Fuzzy Multiobjective Model for Supplier Selection and Production-Distribution Planning Considering Consumer Safety. *Math. Probl Eng.* **2018**, *2018*, 5259876. [CrossRef]
30. Holling, C.S. Resilience and Stability of Ecological Systems. *Annu. Rev. Ecol. Syst.* **1973**, *4*, 1–23. [CrossRef]
31. Hamel, G.; Valikangas, L. The Quest for Resilience. *Revista Icade Revista de las Facultades de Derecho y Ciencias Económicas y Empresariales* **2004**, *62*, 355–358.
32. Sheffi, Y. *The Resilient Enterprise: Overcoming Vulnerability for Competitive Advantage*; MIT Press: Cambridge, MA, USA, 2005.
33. Sutcliffe, M.K.; Vogus, T.J. Organizing for Resilience. *Posit. Organ. Scholarsh. Found. New Discip.* **2003**, *94*, 110.
34. Woods, D.D. Four Concepts for Resilience and the Implications for the Future of Resilience Engineering. *Reliab. Eng. Syst. Saf.* **2015**, *141*, 5–9. [CrossRef]
35. Zhalechian, M.; Torabi, S.A.; Mohammadi, M. Hub-and-Spoke Network Design under Operational and Disruption Risks. *Transp. Res. Part E Logist. Transp. Rev.* **2018**, *109*, 20–43. [CrossRef]

36. Falasca, M.; Zobel, C.W.; Cook, D. A Decision Support Framework to Assess Supply Chain Resilience. Presented at the 5th International ISCRAM Conference, Washington, DC, USA, 4–7 May 2008.
37. Mari, S.I.; Lee, Y.H.; Memon, M.S. Complex Network Theory-Based Approach for Designing Resilient Supply Chain Networks. *Int. J. Logist. Syst. Manag.* **2015**, *21*, 365–384. [CrossRef]
38. Priya Datta, P.; Christopher, M.; Allen, P. Agent-Based Modelling of Complex Production/Distribution Systems to Improve Resilience. *Int. J. Logist. Res. Appl.* **2007**, *10*, 187–203. [CrossRef]
39. Azevedo, S.G.; Govindan, K.; Carvalho, H.; Cruz-Machado, V. Gresilient Index to Assess the Greenness and Resilience of the Automotive Supply Chain. *Discuss. Pap. Bus. Econ.* **2011**, *8*. [CrossRef]
40. Miller-Hooks, E.; Zhang, X.; Faturechi, R. Measuring and Maximizing Resilience of Freight Transportation Networks. *Comput. Oper. Res.* **2012**, *39*, 1633–1643. [CrossRef]
41. Tang, C.S. Robust Strategies for Mitigating Supply Chain Disruptions. *Int. J. Logist. Res. Appl.* **2006**, *9*, 33–45. [CrossRef]
42. Sheffi, Y.; Rice, J.B., Jr. A Supply Chain View of the Resilient Enterprise. *Mit Sloan Manag. Rev.* **2005**, *47*, 41.
43. Mari, S.I.; Lee, Y.H.; Memon, M.S. Sustainable and Resilient Supply Chain Network Design under Disruption Risks. *Sustainability* **2014**, *6*, 6666–6686. [CrossRef]
44. Pramanik, D.; Haldar, A.; Mondal, S.C.; Naskar, S.K.; Ray, A. Resilient Supplier Selection Using Ahp-Topsis-Qfd under a Fuzzy Environment. *Int. J. Manag. Sci. Eng. Manag.* **2017**, *12*, 45–54. [CrossRef]
45. Sen, D.K.; Datta, S.; Mahapatra, S.S. Dominance Based Fuzzy Decision Support Framework for G-Resilient (Ecosilient) Supplier Selection: An Empirical Modelling. *Int. J. Sustain. Eng.* **2017**, *10*, 338–357. [CrossRef]
46. Parkouhi, S.V.; Ghadikolaei, A.S.; Lajimi, H.F. Resilient Supplier Selection and Segmentation in Grey Environment. *J. Clean. Prod.* **2019**, *207*, 1123–1137. [CrossRef]
47. Papadakis, I.S. Financial Performance of Supply Chains after Disruptions: An Event Study. *Supply Chain Manag. Int. J.* **2006**, *11*, 25–33. [CrossRef]
48. FMGlobal. The 2015 FM Global Resilience Index Annual Report. Available online: https://www.fmglobal.com/assets/pdf/Resilience_Methodology.pdf (accessed on 3 July 2018).
49. Mari, S.I.; Lee, Y.H.; Memon, M.S. Sustainable and Resilient Garment Supply Chain Network Design with Fuzzy Multi-Objectives under Uncertainty. *Sustainability* **2016**, *8*, 1038. [CrossRef]
50. Min, H.; Zhou, G. Supply Chain Modeling: Past, Present and Future. *Comput. Ind. Eng.* **2002**, *43*, 231–249. [CrossRef]
51. Jiménez, M.; Arenas, M.; Bilbao, A.; Rodrı, M.V. Linear Programming with Fuzzy Parameters: An Interactive Method Resolution. *Eur. J. Oper. Res.* **2007**, *177*, 1599–1609. [CrossRef]
52. Tayyab, M.; Sarkar, B.; Yahya, B. Imperfect Multi-Stage Lean Manufacturing System with Rework under Fuzzy Demand. *Mathematics* **2019**, *7*, 13. [CrossRef]
53. Li, X.; Chien, C.F.; Yang, L.; Gao, Z. The Train Fueling Cost Minimization Problem with Fuzzy Fuel Prices. *Flex. Serv. Manuf. J.* **2014**, *26*, 249–267. [CrossRef]
54. Selim, H.; Ozkarahan, I. A Supply Chain Distribution Network Design Model: An Interactive Fuzzy Goal Programming-Based Solution Approach. *Int. J. Adv. Manuf. Technol.* **2008**, *36*, 401–418. [CrossRef]
55. Babar Ramzan, M.; Qureshi, S.M.; Mari, S.I.; Memon, M.S.; Mittal, M.; Imran, M.; Iqbal, M.W. Effect of Time-Varying Factors on Optimal Combination of Quality Inspectors for Offline Inspection Station. *Mathematics* **2019**, *7*, 51. [CrossRef]
56. Sarkar, B.; Tayyab, M.; Choi, S.K. Product Channeling in an O2O Supply Chain Management as Power Transmission in Electric Power Distribution Systems. *Mathematics* **2019**, *7*, 4. [CrossRef]

5

Constrained FC 4D MITPs for Damageable Substitutable and Complementary Items in Rough Environments

Sharmistha Halder Jana [1,*], Biswapati Jana [2], Barun Das [3] , Goutam Panigrahi [4] and Manoranjan Maiti [5]

1 Department of Mathematics, Midnapore College (Autonomous), Midnapore 721101, India
2 Department of Computer Science, Vidyasagar University, Midnapore 721102, India; biswapatijana@gmail.com
3 Department of Mathematics, Sidho Kanho Birsha University, Purulia 723104, India; bdasskbu@gmail.com
4 Department of Mathematics, National Institute of Technology, Durgapur 713209, India; panigrahi_goutam@rediffmail.com
5 Department of Mathematics, Vidyasagar University, Midnapore 721102, India; mmaiti2005@yahoo.co.in
* Correspondence: sharmistha792010@gmail.com.

Abstract: Very often items that are substitutable and complementary to each other are sent from suppliers to retailers for business. In this paper, for these types of items, fixed charge (FC) four-dimensional (4D) multi-item transportation problems (MITPs) are formulated with both space and budget constraints under crisp and rough environments. These items are damageable/breakable. The rates of damageability of the items depend on the quantity transported and the distance of travel i.e., path. A fixed charge is applied to each of the routes (independent of items). There are some depots/warehouses (origins) from which the items are transported to the sales counters (destinations) through different conveyances and routes. In proposed FC 4D-MITP models, per unit selling prices, per unit purchasing prices, per unit transportation expenditures, fixed charges, availabilities at the sources, demands at the destinations, conveyance capacities, total available space and budget are expressed by rough intervals, where the transported items are substitutable and complementary in nature. In this business, the demands for the items at the destinations are directly related to their substitutability and complementary natures and prices. The suggested rough model is converted into a deterministic one using lower and upper approximation intervals following Hamzehee et al. as well as Expected Value Techniques. The converted model is optimized through the Generalized Reduced Gradient (GRG) techniques using LINGO 14 software. Finally, numerical examples are presented to illustrate the preciseness of the proposed model. As particular cases, different models such as 2D, 3D FCMITPs for two substitute items, one item with its complement and two non substitute non complementary items are derived and results are presented.

Keywords: four-dimensional transportation problem; fixed charge; space constraint; budget constraint; substitute and complimentary items; rough interval

1. Introduction

Due to globalization, nowadays, transportation of commodities from sources to the destinations by road is getting more important. The advent of transportation problem (TP) was mainly based on real-life problems. Many numbers of real-life goods carrying problems are easily framed as transportation problems. The credit of first transportation problem goes to Hitchcock [1], which is a particular case of Linear Programming Problem (LPP). This optimizing problem consists of two main constraints i.e., source and destination constraints. However, it is very common knowledge that, in practical situations, these two constraints are not enough to formulate the problem perfectly since there exist other constraints, namely mode of transport, type of products, the distance of path traveled, etc. Due to the presence of these additional real-life constraints, the conventional transportation problems (2D-TPs) are modified to a solid transportation problem (STP-3D-TPs), which was first developed by Schell [2]. After the advent of STP, it has swiftly gained a very important place for research and development in transportation. Researchers like Yang et al. [3], Kocken et al. [4], etc. were on the development of STP, in which conveyances at different sources are considered.

In general circumstances, the length of a route remains unchanged in a normal transportation problem, so it does not make any significant change in the minimization of cost or time. However, reality dictates that there may be a number of available option/paths for the transportation of an item from source to destination. The common knowledge suggests that the cost per unit for transportation or other fixed charges related to routes are of different values. Thus, it can be seen that a three-dimensional transportation problem (3D-TP) is converted to a 4D-TP. Such a 4D-TP was considered by Halder et al. [5].

Nowadays, the business of multi -items is quite popular. Normally, in business, it is thought that if one item brings loss, another item may produce the profit. Moreover, due to the constantly changing business environment, the parameters related to transportation are imprecise. This situation has been recently dealt with in two recent investigations Das et al. [6] and Bera et al. [7]. Again, the transportation of multi-items may be substitutable and/or complementary because more options/choices of items bring more customers. This is true for supply chain models. Recently, Sarkar and Lee [8] and sarkar et al. [9] presented optimum pricing strategies for complementary products using game the orectic approach. Khanna et al. [10] suggested supply chain with customer-based two-level credit policies under an imperfect quality environment. Recently, Halder et al. [5] also solved an FCSTP for substitute and breakable items in crisp and fuzzy environments.

1.1. *Scope of the Paper*

Considering the above facts, we formulate the transportation policies of some damageable substitute and/or complementary items from some sources to destinations using some convenient vehicles and appropriate routes where the parameters of the problem are deterministic or imprecise and some realistic resource constraints are imposed.

In this investigation, there are different types of conveyance at different sources and different routes/paths available to travel from a source to a destination. Along some paths, fixed charges such as toll taxes, festival collections, etc. are collected. At each destination, retailers have some limitations on space and budget. The transportation costs, fixed charges, availabilities, demands, resources are imprecise and expressed by rough intervals. As the items are substitutable and complementary, items' demands are influenced by each other's prices. The demand for an item is reduced due to its own and complementary item's prices but increases due to its substitute's price. With these features, the model is formulated as a constrained FC 4DMITP problem and solved using Generalised Reduced Gradient (GRG) method (using LINGO 14.0). As particular cases, transportation policies for several models (i.e., 2D, 3D TPs for different combinations of items) are derived. The model are illustrated numerically. Some sensitivity analyses are also performed. Results of an earlier investigation (Bera et al. [7]) are obtained as a particular case. The novelties of the present investigation are as follows:

- As earlier discussed, TP and STP with various types of constraints are considered by several researchers. However, few researchers have considered 4D-TPs and 4D-MITPs. Moreover, 4D-MITP with rough parameters is an updated contribution.
- The items are complementary and substitutable in nature, that is, demands of the items are appropriately affected by their selling prices.
- The most important issues of this paper are to analyze how the travel distances are related to profit maximization when manufacturing companies transport both complementary and substitute

items that are differentiated by distance including the fixed charge of the path and damageability. The importance of route on profit is illustrated.

- Until now, in transportation, no one has considered the space constraint at the destinations along with the budget constant. The idea of space constraint is introduced here.
- The earlier researchers gave attention to minimization of aggregate transportation expenditure and very few have realized the importance of consideration of total profit instead of total cost/expanses.
- As particular cases, several earlier transportation models are deduced from the present model.

1.2. Structure of the Paper

The structure/organization of the paper is split as:
Section 1: Introduction
Section 2: Notations and Assumptions for the proposed model are given.
Section 3: Model description and formulation
Section 4: Numerical Experiments
Section 5: Particular Cases
Section 6: Sensitivity Analyses
Section 7: A discussion of the models on the basis of numerical results are presented.
Section 8: Practical implication is described.
Section 9: Conclusions drawn.

1.3. Literature Review

Though the transportation problem is quite an age-old one, several researchers are still working in this area as it is important due to national highways within the country and international transportation using cargos and ships. Hirsch and Dantzig [11] first introduced the concept of a fixed charge transportation problem in 1968. In each country, there are different major highways connecting the cities/ports and, for the maintenance, some charges (fixed) are collected from the vehicles plying on these roads. This real-life scenery has been depicted in the fixed charge TPs (FCTP). Kowalski and Lev [12] developed the FCTP as a nonlinear programming problem of practical interest in business and industry. Gen et al. [13] discussed the bi-criteria solid transportation problem with equality constraints solved by a genetic algorithm. Several researchers like Verma et al. [14], Shafaat and Goyal [15], Saad and Abbas [16] and others worked on transportation problems.

Though there are a lot of works on damageable/deteriorating items in inventory (cf. Sarkar and Iqbal [17] and Sarkar et al. [18]), only limited works have been done on damageable/breakable items in transportation. This is because transportation time is not normally considered in TP. However, the damageability may also depend on the route lengths which are different for different routes. Pramanik et al. [19] presented a multi-objective solid transportation problem with reliability for damageable items in a random-fuzzy environment.

Though Sarkar et al. [20] considered a power function of the delivery quantity as a unit transportation cost in inventory, normally constant unit transportation cost is assumed in TP to formulate it as an LPP. Ojha et al. [21] presented a solid transportation problem with nonlinear transportation cost and fuzzy resources, demand and conveyances. However, in the present problem, transportation cost is considered constant. Depending on different aspects, unit transportation costs, resources, demands, available budget and storing capacity at destination, etc. fluctuate due to uncertainty in judgement, lack of evidence, insufficient information, etc. Thus, a transportation model becomes more realistic if these parameters are assumed to be flexible /imprecise in nature i.e., uncertain in non-stochastic sense which may be represented by fuzzy, rough [22], type-2-fuzzy, fuzzy stochastic numbers. There are several investigations on TPs with fuzzy parameters. Verma et al. [14], Bit et al. [23], Jimenez and Verdegay [24], Li and Lai [25], Dey et al. [26] and others solved fuzzy multi-objective TP(s) using a Fuzzy compromise programming approach. Yang and Liu [3] investigated a fixed charge STP with fuzzy costs, fuzzy supplies, fuzzy demands and fuzzy conveyance capacities using an expected value method. There are also many investigations on TP with rough parameters. Recently, Kundu et al. [27] presented an STP considering crisp and rough costs. Tao et al. [28] applied rough multiple objective programming in STP. Das et al. [6] developed a profit-maximizing STP with parameters represented by rough intervals. Recently, Bera et al. [7] solved multi-item 4D TPs under budget constraints using rough interval values for TP parameters. The year-wise investigations of various TPs and STPs with different variations are recorded in the following table.

In spite of all these investigations, there are still some lacunas in making the investigated problem more realistic ones. Some of these gaps such as substitutable and complementary items, both budget and space constraint, damageable units, 4D MITP, fixed charge, profit maximization, etc. have been taken into account in the present investigation.

1.4. Motivation

Although there are so many existing kinds of literature available in the field of transportation problem (cf in Table 1), but still there are some lacunas. Most of the researchers developed two-dimensional or three-dimensional problems that motivated us to form a four-dimensional real transportation problem.

Table 1. Year wise investigations of TP and STP.

References'	Different Kind of TP	Item	Fixed Charge	Space Constraint	Budget Constraint	Different Kind of Environment
Hitchcock et al. [1]	2-dimension	one	×	×	-	crisp
Schell [2]	3-dimension	one	×	×	-	crisp
Haley [29]	3-dimension	one	×	×	-	crisp
Hirsch and Dantzig [11]	2-dimension	one	✓	×	×	crisp
Verma et al. [14]	2-dimension	one	×	×	×	fuzzy
Shafaat and Goyal [15]	2-dimension	one	×	×	×	crisp
Saad and Abbas [16]	2-dimension	one	×	×	×	fuzzy
Jimenez et al. [24]	3-dimension	one	×	×	×	fuzzy
Tao et al. [28]	3-dimension	one	×	×	×	rough
Ojha et al. [30]	2-dimension	multi-item	×	×	✓	fuzzy
Liu et al. [31]	3-dimension	one	×	×	×	type-2 fuzzy
Kundu et al. [27]	3-dimension	multi-item	×	×	×	type-2 fuzzy
Yang et al. [3]	2-dimension	one	✓	×	×	fuzzy
Giri et al. [32]	3-dimension	multi-item	✓	×	×	fuzzy
Kocken et al. [4]	3-dimension	one	×	×	×	fuzzy
Das et al. [6]	3-dimension	one	×	×	×	rough
Present Paper	4 dimensional	multi-item	✓	✓	✓	rough

In modern business, multifarious items are preferable for transportation rather than the single item. In the long term, transportation system fixed charges like (road tax etc.) are very common to real life; this is also a motivation which leads to consideration of fixed charge. Finally, in today's constantly fluctuating life, uncertainty is the only certain factor in our life. In this concept, the different coefficients are not fixed to its deterministic values. It must take an interval. For example, demand of an item is not fixed, but it is represented by its lower and upper limits b_l, b_u, respectively. Again, this limit also may fluctuate, so the necessity of rough interval arise. In this regard, the demand for an item roughly lies on $([\underline{b}_l, \underline{b}_u], [\overline{b}_l, \overline{b}_u])$, respectively.

In reality, most of the problems are not described specifically but described with uncertainty. This may due to the fluctuation of daily life, lack of information, etc. Until now, most of the impreciseness are described by interval values.

In interval analysis, a parameter is represented by lower and upper limits a, b and the possibility of existence of every point within this limits $[a, b]$ are same. However, in reality, it is not so, where the lower 'a' and upper limits 'b' occurred very occasionally. Most of the time, there exists a normal interval $[c, d]$, where $a \leqslant c \leqslant d \leqslant b$. Thus, vagueness here is described by two nested intervals $([a, b], [c, d])$. Such representation is well known as a rough interval.

For example, if we go through "Air Quality Index (AQI)" measurement of the city Ananda Vihar, Delhi in the year 2017, the lower and the upper intervals of this AQI (measured by PM2.5) were $[78, 404]$. The AQI 78 $\mu g/m^3$ found in the day of July 2017 (during rainy season) and the AQI is 404 $\mu g/m^3$ is found on the day of Dewali (November 2017), whereas the normal range of AQI (PM2.5) is [150, 290]. In this regard, the rough representation of AQI (PM2.5) of Ananda Vihar New Delhi is ([78, 404], [150, 290]).

2. Notation and Assumptions

2.1. Notations

2.1.1. Parameters

For r-th item,

- A_i^r: quantity of homogeneous merchandise available at i-th source.
- D_{0j}^r: market demand at j-th goal.
- D_j^r: actual demand at j-th goal.
- e_{kp}: quantity of the merchandise which can be carried by k-th conveyance along p-th route.
- C_{ijpk}^r: per unit transportation price from i-th origin to j-th goal by k-th vehicle via p-th route.
- S_j^r: selling expenses at the j-th destination.
- P_i^r: purchasing price at the i-th origin.
- f_{ijpk}: fixed transportation cost for shipping units from i-th source/origin to j-th goal/destination by k-th vehicles along p-th route.
- λ_{ijpk}^r: rate of breakability per unit distance from i-th source to j-th goal via p-th route and k-th conveyance.
- $Budj$: total budget at the j-th goal point.
- ds_{ijp}: distance from i-th origin to j-th goal along p-th route.
- SP^r: required space for r-th item.
- SPC_j: available space to the j-th retailer.
- α: power of the route length, related with the frangibility.
- $\beta_j, \theta_j^1, \theta_j^{11}$ and τ: Price sensitivity of products.

2.1.2. Decision Variable

- x_{ijpk}^r: the transported quantity from i-th source/origin to j-th goal/destination by k-th vehicle along p-th route (decision variable).

2.1.3. Indices

- R: total number of items.
- M: total number of origin/sources.
- N: total number of goal/destinations.
- L: total number of route/paths.
- K: total number of vehicles/conveyance.

2.2. Assumptions

(i) Particulars are breakable and carried from sources to goals using a vehicle through a path. Broken/damaged amounts depend on conveyance and path.

(ii) Particulars are substitutable and complementary to each other. In case of a substitute item, the demand is negative and is positive when the items are complementary nature:

$$\left.\begin{array}{l} D_j^1 = D_{0j}^1 - \beta_j S_j^1 + \tau\beta_j S_j^2 - \theta_j^1 \beta_j S_j^3 \\ D_j^2 = D_{0j}^2 + \tau\beta_j S_j^1 - \beta_j S_j^2 - \theta_j^{11} \beta_j S_j^3 \\ D_j^3 = D_{0j}^3 - \theta_j^1 \beta_j S_j^1 - \theta_j^{11} \beta_j S_j^2 - \beta_j S_j^3 \end{array}\right\}.$$

3. Model Description and Formulation

Let us define a four-dimensional transportation problem in this way. A merchant/wholesaler having different M godowns/storing houses (sources) at different locations and different N sales counters/shops (destinations) in different cities. The merchant buys R number of items of which some are substitutes and others complementary to each other and stores at the godowns which are of finite capacities. There are different K conveyances and L routes for transportation of goods from sources to destinations.

The schematic diagram of this 4D-MITP is given in Figure 1. The conveyances are of finite capacities and charge different charges for transportation along different routes. There are some fixed costs/charges (toll taxes, etc.) along each route and items are damageable. This damageability depends on type of conveyance and the distance of traveled by that conveyance. At each destination, the items are known and, for that, some amount of items are stored as per the capacities of the go-downs. The merchant can use any conveyance and route for transportation and the above mentioned transportation parameters are crisp or rough interval numbers. At the destination, the merchant sells the good items at some prices. The merchant has a limitation on his initial expenditure. Now, the problem for the merchant is to decide the amount of quantities to be transported from different sources to different destinations by the appropriate conveyance through the appropriate route so that total profit out of this business is at a maximum subject to the constraints.

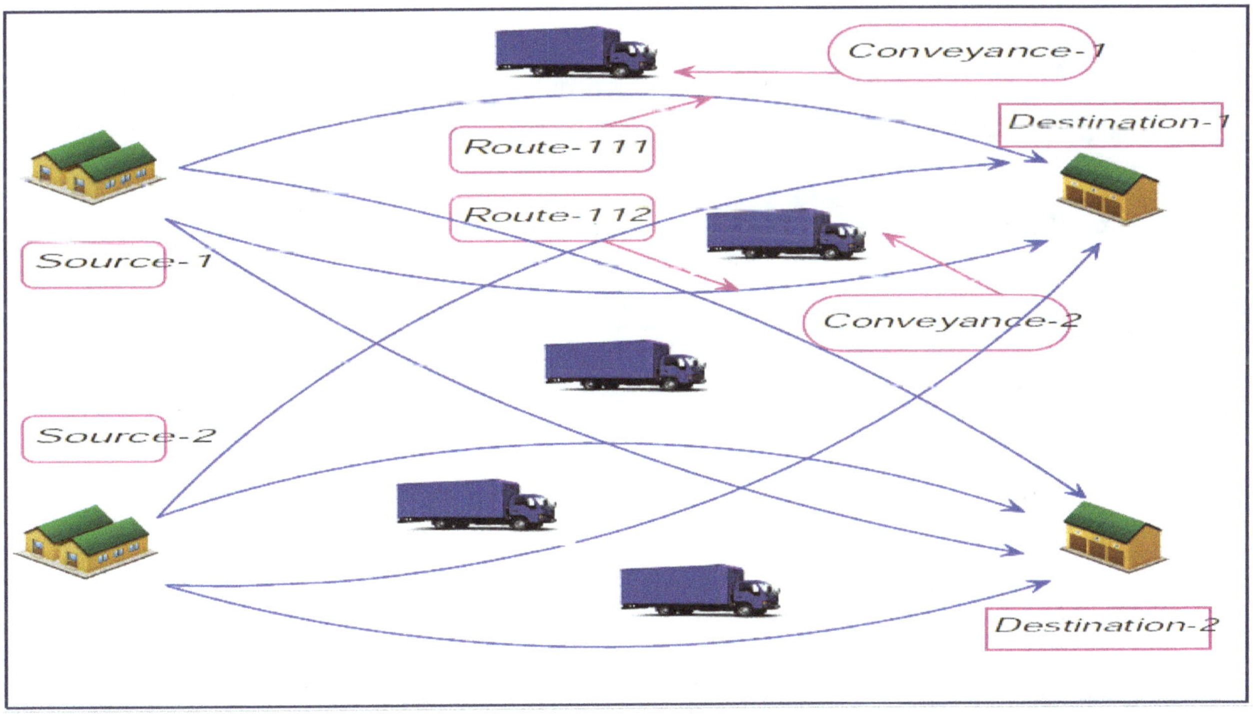

Figure 1. Pictorial representation of a four-dimensional transportation model.

Thus, TP is developed as a profit maximization problem. In this process of business, there are total purchase charge cost, fixed charge cost, transportation cost, revenue in the objective functions and five constraints on source, destination, conveyance capacity, space and budget. Therefore, as per the availability of type of the input parameters with the help of considerable assumptions and notations, we formulate two different models: the crisp model and rough model as given below.

3.1. Model-I: Crisp Model

$$\left.\begin{array}{l}
\text{Maximum Profit} = \text{Total Revenue - Total Purchase Cost - Total Transportation cost - Total fixed charge} \\
\text{Maximize } Z = \sum_{r=1}^{R}\sum_{i=1}^{M}\sum_{j=1}^{N}\sum_{p=1}^{L}\sum_{k=1}^{K}\{S_j^r(1-\lambda_{ijpk}^r ds_{ijp}^{\alpha}) - P_i^r - C_{ijpk}^r ds_{ijp}\}x_{ijpk}^r - \sum_{i=1}^{M}\sum_{j=1}^{N}\sum_{p=1}^{L}\sum_{k=1}^{K} w(x_{ijpk}^r) f_{ijpk}^r \\
\text{subject to } \sum_{j=1}^{N}\sum_{p=1}^{L}\sum_{k=1}^{K} x_{ijpk}^r \le A_i^r, \ \forall\ i = 1,2,...,M, r = 1,2,...,R \\
\sum_{i=1}^{M}\sum_{p=1}^{L}\sum_{k=1}^{K}(1-\lambda_{ijpk}^1 ds_{ijp}^{\alpha})x_{ijpk}^1 \ge D_{0j}^1 - \beta_j s_j^1 + \tau\beta_j s_j^2 - \theta_j^1\beta_j s_j^3 \quad \forall\ j = 1,2,...,N, r = 1,2,...,R \\
\sum_{i=1}^{M}\sum_{p=1}^{L}\sum_{k=1}^{K}(1-\lambda_{ijpk}^2 ds_{ijp}^{\alpha})x_{ijpk}^2 \ge D_{0j}^2 + \tau\beta_j S_j^1 - \beta_j S_j^2 - \theta_j^{11}\beta_j S_j^3 \quad \forall\ j = 1,2,...,N, r = 1,2,...,R \\
\sum_{i=1}^{M}\sum_{p=1}^{L}\sum_{k=1}^{K}(1-\lambda_{ijpk}^3 ds_{ijp}^{\alpha})x_{ijpk}^3 \ge D_{0j}^3 - \theta_j^1\beta_j S_j^1 - \theta_j^{11}\beta_j S_j^2 - \beta_j S_j^3 \quad \forall\ j = 1,2,...,N, r = 1,2,...,R \\
\sum_{r=1}^{R}\sum_{i=1}^{M}\sum_{j=1}^{N} x_{ijpk}^r \le e_{kp} \qquad \forall\ k = 1,2,...,K, p = 1,2,...,L \\
\sum_{r=1}^{R}\sum_{i=1}^{M}\sum_{p=1}^{L}\sum_{k=1}^{K}\left[1-\lambda_{ijpk}^r ds_{ijp}^{\alpha}\right] x_{ijpk}^r * SP^r \le SPC_j^r \quad \forall\ j = 1,2,...,N, \\
\sum_{r=1}^{R}\sum_{i=1}^{M}\sum_{p=1}^{L}\sum_{k=1}^{K}\left[P_i^r + C_{ijpk}^r ds_{ijp}\right] x_{ijpk}^r + \sum_{i=1}^{M}\sum_{p=1}^{L}\sum_{k=1}^{K} w(x_{ijpk}^r) f_{ijpk} \le Bud_j \quad \forall\ j = 1,2,...,N, \\
\text{and } x_{ijpk}^r \ge 0\ \forall\ i = 1,2,...,M, j = 1,2,...,N, p = 1,2,...,L, k = 1,2,...,K, r = 1,2,...,R
\end{array}\right\}. \qquad (1)$$

3.2. Model-II: Rough Model

If the parameters, selling expenses, cost of transportation, purchasing prices, fixed charges, availability, demand, capacities, available space and permittable budget are found as rough intervals i.e.,

$$\left.\begin{array}{l}
\check{S}_j^1 = ([\underline{S}_{jl}^1, \underline{S}_{ju}^1], [\overline{S}_{jl}^1, \overline{S}_{ju}^1]), \check{S}_j^2 = ([\underline{S}_{jl}^2, \underline{S}_{ju}^2], [\overline{S}_{jl}^2, \overline{S}_{ju}^2]), \check{S}_j^3 = ([\underline{S}_{jl}^3, \underline{S}_{ju}^3], [\overline{S}_{jl}^3, \overline{S}_{ju}^3]), \check{P}_i^r = ([\underline{P}_{il}^r, \underline{P}_{iu}^r], [\overline{P}_{il}^r, \overline{P}_{iu}^r]) \check{C}_{ijpk}^r = ([\underline{C}_{ijpkl}, \underline{C}_{ijpku}], [\overline{C}_{ijpkl}, \overline{C}_{ijpku}]), \\
\check{f}_{ijpk} = ([\underline{f}_{ijpkl}, \underline{f}_{ijpku}], [\overline{f}_{ijpkl}, \overline{f}_{ijpku}]), \check{A}_i^r = ([\underline{A}_{il}^r, \underline{A}_{iu}^r], [\overline{A}_{il}^r, \overline{A}_{iu}^r]) \check{D}_j^1 = ([\underline{D}_{0jl}^1, \underline{D}_{0ju}^1], [\overline{D}_{0jl}^1, \overline{D}_{0ju}^1]) \check{D}_j^2 = ([\underline{D}_{0jl}^2, \underline{D}_{0ju}^2], [\overline{D}_{0jl}^2, \overline{D}_{0ju}^2]), \check{D}_j^3 = ([\underline{D}_{0jl}^3, \\
\underline{D}_{0ju}^3], [\overline{D}_{0jl}^3, \overline{D}_{0ju}^3]) \check{e}_{kp} = ([\underline{e}_{kpl}, \underline{e}_{kpu}], [\overline{e}_{kpl}, \overline{e}_{kpu}]), S\check{P}C_j = ([\underline{SPC}_{jl}, \underline{SPC}_{ju}], [\overline{SPC}_{jl}, \overline{SPC}_{ju}]) B\check{u}d_j = ([\underline{Bud}_{jl}, \underline{Bud}_{ju}], [\overline{Bud}_{jl}, \overline{Bud}_{ju}]) \\
\forall\, i = 1,2,...,M,\ j = 1,2,...,N,\ k = 1,2,...,K,\ r = 1,2,...,L.
\end{array}\right\} \qquad (2)$$

Thus, after the introduction of these rough parameters, the above crisp model (1) is translated to the following rough model:

$$\left.\begin{array}{l}
\text{Maximize } Z = \sum_{r=1}^{R}\sum_{i=1}^{M}\sum_{j=1}^{N}\sum_{p=1}^{L}\sum_{k=1}^{K}\left\{\left([\underline{S}_{jl}^r, \underline{S}_{ju}^r], [\overline{S}_{jl}^r, \overline{S}_{ju}^r]\right)(1-\lambda_{ijpk}^r ds_{ijp}^{\alpha}) - \left([\underline{P}_{il}^r, \underline{P}_{iu}^r], [\overline{P}_{il}^r, \overline{P}_{iu}^r]\right)\right. \\
\left. -\left([\underline{C}_{ijpkl}^r, \underline{C}_{ijpku}^r], [\overline{C}_{ijpkl}^r, \overline{C}_{ijpku}^r]\right) ds_{ijp}\right\} x_{ijpk}^r - \sum_{i=1}^{M}\sum_{j=1}^{N}\sum_{p=1}^{L}\sum_{k=1}^{K} w(x_{ijpk}^r)\left([\underline{f}_{ijpkl}, \underline{f}_{ijpku}], [\overline{f}_{ijpkl}, \overline{f}_{ijpku}]\right) \\
\text{subject to } \sum_{j=1}^{N}\sum_{p=1}^{L}\sum_{k=1}^{K} x_{ijpk}^r \le \left([\underline{A}_{il}^r, \underline{A}_{iu}^r], [\overline{A}_{il}^r, \overline{A}_{iu}^r]\right) \quad \forall\ i = 1,2,...,M, r = 1,2,...,R \\
\sum_{i=1}^{M}\sum_{p=1}^{L}\sum_{k=1}^{K}(1-\lambda_{ijpk}^1 ds_{ijp}^{\alpha})x_{ijpk}^1 \ge \left\{\left([\underline{D}_{0jl}^1, \underline{D}_{0ju}^1], [\overline{D}_{0jl}^1, \overline{D}_{0ju}^1]\right) - \beta_j\left([\underline{S}_{jl}^1, \underline{S}_{ju}^1], [\overline{S}_{jl}^1, \overline{S}_{ju}^1]\right)\right. \\
\quad \left. +\tau\beta_j\left([\underline{S}_{jl}^2, \underline{S}_{ju}^2], [\overline{S}_{jl}^2, \overline{S}_{ju}^2]\right) - \theta_j^1\beta_j\left([\underline{S}_{jl}^3, \underline{S}_{ju}^3], [\overline{S}_{jl}^3, \overline{S}_{ju}^3]\right)\right\}, \quad \forall\ j = 1,2,...,N, r = 1,2,...,R \\
\sum_{i=1}^{M}\sum_{p=1}^{L}\sum_{k=1}^{K}(1-\lambda_{ijpk}^2 ds_{ijp}^{\alpha})x_{ijpk}^2 \ge \left\{\left([\underline{D}_{0jl}^2, \underline{D}_{0ju}^2], [\overline{D}_{0jl}^2, \overline{D}_{0ju}^2]\right) + \tau\beta_j\left([\underline{S}_{jl}^1, \underline{S}_{ju}^1], [\overline{S}_{jl}^1, \overline{S}_{ju}^1]\right)\right. \\
\quad \left. -\beta_j\left([\underline{S}_{jl}^2, \underline{S}_{ju}^2], [\overline{S}_{jl}^2, \overline{S}_{ju}^2]\right) - \theta_j^{11}\beta_j\left([\underline{S}_{jl}^3, \underline{S}_{ju}^3], [\overline{S}_{jl}^3, \overline{S}_{ju}^3]\right)\right\}, \quad \forall\ j = 1,2,...,N, r = 1,2,...,R \\
\sum_{i=1}^{M}\sum_{p=1}^{L}\sum_{k=1}^{K}(1-\lambda_{ijpk}^3 ds_{ijp}^{\alpha})x_{ijpk}^3 \ge \left\{\left([\underline{D}_{0jl}^3, \underline{D}_{0ju}^3], [\overline{D}_{0jl}^3, \overline{D}_{0ju}^3]\right) - \beta_j\theta_j^1\left([\underline{S}_{jl}^1, \underline{S}_{ju}^1], [\overline{S}_{jl}^1, \overline{S}_{ju}^1]\right)\right. \\
\quad \left. -\beta_j\theta_j^{11}\left([\underline{S}_{jl}^2, \underline{S}_{ju}^2], [\overline{S}_{jl}^2, \overline{S}_{ju}^2]\right) - \beta_j\left([\underline{S}_{jl}^3, \underline{S}_{ju}^3], [\overline{S}_{jl}^3, \overline{S}_{ju}^3]\right)\right\}, \quad \forall\ j = 1,2,...,N, r = 1,2,...,R \\
\sum_{r=1}^{R}\sum_{i=1}^{M}\sum_{j=1}^{N} x_{ijpk}^r \le \left([\underline{e}_{kpl}, \underline{e}_{kpu}], [\overline{e}_{kpl}, \overline{e}_{kpu}]\right) \quad \forall\ k = 1,2,...,K, p = 1,2,...,L \\
\sum_{r=1}^{R}\sum_{i=1}^{M}\sum_{p=1}^{L}\sum_{k=1}^{K}\left[1-\lambda_{ijpk}^r ds_{ijp}^{\alpha}\right] x_{ijpk}^r * SP^r \le \left([\underline{SPC}_{jl}^r, \underline{SPC}_{ju}^r], [\overline{SPC}_{jl}^r, \overline{SPC}_{ju}^r]\right) \quad \forall\ j = 1,2,...,N, \\
\sum_{r=1}^{R}\sum_{i=1}^{M}\sum_{p=1}^{L}\sum_{k=1}^{K}\left\{\left([\underline{P}_{il}^r, \underline{P}_{iu}^r], [\overline{P}_{il}^r, \overline{P}_{iu}^r]\right) + \left([\underline{C}_{ijpkl}^r, \underline{C}_{ijpku}^r], [\overline{C}_{ijpkl}^r, \overline{C}_{ijpku}^r]\right) ds_{ijp}\right\} x_{ijpk}^r \\
+\sum_{i=1}^{M}\sum_{p=1}^{L}\sum_{k=1}^{K} w(x_{ijpk}^r)\left([\underline{f}_{ijpkl}, \underline{f}_{ijpku}], [\overline{f}_{ijpkl}, \overline{f}_{ijpku}]\right) \le \left([\underline{Bud}_{jl}, \underline{Bud}_{ju}], [\overline{Bud}_{jl}, \overline{Bud}_{ju}]\right) \\
\quad \forall\ j = 1,2,...,N \\
\text{and } x_{ijpk}^r \ge 0\ \forall\ i = 1,2,...,M, j = 1,2,...,N, p = 1,2,...,L, k = 1,2,...,K, r = 1,2,...,R.
\end{array}\right\} \qquad (3)$$

According to the discussion of Linear Programming issues with Rough Interval Coefficient (LPRIC) in Appendix A.1, at first, we get two different transportation problems (TPIC-1, TPIC-2) with interval coefficients—and again, as discussed, TPIC-1 and TPIC-2 broken as TP-1, TP-2 and TP-3, TP-4. The above stated consideration can be graphically represented in Figure 2.

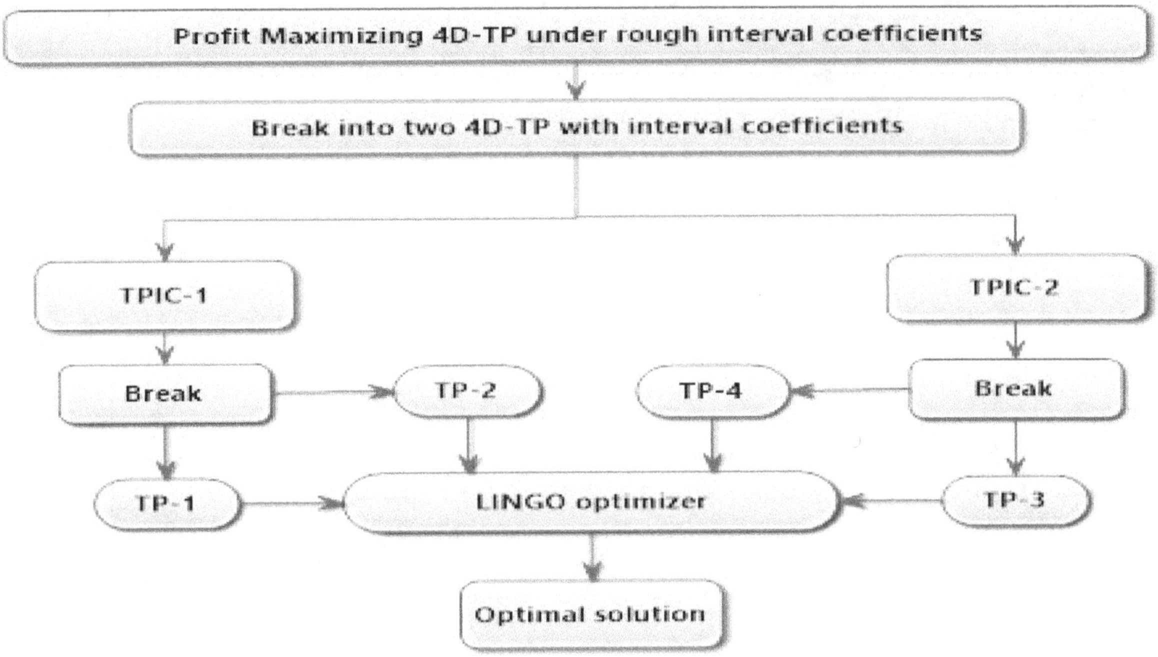

Figure 2. Optimization flowchart of 4D-TP under rough interval coefficients.

3.3. The Mathematical Form of TPIC-1

$$
\left.\begin{aligned}
&\text{Maximize } Z^1 = \sum_{r=1}^{R}\sum_{i=1}^{M}\sum_{j=1}^{N}\sum_{p=1}^{L}\sum_{k=1}^{K}\left\{[\underline{S}^r_{jl},\underline{S}^r_{ju}](1-\lambda^r_{ijpk}ds^{\alpha}_{ijp}) - [\underline{P}^r_{il},\underline{P}^r_{iu}] - [\underline{C}^r_{ijpkl},\underline{C}^r_{ijpku}]ds_{ijp}\right\}x^r_{ijpk}\\
&\qquad - \sum_{i=1}^{M}\sum_{j=1}^{N}\sum_{p=1}^{L}\sum_{k=1}^{K} w(x^r_{ijpk})[\underline{f}_{ijpkl},\underline{f}_{ijpku}]\\
&\text{s.t } \sum_{j=1}^{N}\sum_{p=1}^{L}\sum_{k=1}^{K} x^r_{ijpk} \le [\underline{A}^r_{il},\underline{A}^r_{iu}] \quad \forall\ i=1,2,\ldots,M, r=1,2,\ldots,R\\
&\sum_{i=1}^{M}\sum_{p=1}^{L}\sum_{k=1}^{K}(1-\lambda^1_{ijpk}ds^{\alpha}_{ijp})x^1_{ijpk} \ge [\underline{D}^1_{0jl},\underline{D}^1_{0ju}] - \beta_j[\underline{S}^1_{jl},\underline{S}^1_{ju}] + \tau\beta_j[\underline{S}^2_{jl},\underline{S}^2_{ju}] - \theta^1_j\beta_j[\underline{S}^3_{jl},\underline{S}^3_{ju}],\\
&\qquad \forall\ j=1,2,\ldots,N, r=1,2,\ldots,R\\
&\sum_{i=1}^{M}\sum_{p=1}^{L}\sum_{k=1}^{K}(1-\lambda^2_{ijpk}ds^{\alpha}_{ijp})x^2_{ijpk} \ge [\underline{D}^2_{0jl},\underline{D}^2_{0ju}] + \tau\beta_j[\underline{S}^1_{jl},\underline{S}^1_{ju}] - \beta_j[\underline{S}^2_{jl},\underline{S}^2_{ju}] - \theta^{11}_j\beta_j[\underline{S}^3_{jl},\underline{S}^3_{ju}],\\
&\qquad \forall\ j=1,2,\ldots,N, r=1,2,\ldots,R\\
&\sum_{i=1}^{M}\sum_{p=1}^{L}\sum_{k=1}^{K}(1-\lambda^3_{ijpk}ds^{\alpha}_{ijp})x^3_{ijpk} \ge [\underline{D}^3_{0jl},\underline{D}^3_{0ju}] - \beta_j\theta^1_j[\underline{S}^1_{jl},\underline{S}^1_{ju}] - \beta_j\theta^{11}_j[\underline{S}^2_{jl},\underline{S}^2_{ju}] - \beta_j[\underline{S}^3_{jl},\underline{S}^3_{ju}],\\
&\qquad \forall\ j=1,2,\ldots,N, r=1,2,\ldots,R\\
&\sum_{r=1}^{R}\sum_{i=1}^{M}\sum_{j=1}^{N} x^r_{ijpk} \le [\underline{e}_{kpl},\underline{e}_{kpu}] \ \forall\ k=1,2,\ldots,K, p=1,2,\ldots,L\\
&\sum_{r=1}^{R}\sum_{i=1}^{M}\sum_{p=1}^{L}\sum_{k=1}^{K}\left[1-\lambda^r_{ijpk}ds^{\alpha}_{ijp}\right]x^r_{ijpk} * SP^r \le \left([\underline{SPC}^r_{jl},\underline{SPC}^r_{ju}]\right) \quad \forall\ j=1,2,\ldots,N,\\
&\sum_{r=1}^{R}\sum_{i=1}^{M}\sum_{p=1}^{L}\sum_{k=1}^{K}\{[\underline{P}^r_{il},\underline{P}^r_{iu}] + [\underline{C}^r_{ijpkl},\underline{C}^r_{ijpku}]ds_{ijp}\}x^r_{ijpk} + \sum_{i=1}^{M}\sum_{p=1}^{L}\sum_{k=1}^{K} w(x^r_{ijpk})[\underline{f}_{ijpkl},\underline{f}_{ijpku}]\\
&\qquad \le [\underline{Bud}_{jl},\underline{Bud}_{ju}] \ \forall\ j=1,2,\ldots,N\\
&\text{and } x^r_{ijpk} \ge 0\ \forall\ i=1,2,\ldots,M, j=1,2,\ldots,N, p=1,2,\ldots,L, k=1,2,\ldots,K, r=1,2,\ldots,R.
\end{aligned}\right\} \quad (4)
$$

3.4. The Mathematical Form of TPIC-2

$$\left.\begin{array}{l}
\text{Maximize } Z^2 = \sum\limits_{r=1}^{R}\sum\limits_{i=1}^{M}\sum\limits_{j=1}^{N}\sum\limits_{p=1}^{L}\sum\limits_{k=1}^{K}\{[\overline{S}^r_{jl},\overline{S}^r_{ju}](1-\lambda^r_{ijpk}ds^{\alpha}_{ijp})-[\overline{P}^r_{il},\overline{P}^r_{iu}]-[\overline{C}^r_{ijpkl},\overline{C}^r_{ijpku}]ds_{ijp}\}x^r_{ijpk} \\
\qquad -\sum\limits_{i=1}^{M}\sum\limits_{j=1}^{N}\sum\limits_{p=1}^{L}\sum\limits_{k=1}^{K} w(x^r_{ijpk})[\overline{f}_{ijpkl},\overline{f}_{ijpku}] \\
\text{s.t } \sum\limits_{j=1}^{N}\sum\limits_{p=1}^{L}\sum\limits_{k=1}^{K} x^r_{ijpk} \le [\overline{A}^r_{il},\overline{A}^r_{iu}] \quad \forall\ i=1,2,\dots,M, r=1,2,\dots,R \\
\sum\limits_{i=1}^{M}\sum\limits_{p=1}^{L}\sum\limits_{k=1}^{K}(1-\lambda^1_{ijpk}ds^{\alpha}_{ijp})x^1_{ijpk} \ge [\overline{D}^1_{0jl},\overline{D}^1_{0ju}]-\beta_j[\overline{S}^1_{jl},\overline{S}^1_{ju}]+\tau\beta_j[\overline{S}^2_{jl},\overline{S}^2_{ju}]-\theta^1_j\beta_j[\overline{S}^3_{jl},\overline{S}^3_{ju}], \\
\qquad \forall\ j=1,2,\dots,N, r=1,2,\dots,R \\
\sum\limits_{i=1}^{M}\sum\limits_{p=1}^{L}\sum\limits_{k=1}^{K}(1-\lambda^2_{ijpk}ds^{\alpha}_{ijp})x^2_{ijpk} \ge [\overline{D}^2_{0jl},\overline{D}^2_{0ju}]+\tau\beta_j[\overline{S}^1_{jl},\overline{S}^1_{ju}]-\beta_j[\overline{S}^2_{jl},\overline{S}^2_{ju}]-\theta^{11}_j\beta_j([\overline{S}^3_{jl},\overline{S}^3_{ju}], \\
\qquad \forall\ j=1,2,\dots,N, r=1,2,\dots,R \\
\sum\limits_{i=1}^{M}\sum\limits_{p=1}^{L}\sum\limits_{k=1}^{K}(1-\lambda^3_{ijpk}ds^{\alpha}_{ijp})x^3_{ijpk} \ge [\overline{D}^3_{0jl},\overline{D}^3_{0ju}]-\beta_j\theta^1_j[\overline{S}^1_{jl},\overline{S}^1_{ju}]-\beta_j\theta^{11}_j[\overline{S}^2_{jl},\overline{S}^2_{ju}]-\beta_j[\overline{S}^3_{jl},\overline{S}^3_{ju}], \\
\qquad \forall\ j=1,2,\dots,N, r=1,2,\dots,R \\
\sum\limits_{r=1}^{R}\sum\limits_{i=1}^{M}\sum\limits_{j=1}^{N} x^r_{ijpk} \le [\overline{e}_{kpl},\overline{e}_{kpu}] \quad \forall\ k=1,2,\dots,K, p=1,2,\dots,L \\
\sum\limits_{r=1}^{R}\sum\limits_{i=1}^{M}\sum\limits_{p=1}^{L}\sum\limits_{k=1}^{K}\left[1-\lambda^r_{ijpk}ds^{\alpha}_{ijp}\right]x^r_{ijpk} * SP^r \le \left([\overline{SPC}^r_{jl},\overline{SPC}^r_{ju}]\right) \quad \forall\ j=1,2,\dots,N, \\
\sum\limits_{r=1}^{R}\sum\limits_{i=1}^{M}\sum\limits_{p=1}^{L}\sum\limits_{k=1}^{K}\{[\overline{P}^r_{il},\overline{P}^r_{iu}]+[\overline{C}^r_{ijpkl},\overline{C}^r_{ijpku}]ds_{ijp}\}x^r_{ijpk}+\sum\limits_{i=1}^{M}\sum\limits_{p=1}^{L}\sum\limits_{k=1}^{K} w(x^r_{ijpk})[\overline{f}_{ijpkl},\overline{f}_{ijpku}] \\
\qquad \le [\overline{Bud}_{jl},\overline{Bud}_{ju}] \ \forall\ j=1,2,\dots,N \\
\text{and } x^r_{ijpk} \ge 0 \ \forall\ i=1,2,\dots,M, j=1,2,\dots,N, p=1,2,\dots,L, k=1,2,\dots,K, r=1,2,\dots,R.
\end{array}\right\} \quad (5)$$

3.4.1. TP-1

$$\left.\begin{array}{l}
\text{Maximize } \underline{Z}^l = \sum\limits_{r=1}^{R}\sum\limits_{i=1}^{M}\sum\limits_{j=1}^{N}\sum\limits_{p=1}^{L}\sum\limits_{k=1}^{K}\left\{\underline{S}^r_{jl}(1-\lambda^r_{ijpk}ds^{\alpha}_{ijp})-\underline{P}^r_{iu}-\underline{C}^r_{ijpku}ds_{ijp}\right\}x^r_{ijpk}-\sum\limits_{i=1}^{M}\sum\limits_{j=1}^{N}\sum\limits_{p=1}^{L}\sum\limits_{k=1}^{K} w(x^r_{ijpk})\underline{f}_{ijpku} \\
\text{s.t } \sum\limits_{j=1}^{N}\sum\limits_{p=1}^{L}\sum\limits_{k=1}^{K} x^r_{ijpk} \le \underline{A}^r_{il} \quad \forall\ i=1,2,\dots,M, r=1,2,\dots,R \\
\sum\limits_{i=1}^{M}\sum\limits_{p=1}^{L}\sum\limits_{k=1}^{K}(1-\lambda^1_{ijpk}ds^{\alpha}_{ijp})x^1_{ijpk} \ge \left\{\underline{D}^1_{0ju}-\beta_j\underline{S}^1_{jl}+\tau\beta_j\underline{S}^2_{ju}-\theta^1_j\beta_j\underline{S}^3_{jl}\right\}, \quad \forall\ j=1,2,\dots,N, r=1,2,\dots,R \\
\sum\limits_{i=1}^{M}\sum\limits_{p=1}^{L}\sum\limits_{k=1}^{K}(1-\lambda^2_{ijpk}ds^{\alpha}_{ijp})x^2_{ijpk} \ge \left\{\underline{D}^2_{0ju}+\tau\beta_j\underline{S}^1_{ju}-\beta_j\underline{S}^2_{jl}-\theta^{11}_j\beta_j\underline{S}^3_{jl}\right\}, \quad \forall\ j=1,2,\dots,N, r=1,2,\dots,R \\
\sum\limits_{i=1}^{M}\sum\limits_{p=1}^{L}\sum\limits_{k=1}^{K}(1-\lambda^3_{ijpk}ds^{\alpha}_{ijp})x^3_{ijpk} \ge \left\{\underline{D}^3_{0ju}-\beta_j\theta^1_j\underline{S}^1_{jl}-\beta_j\theta^{11}_j\underline{S}^2_{jl}-\beta_j\underline{S}^3_{jl}\right\}, \quad \forall\ j=1,2,\dots,N, r=1,2,\dots,R \\
\sum\limits_{r=1}^{R}\sum\limits_{i=1}^{M}\sum\limits_{j=1}^{N} x^r_{ijpk} \le \underline{e}_{kpl} \quad k=1,2,\dots,K, p=1,2,\dots,L \\
\sum\limits_{r=1}^{R}\sum\limits_{i=1}^{M}\sum\limits_{p=1}^{L}\sum\limits_{k=1}^{K}\left[1-\lambda^r_{ijpk}ds^{\alpha}_{ijp}\right]x^r_{ijpk} * SP^r \le \underline{SPC}^r_{jl} \quad \forall\ j=1,2,\dots,N, \\
\sum\limits_{r=1}^{R}\sum\limits_{i=1}^{M}\sum\limits_{p=1}^{L}\sum\limits_{k=1}^{K}\left\{\underline{P}^r_{iu}+\underline{C}^r_{ijpku}ds_{ijp}\right\}x^r_{ijpk}+\sum\limits_{i=1}^{M}\sum\limits_{p=1}^{L}\sum\limits_{k=1}^{K} w(x^r_{ijpk})\underline{f}_{ijpku}\Big) \le \underline{Bud}_{jl}, \ \forall\ j=1,2,\dots,N \\
\text{and } x^r_{ijpk} \ge 0 \ \forall\ i=1,2,\dots,M, j=1,2,\dots,N, p=1,2,\dots,L, k=1,2,\dots,K, r=1,2,\dots,R.
\end{array}\right\} \quad (6)$$

3.4.2. TP-2

$$
\left.\begin{aligned}
&\text{Maximize } \underline{Z}^u = \sum_{r=1}^{R}\sum_{i=1}^{M}\sum_{j=1}^{N}\sum_{p=1}^{L}\sum_{k=1}^{K}\left\{\underline{S}^r_{ju}(1-\lambda^r_{ijpk}ds^{\alpha}_{ijp}) - \underline{P}^r_{il} - \underline{C}^r_{ijpkl}ds_{ijp}\right\}x^r_{ijpk} - \sum_{i=1}^{M}\sum_{j=1}^{N}\sum_{p=1}^{L}\sum_{k=1}^{K} w(x^r_{ijpk})\underline{f}_{ijpkl}\\
&\text{s.t } \sum_{j=1}^{N}\sum_{p=1}^{L}\sum_{k=1}^{K} x^r_{ijpk} \le \underline{A}^r_{iu} \quad \forall\ i = 1,2,\ldots,M, r = 1,2,\ldots,R\\
&\sum_{i=1}^{M}\sum_{p=1}^{L}\sum_{k=1}^{K}(1-\lambda^1_{ijpk}ds^{\alpha}_{ijp})x^1_{ijpk} \ge \left\{\underline{D}^1_{0jl} - \beta_j\underline{S}^1_{ju} + \tau\beta_j\underline{S}^2_{jl} - \theta^1_j\beta_j\underline{S}^3_{ju}\right\}, \quad \forall\ j = 1,2,\ldots,N, r = 1,2,\ldots,R\\
&\sum_{i=1}^{M}\sum_{p=1}^{L}\sum_{k=1}^{K}(1-\lambda^2_{ijpk}ds^{\alpha}_{ijp})x^2_{ijpk} \ge \left\{\underline{D}^2_{0jl} + \tau\beta_j\underline{S}^1_{jl} - \beta_j\underline{S}^2_{ju} - \theta^{11}_j\beta_j\underline{S}^3_{ju}\right\}, \quad \forall\ j = 1,2,\ldots,N, r = 1,2,\ldots,R\\
&\sum_{i=1}^{M}\sum_{p=1}^{L}\sum_{k=1}^{K}(1-\lambda^3_{ijpk}ds^{\alpha}_{ijp})x^3_{ijpk} \ge \left\{\underline{D}^3_{0jl} - \beta_j\theta^1_j\underline{S}^1_{ju} - \beta_j\theta^{11}_j\underline{S}^2_{ju} - \beta_j\underline{S}^3_{ju}\right\}, \quad \forall\ j = 1,2,\ldots,N, r = 1,2,\ldots,R\\
&\sum_{r=1}^{R}\sum_{i=1}^{M}\sum_{j=1}^{N} x^r_{ijpk} \le \underline{e}_{kpu} \quad \forall\ k = 1,2,\ldots,K, p = 1,2,\ldots,L\\
&\sum_{r=1}^{R}\sum_{i=1}^{M}\sum_{p=1}^{L}\sum_{k=1}^{K}\left[1-\lambda^r_{ijpk}ds^{\alpha}_{ijp}\right]x^r_{ijpk} * SP^r \le \underline{SPC}^r_{ju} \quad \forall\ j = 1,2,\ldots,N,\\
&\sum_{r=1}^{R}\sum_{i=1}^{M}\sum_{p=1}^{L}\sum_{k=1}^{K}\left\{\underline{P}^r_{il} + \underline{C}^r_{ijpkl}ds_{ijp}\right\}x^r_{ijpk} + \sum_{i=1}^{M}\sum_{p=1}^{L}\sum_{k=1}^{K} w(x^r_{ijpk})\underline{f}_{ijpkl} \le \underline{Bud}_{ju} \quad \forall\ j = 1,2,\ldots,N\\
&\text{and } x^r_{ijpk} \ge 0\ \forall\ i = 1,2,\ldots,M, j = 1,2,\ldots,N, p = 1,2,\ldots,L, k = 1,2,\ldots,K, r = 1,2,\ldots,R.
\end{aligned}\right\} \quad (7)
$$

3.4.3. TP-3

$$
\left.\begin{aligned}
&\text{Maximize } \overline{Z}^l = \sum_{r=1}^{R}\sum_{i=1}^{M}\sum_{j=1}^{N}\sum_{p=1}^{L}\sum_{k=1}^{K}\left\{\overline{S}^r_{jl}(1-\lambda^r_{ijpk}ds^{\alpha}_{ijp}) - \overline{P}^r_{iu} - \overline{C}^r_{ijpku}ds_{ijp}\right\}x^r_{ijpk} - \sum_{i=1}^{M}\sum_{j=1}^{N}\sum_{p=1}^{L}\sum_{k=1}^{K} w(x^r_{ijpk})\overline{f}_{ijpku}\\
&\text{subject to } \sum_{j=1}^{N}\sum_{p=1}^{L}\sum_{k=1}^{K} x^r_{ijpk} \le \overline{A}^r_{il} \quad \forall\ i = 1,2,\ldots,M, r = 1,2,\ldots,R\\
&\sum_{i=1}^{M}\sum_{p=1}^{L}\sum_{k=1}^{K}(1-\lambda^1_{ijpk}ds^{\alpha}_{ijp})x^1_{ijpk} \ge \left\{\overline{D}^1_{0ju} - \beta_j\overline{S}^1_{jl} + \tau\beta_j\overline{S}^2_{ju} - \theta^1_j\beta_j\overline{S}^3_{jl}\right\}, \quad j = 1,2,\ldots,N, r = 1,2,\ldots,R\\
&\sum_{i=1}^{M}\sum_{p=1}^{L}\sum_{k=1}^{K}(1-\lambda^2_{ijpk}ds^{\alpha}_{ijp})x^2_{ijpk} \ge \left\{\overline{D}^2_{0ju} + \tau\beta_j\overline{S}^1_{ju} - \beta_j\overline{S}^2_{jl} - \theta^{11}_j\beta_j\overline{S}^3_{jl}\right\}, \quad \forall\ j = 1,2,\ldots,N, r = 1,2,\ldots,R\\
&\sum_{i=1}^{M}\sum_{p=1}^{L}\sum_{k=1}^{K}(1-\lambda^3_{ijpk}ds^{\alpha}_{ijp})x^3_{ijpk} \ge \left\{\overline{D}^3_{0ju} - \beta_j\theta^1_j\overline{S}^1_{jl} - \beta_j\theta^{11}_j\overline{S}^2_{jl} - \beta_j\overline{S}^3_{jl}\right\}, \quad \vee\ j - 1,2,\ldots,N, r = 1,2,\ldots,R\\
&\sum_{r=1}^{R}\sum_{i=1}^{M}\sum_{j=1}^{N} x^r_{ijpk} \le \overline{e}_{kpl} \quad k = 1,2,\ldots,K, p = 1,2,\ldots,L\\
&\sum_{r=1}^{R}\sum_{i=1}^{M}\sum_{p=1}^{L}\sum_{k=1}^{K}\left[1-\lambda^r_{ijpk}ds^{\alpha}_{ijp}\right]x^r_{ijpk} * SP^r \le \overline{SPC}^r_{jl} \quad \forall\ j = 1,2,\ldots,N,\\
&\sum_{r=1}^{R}\sum_{i=1}^{M}\sum_{p=1}^{L}\sum_{k=1}^{K}\left\{\overline{P}^r_{iu} + \overline{C}^r_{ijpku}ds_{ijp}\right\}x^r_{ijpk} + \sum_{i=1}^{M}\sum_{p=1}^{L}\sum_{k=1}^{K} w(x^r_{ijpk})\overline{f}_{ijpku} \le \overline{Bud}_{jl} \quad \forall \quad j = 1,2,\ldots,N\\
&\text{and } x^r_{ijpk} \ge 0\ \forall\ i = 1,2,\ldots,M, j = 1,2,\ldots,N, p = 1,2,\ldots,L, k = 1,2,\ldots,K, r = 1,2,\ldots,R.
\end{aligned}\right\} \quad (8)
$$

3.4.4. TP-4

$$
\left.\begin{aligned}
&\text{Maximize } \overline{Z}^u = \sum_{r=1}^{R}\sum_{i=1}^{M}\sum_{j=1}^{N}\sum_{p=1}^{L}\sum_{k=1}^{K}\left\{\overline{S}^r_{ju}(1-\lambda^r_{ijpk}ds^{\alpha}_{ijp}) - \overline{P}^r_{il} - \overline{C}^r_{ijpkl}ds_{ijp}\right\}x^r_{ijpk} - \sum_{i=1}^{M}\sum_{j=1}^{N}\sum_{p=1}^{L}\sum_{k=1}^{K} w(x^r_{ijpk})\overline{f}_{ijpkl}\\
&\text{subject to } \sum_{j=1}^{N}\sum_{p=1}^{L}\sum_{k=1}^{K} x^r_{ijpk} \le \overline{A}^r_{iu} \quad \forall\ i = 1,2,\ldots,M, r = 1,2,\ldots,R\\
&\sum_{i=1}^{M}\sum_{p=1}^{L}\sum_{k=1}^{K}(1-\lambda^1_{ijpk}ds^{\alpha}_{ijp})x^1_{ijpk} \ge \left\{\overline{D}^1_{0jl} - \beta_j\overline{S}^1_{ju} + \tau\beta_j\overline{S}^2_{jl} - \theta^1_j\beta_j\overline{S}^3_{ju}\right\}, \quad \forall\ j = 1,2,\ldots,N, r = 1,2,\ldots,R\\
&\sum_{i=1}^{M}\sum_{p=1}^{L}\sum_{k=1}^{K}(1-\lambda^2_{ijpk}ds^{\alpha}_{ijp})x^2_{ijpk} \ge \left\{\overline{D}^2_{0jl} + \tau\beta_j\overline{S}^1_{jl} - \beta_j\overline{S}^2_{ju} - \theta^{11}_j\beta_j\overline{S}^3_{ju}\right\}, \quad \forall\ j = 1,2,\ldots,N, r = 1,2,\ldots,R\\
&\sum_{i=1}^{M}\sum_{p=1}^{L}\sum_{k=1}^{K}(1-\lambda^3_{ijpk}ds^{\alpha}_{ijp})x^3_{ijpk} \ge \left\{\overline{D}^3_{0jl} - \beta_j\theta^1_j\overline{S}^1_{ju} - \beta_j\theta^{11}_j\overline{S}^2_{ju} - \beta_j\overline{S}^3_{ju}\right\}, \quad \forall\ j = 1,2,\ldots,N, r = 1,2,\ldots,R\\
&\sum_{r=1}^{R}\sum_{i=1}^{M}\sum_{j=1}^{N} x^r_{ijpk} \le \overline{e}_{kpu} \quad k = 1,2,\ldots,K, p = 1,2,\ldots,L\\
&\sum_{r=1}^{R}\sum_{i=1}^{M}\sum_{p=1}^{L}\sum_{k=1}^{K}\left[1-\lambda^r_{ijpk}ds^{\alpha}_{ijp}\right]x^r_{ijpk} * SP^r \le \overline{SPC}^r_{ju} \quad \forall\ j = 1,2,\ldots,N,\\
&\sum_{r=1}^{R}\sum_{i=1}^{M}\sum_{p=1}^{L}\sum_{k=1}^{K}\left\{\overline{P}^r_{il} + \overline{C}^r_{ijpkl}ds_{ijp}\right\}x^r_{ijpk} + \sum_{i=1}^{M}\sum_{p=1}^{L}\sum_{k=1}^{K} w(x^r_{ijpk})\overline{f}_{ijpkl} \le \overline{Bud}_{ju} \quad \forall\ j = 1,2,\ldots,N\\
&\text{and } x^r_{ijpk} \ge 0\ \forall\ i = 1,2,\ldots,M, j = 1,2,\ldots,N, p = 1,2,\ldots,L, k = 1,2,\ldots,K, r = 1,2,\ldots,R.
\end{aligned}\right\} \quad (9)
$$

3.4.5. Approach-2

After use of the Expected Value Method, the Rough transportation problem (Model-II) mathematically reduces to:

$$
\left.\begin{aligned}
&\text{Maximize } E\Big[Z\Big] = E\Big[\sum_{r=1}^{R}\sum_{i=1}^{M}\sum_{j=1}^{N}\sum_{p=1}^{L}\sum_{k=1}^{K}\Big\{\Big([\underline{S}^r_{jl},\underline{S}^r_{ju}],[\overline{S}^r_{jl},\overline{S}^r_{ju}]\Big)(1-\lambda^r_{ijpk}ds^\alpha_{ijp}) - \Big([\underline{P}^r_{il},\underline{P}^r_{iu}],[\overline{P}^r_{il},\overline{P}^r_{iu}]\Big)\\
&-\Big([\underline{C}^r_{ijpkl},\underline{C}^r_{ijpku}],[\overline{C}^r_{ijpkl},\overline{C}^r_{ijpku}]\Big)ds_{ijp}\Big\}x^r_{ijpk} - \sum_{i=1}^{M}\sum_{j=1}^{N}\sum_{p=1}^{L}\sum_{k=1}^{K} w(x^r_{ijpk})\Big([\underline{f}_{ijpkl},\underline{f}_{ijpku}],[\overline{f}_{ijpkl},\overline{f}_{ijpku}]\Big)\Big)\Big]\\
&\text{subject to } \sum_{j=1}^{N}\sum_{p=1}^{L}\sum_{k=1}^{K} x^r_{ijpk} \le E\Big[\Big([\underline{A}^r_{il},\underline{A}^r_{iu}],[\overline{A}^r_{il},\overline{A}^r_{iu}]\Big)\Big] \quad \forall\ i=1,2,\ldots,M, r=1,2,\ldots,R\\
&\sum_{i=1}^{M}\sum_{p=1}^{L}\sum_{k=1}^{K}(1-\lambda^1_{ijpk}ds^\alpha_{ijp})x^1_{ijpk} \ge E\Big[\Big\{\Big([\underline{D}^1_{0jl},\underline{D}^1_{0ju}],[\overline{D}^1_{0jl},\overline{D}^1_{0ju}]\Big) - \beta_j\Big([\underline{S}^1_{jl},\underline{S}^1_{ju}],[\overline{S}^1_{jl},\overline{S}^1_{ju}]\Big)\\
&\qquad +\tau\beta_j\Big([\underline{S}^2_{jl},\underline{S}^2_{ju}],[\overline{S}^2_{jl},\overline{S}^2_{ju}]\Big) - \theta^1_j\beta_j\Big([\underline{S}^3_{jl},\underline{S}^3_{ju}],[\overline{S}^3_{jl},\overline{S}^3_{ju}]\Big)\Big\}\Big], \quad \forall\ j=1,2,\ldots,N, r=1,2,\ldots,R\\
&\sum_{i=1}^{M}\sum_{p=1}^{L}\sum_{k=1}^{K}(1-\lambda^2_{ijpk}ds^\alpha_{ijp})x^2_{ijpk} \ge E\Big[\Big\{\Big([\underline{D}^2_{0jl},\underline{D}^2_{0ju}],[\overline{D}^2_{0jl},\overline{D}^2_{0ju}]\Big) + \tau\beta_j\Big([\underline{S}^1_{jl},\underline{S}^1_{ju}],[\overline{S}^1_{jl},\overline{S}^1_{ju}]\Big)\\
&\qquad -\beta_j\Big([\underline{S}^2_{jl},\underline{S}^2_{ju}],[\overline{S}^2_{jl},\overline{S}^2_{ju}]\Big) - \theta^{11}_j\beta_j\Big([\underline{S}^3_{jl},\underline{S}^3_{ju}],[\overline{S}^3_{jl},\overline{S}^3_{ju}]\Big)\Big\}\Big], \quad \forall\ j=1,2,\ldots,N, r=1,2,\ldots,R\\
&\sum_{i=1}^{M}\sum_{p=1}^{L}\sum_{k=1}^{K}(1-\lambda^3_{ijpk}ds^\alpha_{ijp})x^3_{ijpk} \ge E\Big[\Big\{\Big([\underline{D}^3_{0jl},\underline{D}^3_{0ju}],[\overline{D}^3_{0jl},\overline{D}^3_{0ju}]\Big) - \beta_j\theta^1_j\Big([\underline{S}^1_{jl},\underline{S}^1_{ju}],[\overline{S}^1_{jl},\overline{S}^1_{ju}]\Big)\\
&\qquad -\beta_j\theta^{11}_j\Big([\underline{S}^2_{jl},\underline{S}^2_{ju}],[\overline{S}^2_{jl},\overline{S}^2_{ju}]\Big) - \beta_j\Big([\underline{S}^3_{jl},\underline{S}^3_{ju}],[\overline{S}^3_{jl},\overline{S}^3_{ju}]\Big)\Big\}\Big], \quad \forall\ j=1,2,\ldots,N, r=1,2,\ldots,R\\
&\sum_{r=1}^{R}\sum_{i=1}^{M}\sum_{j=1}^{N} x^r_{ijpk} \le E\Big[\Big([\underline{e}_{kpl},\underline{e}_{kpu}],[\overline{e}_{kpl},\overline{e}_{kpu}]\Big)\Big] \quad \forall\ k=1,2,\ldots,K, p=1,2,\ldots,L\\
&\sum_{r=1}^{R}\sum_{i=1}^{M}\sum_{p=1}^{L}\sum_{k=1}^{K}\Big[1-\lambda^r_{ijpk}ds^\alpha_{ijp}\Big]x^r_{ijpk} * SP^r \le E\Big[\Big([\underline{SPC}^r_{jl},\underline{SPC}^r_{ju}],[\overline{SPC}^r_{jl},\overline{SPC}^r_{ju}]\Big)\Big] \quad \forall\ j=1,2,\ldots,N,\\
&\sum_{r=1}^{R}\sum_{i=1}^{M}\sum_{p=1}^{L}\sum_{k=1}^{K} E\Big[\Big\{\Big([\underline{P}^r_{il},\underline{P}^r_{iu}],[\overline{P}^r_{il},\overline{P}^r_{iu}]\Big) + \Big([\underline{C}^r_{ijpkl},\underline{C}^r_{ijpku}],[\overline{C}^r_{ijpkl},\overline{C}^r_{ijpku}]\Big)ds_{ijp}\Big\}x^r_{ijpk}\\
&+\sum_{i=1}^{M}\sum_{p=1}^{L}\sum_{k=1}^{K} w(x^r_{ijpk})\Big([\underline{f}_{ijpkl},\underline{f}_{ijpku}],[\overline{f}_{ijpkl},\overline{f}_{ijpku}]\Big)\Big] \le E\Big[\Big([\underline{Bud}_{jl},\underline{Bud}_{ju}],[\overline{Bud}_{jl},\overline{Bud}_{ju}]\Big)\Big]\\
&\qquad \forall\ j=1,2,\ldots,N\\
&\text{and } x^r_{ijpk} \ge 0\ \forall\ i=1,2,\ldots,M, j=1,2,\ldots,N, p=1,2,\ldots,L, k=1,2,\ldots,K, r=1,2,\ldots,R.
\end{aligned}\right\} \tag{10}
$$

The problem (16) can be written as:

$$
\left.\begin{aligned}
&\text{Maximize } Z = \sum_{r=1}^{R}\sum_{i=1}^{M}\sum_{j=1}^{N}\sum_{p=1}^{L}\sum_{k=1}^{K}\Big[\{\eta(\underline{S}^r_{jl}+\underline{S}^r_{ju}) + (1-\eta)(\overline{S}^r_{jl}+\overline{S}^r_{ju})\}/2 * (1-\lambda^r_{ijpk}ds^\alpha_{ijp})\\
&-\{\eta(\underline{P}^r_{il}+\underline{P}^r_{iu}) + (1-\eta)(\overline{P}^r_{il}+\overline{P}^r_{iu})\}/2 - \{\eta(\underline{C}^r_{ijpkl}+\underline{C}^r_{ijpku}) + (1-\eta)(\overline{C}^r_{ijpkl}+\overline{C}^r_{ijpku})\}/2 * ds_{ijp}\Big]x^r_{ijpk}\\
&\qquad - \sum_{i=1}^{M}\sum_{j=1}^{N}\sum_{p=1}^{L}\sum_{k=1}^{K} w(x^r_{ijpk})\Big[\{\eta(\underline{f}_{ijpkl}+\underline{f}_{ijpku}) + (1-\eta)(\overline{f}_{ijpkl}+\overline{f}_{ijpku})\}/2\Big]\\
&\text{subject to } \sum_{j=1}^{N}\sum_{p=1}^{L}\sum_{k=1}^{K} x^r_{ijpk} \le \Big[\{\eta(\underline{A}^r_{il}+\underline{A}^r_{iu}) + (1-\eta)(\overline{A}^r_{il}+\overline{A}^r_{iu})\}/2\Big] \quad \forall\ i=1,2,\ldots,M, r=1,2,\ldots,R\\
&\sum_{i=1}^{M}\sum_{p=1}^{L}\sum_{k=1}^{K}(1-\lambda^1_{ijpk}ds^\alpha_{ijp})x^1_{ijpk} \ge \Big[\{\eta(\underline{D}^1_{0jl}+\underline{D}^1_{0ju}) + (1-\eta)(\overline{D}^1_{0jl}+\overline{D}^1_{0ju})\}/2 - \beta_j\{\eta(\underline{S}^1_{jl}+\underline{S}^1_{ju}) + (1-\eta)(\overline{S}^1_{jl}+\overline{S}^1_{ju})\}/2\\
&+\tau\beta_j\{\eta(\underline{S}^2_{jl}+\underline{S}^2_{ju}) + (1-\eta)(\overline{S}^2_{jl}+\overline{S}^2_{ju})\}/2 - \theta^1_j\beta_j\{(\underline{S}^3_{jl}+\underline{S}^3_{ju}) + (1-\eta)(\overline{S}^3_{jl}+\overline{S}^3_{ju})\}/2\Big], \quad \forall\ j=1,2,\ldots,N, r=1,2,\ldots,R\\
&\sum_{i=1}^{M}\sum_{p=1}^{L}\sum_{k=1}^{K}(1-\lambda^2_{ijpk}ds^\alpha_{ijp})x^2_{ijpk} \ge \Big[\{\eta(\underline{D}^2_{0jl}+\underline{D}^2_{0ju}) + (1-\eta)(\overline{D}^2_{0jl}+\overline{D}^2_{0ju})\}/2 + \tau\beta_j\{\eta(\underline{S}^1_{jl}+\underline{S}^1_{ju}) + (1-\eta)(\overline{S}^1_{jl}+\overline{S}^1_{ju})\}/2\\
&-\beta_j\{\eta(\underline{S}^2_{jl}+\underline{S}^2_{ju}) + (1-\eta)(\overline{S}^2_{jl}+\overline{S}^2_{ju})/2 - \theta^{11}_j\beta_j\{\eta(\underline{S}^3_{jl}+\underline{S}^3_{ju}) + (1-\eta)(\overline{S}^3_{jl}+\overline{S}^3_{ju})\}/2\Big], \quad \forall\ j=1,2,\ldots,N, r=1,2,\ldots,R\\
&\sum_{i=1}^{M}\sum_{p=1}^{L}\sum_{k=1}^{K}(1-\lambda^3_{ijpk}ds^\alpha_{ijp})x^3_{ijpk} \ge \Big[\{\eta(\underline{D}^3_{0jl}+\underline{D}^3_{0ju}) + (1-\eta)(\overline{D}^3_{0jl}+\overline{D}^3_{0ju})\}/2 - \beta_j\theta^1_j\{\eta(\underline{S}^1_{jl}+\underline{S}^1_{ju}) + (1-\eta)(\overline{S}^1_{jl}+\overline{S}^1_{ju})\}/2\\
&-\beta_j\theta^{11}_j\{\eta(\underline{S}^2_{jl}+\underline{S}^2_{ju}) + (1-\eta)(\overline{S}^2_{jl}+\overline{S}^2_{ju})\}/2 - \beta_j\{\eta(\underline{S}^3_{jl}+\underline{S}^3_{ju}) + (1-\eta)(\overline{S}^3_{jl}+\overline{S}^3_{ju})\}/2\Big], j=1,2,\ldots,N, r=1,2,\ldots,R\\
&\sum_{r=1}^{R}\sum_{i=1}^{M}\sum_{j=1}^{N} x^r_{ijpk} \le \Big[\eta(\underline{e}_{kpl}+\underline{e}_{kpu}) + (1-\eta)(\overline{e}_{kpl}+\overline{e}_{kpu})]/2)\Big] \quad \forall\ k=1,2,\ldots,K, p=1,2,\ldots,L\\
&\sum_{r=1}^{R}\sum_{i=1}^{M}\sum_{p=1}^{L}\sum_{k=1}^{K}\Big[1-\lambda^r_{ijpk}ds^\alpha_{ijp}\Big]x^r_{ijpk} * SP^r \le \Big[\{\eta(\underline{SPC}^r_{jl}+\underline{SPC}^r_{ju}) + (1-\eta)(\overline{SPC}^r_{jl}+\overline{SPC}^r_{ju})\}/2\Big] \quad \forall\ j=1,2,\ldots,N,\\
&\sum_{r=1}^{R}\sum_{i=1}^{M}\sum_{p=1}^{L}\sum_{k=1}^{K}\Big[\{\eta(\underline{P}^r_{il} + (1-\eta)\underline{P}^r_{iu}) + (\overline{P}^r_{il}+\overline{P}^r_{iu})\}/2 + \{\eta(\underline{C}^r_{ijpkl}+\underline{C}^r_{ijpku}) + (1-\eta)(\overline{C}^r_{ijpkl}+\overline{C}^r_{ijpku})\}/2 * ds_{ijp}\Big]x^r_{ijpk} +\\
&\sum_{i=1}^{M}\sum_{p=1}^{L}\sum_{k=1}^{K} w(x^r_{ijpk})\Big[\{\eta(\underline{f}_{ijpkl}+\underline{f}_{ijpku}) + (1-\eta)(\overline{f}_{ijpkl}+\overline{f}_{ijpku})\}/2\Big] \le \Big[\{\eta(\underline{Bud}_{jl}+\underline{Bud}_{ju}) + (1-\eta)(\overline{Bud}_{jl}+\overline{Bud}_{ju})\}/2\Big]\\
&\quad \forall\ j=1,2,\ldots,N\\
&\text{and } x^r_{ijpk} \ge 0\ \forall\ i=1,2,\ldots,M, j=1,2,\ldots,N, p=1,2,\ldots,L, k=1,2,\ldots,K, r=1,2,\ldots,R.
\end{aligned}\right\} \tag{11}
$$

4. Numerical Experiments

Let us consider a real life transportation system where number of sources = 2 (i.e., $M = 2$), number of destination = 2 (i.e., $N = 2$), number of vehicles = 2 (i.e., $K = 2$) and number of routes = 2 ($L = 2$), number of items = 3 (i.e., $R = 3$). The input values of unit transportation costs, fixed charges are of crisp and rough interval forms which are presented in Tables 2 and 3. The amount of breakability in a crisp environment is also presented in Table 2. Here, it is assumed that there are three types of products. The 1st and 2nd ones are the substitute, the 1st and 3rd are complementary, and the 2nd and 3rd are also complimentary.

Table 2. Unit cost of transportation (C^r_{ijpk}), fixed charges (f_{ijpk}), breakability (λ^r_{ijpk}) for proposed Models.

	p	1				2			
	k	1		2		1		2	
	i/j	1	2	1	2	1	2	1	2
		Model-I							
		(Crisp Model)							
C^1_{ijkp}	1	0.5	2.1	1.7	1.12	0.85	0.5	1.38	1.28
	2	1.21	1.28	2.05	0.6	1.65	2.72	1.66	1.78
C^2_{ijkp}	1	1.5	2.25	2.0	1.3	0.08	0.7	1.05	1.25
	2	1.05	2.15	2.2	0.9	1.35	2.25	1.85	0.98
C^3_{ijkp}	1	1.3	1.2	1.91	1.2	0.9	0.94	0.73	0.93
	2	1.22	1.3	2	0.7	1.8	3	1.8	1.8
λ^1_{ijkp}	1	0.01	0.01	0.01	0.01	0.02	0.02	0.02	0.02
	2	0.01	0.01	0.01	0.01	0.02	0.02	0.02	0.02
λ^2_{ijkp}	1	0.01	0.01	0.01	0.01	0.02	0.02	0.02	0.02
	2	0.01	0.01	0.01	0.01	0.02	0.02	0.02	0.02
λ^3_{ijkp}	1	0.01	0.01	0.01	0.01	0.02	0.02	0.02	0.02
	2	0.01	0.01	0.01	0.01	0.02	0.02	0.02	0.02
f_{ijkp}	1	1.9	1.7	1.1	1.3	1.2	1.4	0.3	1.8
	2	1.2	1.85	0.9	1.5	1.5	2.1	2.1	1.6
		Model-II							
		(Rough Model)							
$\check{C}^1_{ijkp}$	1	([1, 2], [0.7, 2.3])	([0.5, 1.5], [0.5, 2])	([1.1, 2.3], [1, 2.5])	([0.8, 1.3], [0.5, 2])	([1.4, 2.3], [1.2, 2.8])	([0.7, 1], [0.5, 1.5])	([1.2, 2], [1.2, 2.3])	([1.2, 2.4], [1, 3])
	2	([1.2, 2.5], [1, 3])	([0.9, 2], [0.5, 3.2])	([1.1, 2], [0.7, 2.5])	([1, 2], [0.9, 2.6])	([1.1, 2.2], [1, 2.7])	([1, 2.5], [0.3, 3])	([0.9, 2.1], [0.7, 2.8])	([1.3, 2.5], [1.1, 3.5])
$\check{C}^2_{ijkp}$	1	([1, 2], [0.7, 2.3])	([0.5, 1.5], [0.5, 2])	([1.1, 2.3], [1, 2.5])	([0.8, 1.3], [0.5, 2])	([1.4, 2.3], [1.2, 2.8])	([0.7, 1], [0.5, 1.5])	([1.2, 2], [1.2, 2.3])	([1.2, 2.4], [1, 3])
	2	([1.2, 2.5], [1, 3])	([0.9, 2], [0.5, 3.2])	([1.1, 2], [0.7, 2.5])	([1, 2], [0.9, 2.6])	([1.1, 2.2], [1, 2.7])	([1, 2.5], [0.3, 3])	([0.9, 2.1], [0.7, 2.8])	([1.3, 2.5], [1.1, 3.5])
$\check{C}^3_{ijkp}$	1	([1.3, 2.7], [1, 3])	([1.7, 2.8], [1.2, 3.1])	([1.5, 2.5], [1.4, 3.1])	([1.6, 2.4], [1.5, 3.5])	([1.3, 3], [1.2, 3.5])	([1.4, 2], [1.1, 2.5])	([1.5, 2.6], [1.3, 3])	([1.8, 2.7], [1.6, 3.4])
	2	([1.5, 2.5], [1, 3.5])	([1.3, 3], [1.1, 3.7])	([1.7, 2.7], [1.5, 3])	([1.4, 3], [1.2, 3.2])	([2, 2.7], [1.8, 3])	([1.3, 2.2], [1, 2.5])	([1.6, 2.1], [1.5, 2.8])	([1.6, 2.5], [1.4, 2.8])
$\check{f}_{ijkp}$	1	([0.5, 1.5], [0.5, 2])	([0.7, 1], [0.6, 1.5])	([0.7, 1.3], [0.4, 2])	([0.8, 1.6], [0.3, 1.7])	([0.7, 1.4], [0.6, 2])	([0.8, 1.8], [0.3, 2.3])	([1, 1.5], [0.8, 1.8])	([0.9, 1.8], [0.7, 2.1])
	2	([0.2, 0.5], [0.1, 1])	([0.3, 0.6], [0.2, 1])	([0.9, 1.7], [0.6, 1.8])	([1, 1.5], [0.5, 1.6])	([0.9, 1.5], [0.4, 1.9])	([0.5, 1.6], [0.2, 2])	([0.4, 0.9], [0.2, 1])	([1, 1.7], [0.5, 2])

The other parametric magnitudes such as unit purchasing price and unit selling price, capacities of conveyances, available space, budget, sources and demands of the Models-I and -II are given in Tables 3 and 4. Let the unit space for 1st, 2nd and 3rd items are two units, three units and two units.

Table 3. Parametric values for the Models.

Models	Source	Demand	Capacities. of Conveyance.
	$(A_1^1, A_1^2, A_1^3$ $A_2^1, A_2^2, A_2^3)$	$(D_{01}^1, D_{01}^2, D_{01}^3$ $D_{02}^1, D_{02}^2, D_{02}^3)$	(e_1^1, e_1^2, e_1^3) (e_2^1, e_2^2, e_2^3)
-I	(90, 80, 85 85, 75, 70)	(75, 73, 70, 65, 63, 60)	(90, 85, 80 85, 70, 75)
-II	{([89.5, 90], [88.6, 91]), ([79.6, 80], [78.5, 81.3]), ([84.7, 85], [83.5, 86]), ([84.4, 85], [83.6, 86]), ([74, 75], [73.6, 76]), ([69.4, 70], [68.7, 71])}	{([74, 75], [73.6, 76]), ([72.6, 73], [72, 74]), ([69.4, 70], [68.7, 71]), ([64.5,65], [64, 66]), ([62.4, 63], [62, 64]), ([59.4, 60], [59, 61])}	{([89, 90], [88, 91]), ([84, 85], [83, 86.3]), ([79, 80], [78, 81.3]), ([84, 85], [83, 86.3]), ([69, 70], [68, 71]), ([74, 75], [73, 76])}

Table 4. Parametric values for the Models.

Models	Purchasing Costs	Unit Selling Price	Budget	
	$(P_1^1, P_1^2, P_1^3$ $P_2^1, P_2^2, P_2^3)$	$(S_1^1, S_1^2, S_1^3$ $S_2^1, S_2^2, S_2^3)$	(Bud_1, Bud_2, Bud_3)	(SPC_1, SPC_2, SPC_3)
-I	(9, 8, 10 6, 7, 8.5)	(52, 30, 24, 46, 28, 20)	(3000, 2500, 2700)	(510, 520, 525)
-III	{([8.3, 9], [8, 10]), ([7.5, 8], [7, 9]), ([9, 10], [8, 10.5]), ([5.5, 6], [5, 7]), ([6.5, 7], [6, 7.8]), ([8, 8.5], [7.5, 9])}	{([51.5, 52], [51, 53]), ([29, 30], [28, 31]), ([23.5, 24], [23, 25]), ([45.5,45], [45.3, 46]), ([27.4, 28], [27, 29]), ([19.5, 20], [19.1, 21])}	{([8250,5000], [7500,8600]), ([7500,8600], [4500,4600]), ([4400,4700]), [4400,4700])}	{([509.6,510.3], [509, 511]), ([519.4,520], [519, 521]), ([524, 525]), [523,526])}

By considering the general demand function, it is necessary to define β_j, θ_j^1, θ_j^{11} and τ. These parameters are defined as follows: $\beta_j = 0.1$, $\theta_j^1 = 0.052$, $\theta_j^{11} = 0.051$ and $\tau = 0.05$. The following Table 5 represent the distances from different origins to destinations by route-1 and route-2.

Table 5. Distances (ds_{ijp}) of the routes for the Models.

Route	Destination	Origin-1	Origin-2
1	1	35	30
	2	40	25
2	1	30	35
	2	25	30

Now, the problem is to determine the optimal policy of transportation to maximize the profit. Under the objective of maximization of profit subjected to the traditional constraints, available space, and budget constraints for the above given input data, the crisp model is solved using Generalised Reduced Gradient (GRG) method and rough model by (i) Hamzehee et al and (ii) Expected Value Techniques along with GRG method. The results are presented in Table 6 (Crisp Model and Rough Model by Hamzehee Method) and Table 7 (Rough Model by Expected Value Technique). The convergence of the GRG method is also well established (shown in Appendix A.5) [33].

Table 6. Optimum results for Model-I and Model-II (Hamzehee Method) by Lingo.

Models	Crisp Model	4D Rough Model			
		TP-1	TP-2	TP-3	TP-4
Optimal profit	6966.775	5293.95	8058.61	4109.11	9130.81
		Set-1	Set-2	Set-3	Set-4
	x11111 = 63.07 x11122 = 52.75 x11222 = 8.75 x21121 = 67.15 x21122 = 47.1 x22211 = 27.74 x22212 = 39.9 x22213 = 26.9 x11213 = 7.9 x11123 = 15.91 others are zero	x11221 = 14.04 x12112 = 46 x12211 = 29.96 x21121 = 50.46 x21122 = 21.84 x21211 = 33.04 x21222 = 45.16 x21223 = 26.07 x21123 = 27.16 x11123 = 15.91 others are zero	x11122 = 48.7 x11221 = 65.3 x11222 = 49.6 x12211 = 22.1 x12221 = 42.2 x21111 = 44 x21121 = 9.80 x21123 = 14.80 x22123 = 23.80 x11123 = 15.91 x21122 = 33.4 x22212 = 14.9 other are zero	x11111 = 54.4 x11221 = 32.2 x11222 = 17.9 x12211 = 0.2 x21121 = 20.3 x21122 = 19.9 x22211 = 31.31 x22213 = 2.23 x22113 = 62.04 x22212 = 18.65 other are zero	x11111 = 52.44 x11112 = 25.41 x11222 = 19.9 x12112 = 9.09 x12211 = 62.24 x21121 = 54.71 x21122 = 65.91 x21123 = 27.09 x22123 = 15.09 x21113 = 28.51 x21211 = 8.29 x22212 = 31.21 other are zero

Table 7. Optimum results for Model-II (Expected Value Technique) by Lingo.

η = 0.0	η = 0.2	η = 0.4	η = 0.5	η = 0.6	η = 0.8	η = 1.0
6351.76	6393.95	6412.61	6463.11	6488.81	6504.54	6523.54
x11111 = 32.75 x11222 = 30.33 x12211 = 45.65 x21121 = 52.20 x21122 = 10.8 x21223 = 30.74 x22111 = 23.9 x22213 = 32.9 and all others variables are zero	x11111 = 32.84 x11222 = 30.33 x12211 = 45.576 x21121 = 52.16 x21122 = 18.75 x21223 = 30.44 x22111 = 23.13 x22213 = 32.06 and all others variables are zero	x11111 = 32.86 x22212 = 27.46 x12211 = 45.52 x21121 = 52.13 x21122 = 10.712 x21223 = 30.04 x22111 = 23.80 x22213 = 32.16 x22212 = 18.9 and all others variables are zero	x11111 = 28.85 x11222 = 30.13 x12211 = 45.5 x21111 = 39.2 x21121 = 32.3 x21123 = 14.9 x21222 = 23.14 x22213 = 32.16 x22212 = 27.46 and all others variables are zero	x11111 = 28.14 x11222 = 30.02 x12211 = 45.48 x21113 = 39.09 x21121 = 32.24 x21123 = 14.71 x21222 = 23.37 x22213 = 32.22 x21211 = 27.49 x22212 = 5.61 and all others are zero	x11111 = 28.09 x11222 = 30.33 x12211 = 45.44 x21113 = 42.76 x21121 = 54.05 x21123 = 24.01 x21222 = 23.41 x22113 = 32.8 and all others are zero	x11111 = 28.92 x11222 = 30.33 x12211 = 45.40 x21113 = 42.76 x21121 = 54.21 x21123 = 24.32 x21222 = 33.51 x22113 = 41.05 and all others are zero

To compare the results for existence of different constraints, we present the following Table 8 for $\eta = 0.5$.

Table 8. Optimum results for existence of different constraints (Expected Value Technique) by Lingo.

Results without Space and Budget	Results with Space and without Budget	Results without Space and with Budget
x11111 = 42.75 x11212 = 13.33 x12211 = 25.65 x21121 = 52.20 x21122 = 10.8 x21223 = 57.74 x22112 = 43.9 x22213 = 32.9 x12213 = 12.9 x11123 = 21.17 and all others variables are zero Optimal profit 6579.135	x11221 = 22.84 x11222 = 32.04 x12211 = 45.57 x21121 = 52.16 x21122 = 10.75 x21223 = 30.44 x22112 = 53.16 x22213 = 32.16 x12213 = 12.9 x11123 = 11.17 and all others variables are zero Optimal profit 6529.07	x11211 = 19.06 x21222 = 53.33 x22212 = 14.9 x21121 = 32.13 x21122 = 10.71 x21213 = 24.54 x22112 = 53.80 x22213 = 32.16 x21122 = 27.4 x11123 = 31.12 and all others variables are zero Optimal profit 6513.513

5. Particular Cases

5.1. Three-Dimensional TP Model

If we take the system where there is only one route only, i.e., $L = 1$ in the Model-II (Rough Model), then the converted transportation problem is termed as a solid transportation problem (STP). The corresponding three dimensional TP-1, TP-2, TP-3, TP-4 is obtained by substituting $L = 1$ in the TP-1, TP-2, TP-3, TP-4 of the four-dimensional transportation Model. The results of the 3DTP (i.e., STP) model are given in Table 9.

Table 9. Optimum results of the STP Model ignoring route.

823.76	1263.95	323.61	2623.21
x1111 = 14.22	x1111 = 0.82	x1111 = 16.52	x1111 = 18.04
x1112 = 0.75	x1112 = 4.19	x1112 = 4.43	x1112 = 9.12
x1121 = 0.05	x1121 = 6.26	x1121 = 0.72	x1121 = 0.21
x1122 = 11.225	x1122 = 12.06	x1122 = 0.92	x1122 = 11.42
x1211 = 8.25	x1211 = 0.36	x1211 = 0.26	x1211 = 0.53
x1212 = 0.65	x1212 = 0.16	x1212 = 0.36	x1212 = 0.43
x1222 = 0.25	x1222 = 5.06	x1222 = 0.16	x1221 = 8.12
x2111 = 0.51	x2111 = 0.06	x2111 = 0.91	x1222 = 0.09
x2112 = 0.62	x2112 = 8.03	x2112 = 0.82	x2111 = 5.05
x2121 = 29.74	x2121 = 0.4	x2121 = 4.24	x2112 = 27.3
x2122 = 5.9	x2122 = 0.44	x2122 = 0.80	x2121 = 8.44
x2211 = 0.65	x2211 = 5.26	x2211 = 0.6	x2122 = 5.32
x2213 = 10.25	x2213 = 0.96	x2213 = 0.03	x2213 = 14.76
x2221 = 0.35	x2221 = 4.81	x2221 = 6.6	x2212 = 5.75
x2223 = 5.05	x2223 = 5.76	x2223 = 7.6	x2223 = 9.2
others zero	others zero	others zero	others zero

If we take $K = 1$, then we get another three-dimensional transportation Model. Like above, the corresponding 3D TP-1, TP-2, TP-3, TP-4 are found by substituting $K = 1$ in the TP-1, TP-2, TP-3, TP-4 of the four-dimensional transportation Model. The optimum results of this STP model are given in Table 10.

Table 10. Optimum results of the STP Model ignoring conveyance.

623.76	1103.95	383.61	2056.32
x1111 = 0.24	x1111 = 0.84	x1111 = 0.57	x1111 = 14.17
x1112 = 0.76	x1112 = 0.11	x1112 = 0.33	x111 = 0.42
x1121 = 0.05	x1121 = 0.26	x1121 = 0.62	x1121 = 0.43
x1122 = 0.24	x1122 = 0.26	x1122 = 0.92	x1122 = 0.97
x1211 = 0.23	x1211 = 0.46	x1211 = 0.36	x1211 = 0.64
x1212 = 0.54	x1212 = 0.16	x1212 = 0.46	x1212 = 0.54
x1222 = 0.25	x1222 = 14.06	x1222 = 0.26	x1223 = 0.17
x2111 = 0.51	x2111 = 0.06	x2111 = 0.91	x1333 = 0.65
x2112 = 0.62	x2112 = 42.03	x2112 = 0.82	x1213 = 0.42
x2121 = 32.74	x2121 = 0.4	x2121 = 44.24	x2121 = 41.53
x2122 = 20.9	x2122 = 0.44	x2122 = 0.80	x2122 = 10.42
x2211 = 7.65	x2211 = 10.26	x2211 = 0.6	x2211 = 0.95
x2213 = 0.25	x2213 = 0.96	x2213 = 0.03	x221 = 0.94
x2221 = 0.35	x2221 = 25.11	x2221 = 1.6	x2221 = 0.79
x2223 = 0.05	x2223 = 8.76	x2223 = 12.6	x2222 = 0.57
others zero	others zero	others zero	others zero

5.2. 2D-TP Model

If we take $L = 1$ and $K = 1$ in the Model-II (Rough Model), then we get the 2D-transportation problem. The corresponding 2D TP-1, TP-2, TP-3, TP-4 are found in the following Table 11.

Table 11. Optimum solutions for 2D-TP Model.

4113.06	8093.95	453.61	8243.93
x111 = 63	x111 = 64.82	x111 = 63.32	x111 = 62
x112 = 52	x112 = 19.26	x112 = 2.62	x112 = 17.2
x121 = 50.0	x121 = 16.36	x121 = 0.67	x211 = 8
x122 = 5.0	x122 = 0.70	x122 = 0.32	x223 = 23
x211 = 6	x211 = 23.8	x211 = 2.18	x222 = 26.21
x213 = 5.5	x213 = 5.0	x213 = 2.23	other zero
x221 = 16.5	x221 = 8.0	x221 = 4.15	
x223 = 19.5	x223 = 27.4	x223 = 4.32	
other zero	other zero	other zero	

5.3. 4D-TPs with Different Natures of the Items

Results for the different combinations of substitute and complementary items are shown in Table 12.

Table 12. Results for the different combinations of substitute and complementary items.

Problem	Independent	Substitute	Complementary	Profit for Model-1
1	Item-1, Item-2, Item-3			7388.841
1		Item-1, Item-2	Item-3	6966.775
2		Item-1, Item-2		5497.527
3			Item-1, Item-3	6021.859
4			Item-2, Item-3	5952.067

6. Sensitivity Analyses

Taking $\beta_j = 0.1, 0.2, ..., 0.7$, and keeping θ_j^1, θ_j^{11} and τ as same, the sensitivity analyses for the Model-1 are presented in Table 13.

Table 13. Sensitivity analysis (w.r.t β_j).

Set	τ	$\theta_j^1 = \theta_j^{11}$	β_j	Maximum Profit	Demand
	0.052	0.052	0.1	6966.68	379.542
	0.052	0.052	0.2	6519.775	359.084
	0.052	0.052	0.3	6072.868	338.627
1	0.052	0.052	0.4	5625.962	318.1696
	0.052	0.052	0.5	5179.055	297.712
	0.052	0.052	0.6	4732.149	277.254
	0.052	0.052	0.7	4285.242	256.79

If price of items are sensitives with $\tau = 0.1, 0.2, ..., 0.7$, and $\beta_j = 0.5$, keeping θ_j^1 and θ_j^{11} as unchanged, the sensitivity analyses for the Model-1 are presented in Table 14.

Table 14. Sensitivity Analysis (w.r.t τ).

Set	τ	$\theta_j^1 = \theta_j^{11}$	β_j	Maximum Profit	Demand
	0.1	0.052	0.5	5223.015	298.09
	0.2	0.052	0.5	5314.598	298.89
	0.3	0.052	0.5	5406.181	299.69
1	0.4	0.052	0.5	5497.76	300.49
	0.5	0.052	0.5	5589.34	301.29
	0.6	0.052	0.5	5680.93	302.09
	0.7	0.052	0.5	5772.513	303.89

7. Discussion

Tables 6–10 represent the optimal solutions for given assumed values, which are self explained. From the above observation of TP-1 and TP-2, we can conclude that it yields satisfactory results and that of TP-3 and TP-4 yield almost satisfactory results. From the above observation of Table 7, we can easily infer that, with the increasing value of η , the profit increases. The profit is varied for the substitute and complementary nature of the items observed from Table 11. When items are independent, then profit is maximum that is \$7388.841. When item-1 and item-2 are substituted in nature with each other and item-3 is complementary nature then profit is \$6966.775. However, when the item-1 and item-2 are the substitutes in nature to each other, then profit is minimum that is \$5499.527.

Table 11 shows the optimal profit of model-1 for the nature of the items. This table helps the manager to make the decisions of the type of items he/she should use for his/her business policy. For the current consideration, it is observed that the existence of the complementary item is more profitable than the existence of the substitute item.

From Table 12, it is seen that the demand parameter τ, θ_j^1, θ_j^{11} and β_j have an effect on the amount of demand. As the parameter β_j is the negative indicator of the demand expression, its increase in value causes decreasing demand and optimum profit.

Similarly, Table 13 gives reverse calculation as τ is a positive (+ve) parameter present in the demand expression. The Figures 3 and 4 have also established the same conclusion in Tables 12 and 13. It is to be noticed that the coefficients of selling prices and amounts of gains do not depend on the actual amounts of these factors. Profit depends only on the deviation of the factors.

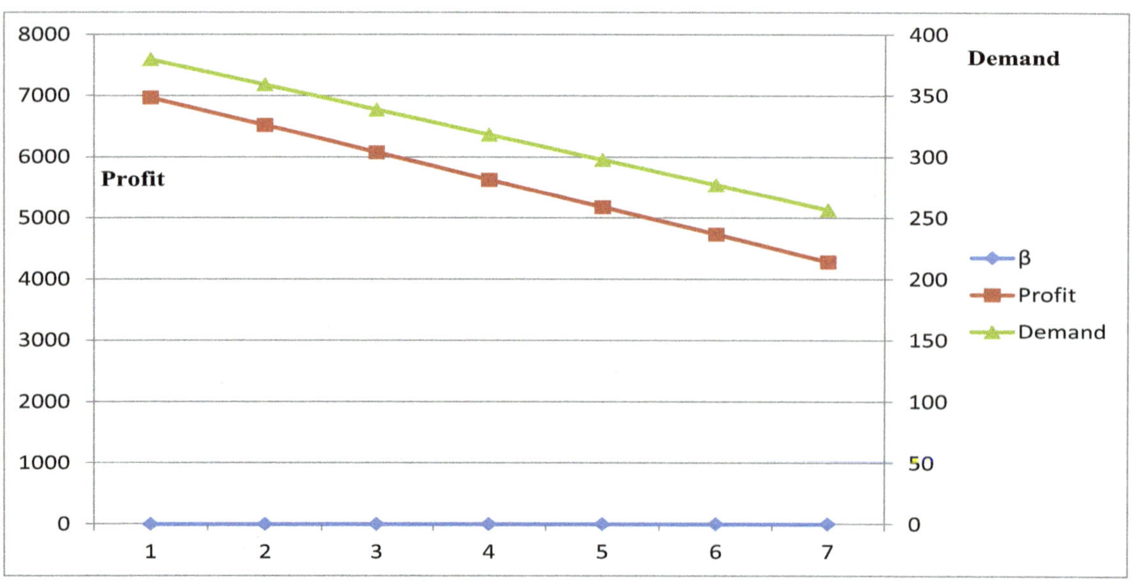

Figure 3. Profit vs. demand with respect to β.

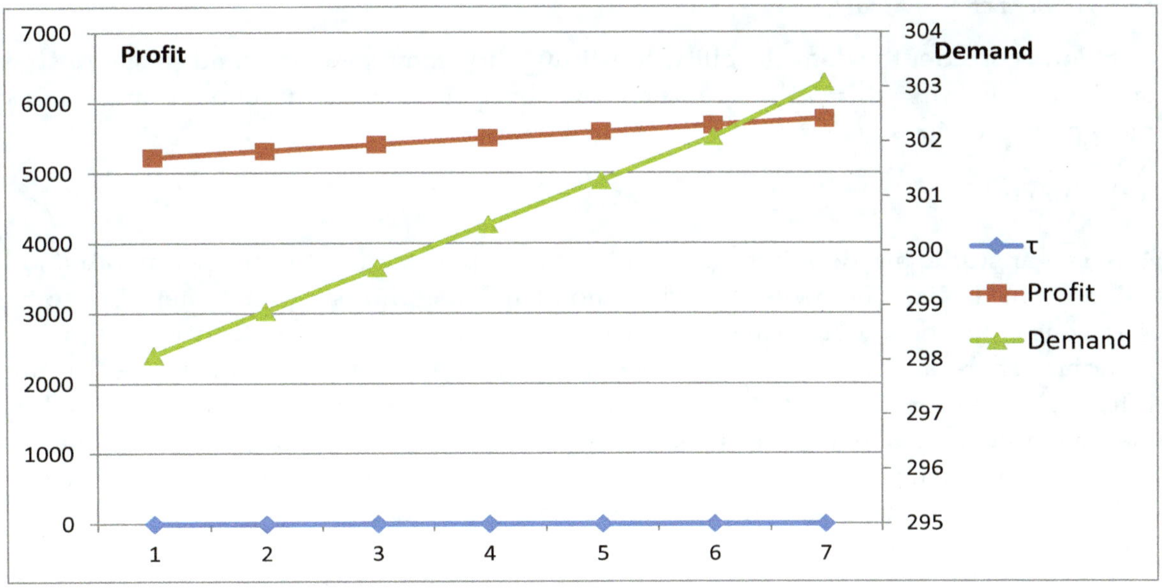

Figure 4. Profit vs. demand with respect to τ.

Thus, with the constant degree of substitution and complementary, profit and demand are constant though the measures of responsiveness of both merchandise are dissimilar for all cases.

Figure 5 draws a comparison between the profit versus price of sensitivities β_j and τ. It is observed that, when the price of sensitivity of product (β_j) is increasing, then profit is decreasing and, when τ is increasing, then profit is increasing. The maximum profit is dependent on both β_j and τ. Thus, it is essential to trade-off between these sensitivities parameters to achieve maximum profit when substitute and complimentary items are transported.

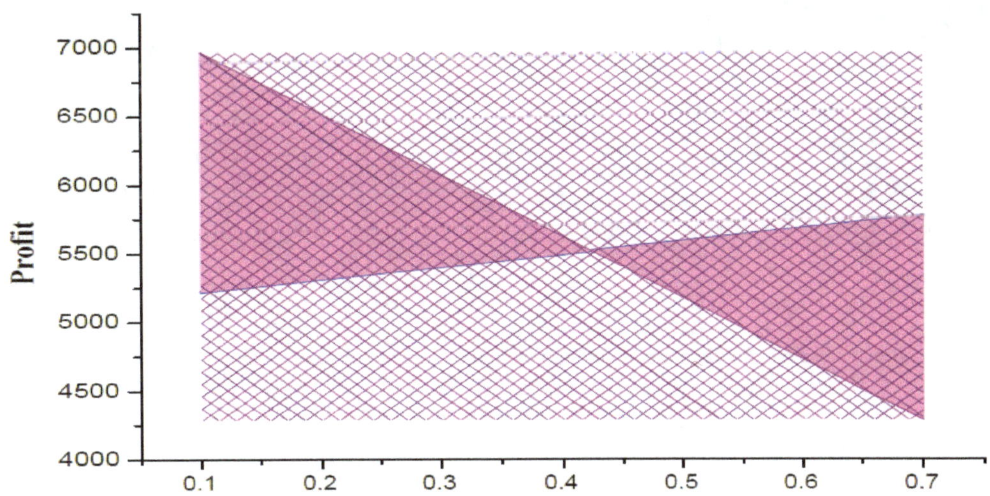

Figure 5. Profit vs. demand with respect to β and τ.

7.1. Discussion for Particular Models

From Table 8, when we consider only the first route of Model-II, the profit range of a corresponding 3D TP is ([823.76, 1263.95], [323.61, 2623.21]) and, from Table 9, when we take only the first conveyance of Model-II, the profit range of corresponding 3D TP is ([623.76, 1103.95], [383.61, 2056.32]). Again when we consider only the first route and first conveyance in the Model-II, the profit range of corresponding 2D TP is ([4113.06, 8093.95], [453.61, 8243.93]).

7.2. Results of Bera et al.'s Model

If we don't consider the damageability, substitutability, complementary and space constraint i.e., if we put $\lambda^r_{ijpk} = 0 = \beta_j = \theta^1_j = \theta^{11}_j = \tau$ and remove the space constraint in Equation (1), it becomes the same type expressions as Bera et al. [7] for three items.

8. A Real-Life Problem

M/S. Sargar studio has two shops (studios) at Midnapore Sadar, Paschim Midnapur and Digha, Purba Medinipur, India. The owner of the studio brings the photographic materials-two types of plastics and one type wooden frame material (by weight) from Kolkata, West Bengal, and Bhubaneswar, Orissa, India. There are two conveyances—bus and lorries—between sources and destinations. Two different routes for transportation are also available. Here, plastic sheets (one high quality and other one little low) are substitutes and the wooden frame is complementary. The relevant data and the results are given in Table 15. The company has sufficient storing spaces at destinations along with the sufficient budget. Here, we take the crisp values for the parameters given by the studio company (Table 16).

Table 15. Unit cost of transportation (C^r_{ijpk}), fixed charges (f_{ijpk}), breakability (λ^r_{ijpk}) for proposed Models.

	p	1				2			
	k	1		2		1		2	
	i/j	1	2	1	2	1	2	1	2
		Model-I							
		(Crisp Model)							
C^1_{ijkp}	1	0.15	1.1	1.2	0.12	0.85	0.5	1.38	1.28
	2	0.21	1.3	2.5	0.6	1.65	1.72	0.66	0.78
C^2_{ijkp}	1	1.5	2.25	2.0	1.3	0.08	0.7	1.05	1.25
	2	0.25	0.25	1.2	1.9	1.35	1.25	1.85	0.98
C^3_{ijkp}	1	1.3	1.2	1.91	1.2	0.9	0.94	0.73	0.93
	2	0.52	1.53	1.2	0.7	1.8	0.53	1.9	0.98
λ^1_{ijkp}	1	0.01	0.01	0.02	0.03	0.01	0.02	0.01	0.02
	2	0.02	0.01	0.03	0.04	0.02	0.01	0.03	0.02
λ^2_{ijkp}	1	0.04	0.03	0.01	0.02	0.05	0.01	0.02	0.03
	2	0.05	0.04	0.02	0.03	0.02	0.01	0.03	0.02
λ^3_{ijkp}	1	0.01	0.01	0.01	0.01	0.02	0.02	0.02	0.02
	2	0.04	0.05	0.02	0.03	0.01	0.02	0.04	0.02
f_{ijkp}	1	1.1	1.4	1.5	1.0	1.2	1.4	0.3	1.8
	2	1.3	1.5	0.9	1.1	1.3	2.1	2.1	1.02

Table 16. Parametric values for the Models.

Models	Source	Demand	Capacities. of Conveyance.	Purchasing Costs	Unit Selling Price
	$(A^1_1, A^2_1, A^3_1$ $A^1_2, A^2_2, A^3_2)$	$(D^1_{01}, D^2_{01}, D^3_{01}$ $D^1_{02}, D^2_{02}, D^2_{02})$	(e^1_1, e^2_1, e^3_1) (e^1_2, e^2_2, e^3_2)	$(P^1_1, P^2_1, P^3_1$ $P^1_2, P^2_2, P^3_2)$	$(S^1_1, S^2_1, S^3_1$ $S^1_2, S^2_2, S^3_2)$
-I	(85, 75, 80 83, 72, 68)	(60, 70, 75, 62, 64, 59)	(86, 89, 90 81, 65, 70)	(45, 48, 51 36, 47, 58.5)	(132, 145, 124, 126, 135, 137)

Out put results are without constraints are x11111 = 67.07, x11122 = 51.75, x21121 = 66.15, x21122 = 57.1, x22211 = 16.54, x22212 = 49.9, x22213 = 16.9, x11213 = 7.9, x11123 = 15.91 others are zero. Max z = 6947.52.

9. Conclusions

Renowned scientific discoveries and Research always had a practical application in the real world. Standing in the 21st century where machines, technology and industries govern the civilization, the need for sophisticated and easy techniques in the industry for optimizing products and profit is of immense importance. To further proceed and bring new ideas, we investigate a four-dimensional multifarious item transportation issue with rough interval coefficient, which maximizes the profit considering substitute and complementary products. The concept of rough interval and its properties are discussed briefly. In recent times, the rough interval has played an important role in different techniques handling vagueness for decision-making problems. The models we have discussed comprises of all earlier factors that are all considered as rough intervals. While solving the proposed model, at first, we discuss a solving procedure of a general linear programming problem with rough intervals and then we solve our model according to the concept of the above model. The rough problem is disintegrated into four classical transportation cases. We derive some special cases like a three-dimensional model and two-dimensional model to realize the problem more practically. Few examples are illustrated using LINGO 14.0 software showcasing the smooth working of the technique. We have considered a single profit maximization problem which can be extended to become a multi-objective optimization problem if we simultaneously maximize profit and minimize time. Moreover, the travelling salesman problem (TSP), Inventory routing problem, Travelling purchase problem, etc. are NP hard problems and are represented as LPPs and formulated as both single and multi-objective problems (cf. Khanra et al. [34], Manbera [35], etc.). In these problems, the travelling cost, stay charge, etc. can be taken as rough intervals and Hamzehee et al. [36] can be applied to convert to crisp models, which can be solved by heuristic methods, Genetic-Algorithms, Ant Colony Optimization, etc.

The drawback of the present methodology is that it provides a wide-range of possible profit values. The average of these possible values may give an approximate single value.

Author Contributions: S.H.J., B.D. and M.M. determined this problem and proposed the method for solutions. B.J. and G.P. did computer programming and mathematical calculations. B.J., S.H.J. and B.D. wrote the paper. M.M. and S.H.J. thoroughly checked the paper and gave interpretation to the results.

Appendix A

In this section, some definitions and properties on Rough Intervals (RI) are given. For more details, see Hamzehee et al. [36]. An RI can be treated as a qualitative amount from a vague concept specified on a variable x in R [37], which is abstracted in the following definition.

Appendix A.1. Rough Intervals and Its Algebra

The algebraic operations applied on Rough Intervals (RIs) are taken from Moore's Interval Arithmetic [14]. Rebolledo [13] discussed the rough interval arithmetic very significantly.

Let $Q = ([\underline{a}_l, \underline{a}_u], [\overline{a}_l, \overline{a}_u])$ and $T = ([\underline{b}_l, \underline{b}_u], [\overline{b}_l, \overline{b}_u])$ represents the rough intervals. Then, the different operations on these are:

Addition: $Q + T = ([\underline{a}_l + \underline{b}_l, \underline{a}_u + \underline{b}_u], [\overline{a}_l + \overline{b}_l, \overline{a}_u + \overline{b}_u])$,
Subtraction: $Q - T = ([\underline{a}_l - \underline{b}_u, \underline{a}_u - \underline{b}_l], [\overline{a}_l - \overline{b}_u, \overline{a}_u - \overline{b}_l])$,
Negation: $-Q = ([-\underline{a}_u, \underline{a}_l], [-\underline{a}_u, \underline{a}_l])$.

Appendix A.2. Expected Value of a Rough Interval (Shu and Edmund [38])

Let an event X be uttered by $\{x \mid \psi(x) \in T\}$, whereas ψ is a mathematical function in universe U in $\mathbb{R}$, $T \subseteq \mathbb{R}$ and X is estimated by $(\underline{X}, \overline{X})$ consorting to the similar relation in $\mathbb{R}$. The corresponding lower expected measure of X is given by

$$\underline{E}[\psi] = \int_0^\infty \underline{Appro}\{\psi \geq y\}dy - \int_{-\infty}^0 \underline{Appro}\{\psi \leq y\}dy \ \ where \ \ \underline{Appro}(X) = \frac{|X \cap \underline{X}|}{\underline{X}}$$

and the corresponding upper expected measure of X is given by

$$\overline{E}[\psi] = \int_0^\infty \overline{Appro}\{\psi \geq y\}dy - \int_{-\infty}^0 \overline{Appro}\{\psi \leq y\}dy \ \ where \ \ \overline{Appro}(X) = \frac{|X|}{\overline{X}}$$

and the corresponding expected value of X is given by

$$E[X] = \int_0^\infty Appro\{\psi \geq y\}dy - \int_{-\infty}^0 Appro\{\psi \leq y\}dy.$$

Proposition (Shu and Edmund [38])

For any predetermined parameter η, chosen by the decision maker's (DM's) preference, $E(X) = \eta\underline{E}(X) + (1-\eta)\overline{E}(X)$.

Appendix A.3. General Linear Programming Problem in Rough Interval Environments

A Linear Programming Problem (LPP) in which the coefficients of the objective function and the constraints are rough intervals:

$$\left.\begin{aligned}
&\text{Maximize } Z = \sum_{j=1}^{N} ([\underline{c}_{jl}, \underline{c}_{ju}], [\overline{c}_{jl}, \overline{c}_{ju}])x_j \\
&\text{s.t } \sum_{j=1}^{N} ([\underline{a}_{ijl}, \underline{a}_{iju}], [\overline{a}_{ijl}, \overline{a}_{iju}])x_j \leq ([\underline{b}_{il}, \underline{b}_{iu}], [\overline{b}_{il}, \overline{b}_{iu}]) \ \forall \ i = 1,2,...m, \\
&\sum_{j=1}^{N} ([\underline{d}_{kjl}, \underline{d}_{kju}], [\overline{d}_{kjl}, \overline{d}_{kju}])x_j \geq ([\underline{e}_{kl}, \underline{e}_{ku}], [\overline{e}_{kl}, \overline{e}_{ku}]) \ \forall \ k = 1,2,...p, \\
&\text{and } x_j \geq 0 \quad \forall \ j = 1,2,...n,
\end{aligned}\right\} \quad \text{(A1)}$$

where $([\underline{c}_{jl}, \underline{c}_{ju}], [\overline{c}_{jl}, \overline{c}_{ju}])$, $([\underline{a}_{ijl}, \underline{a}_{iju}], [\overline{a}_{ijl}, \overline{a}_{iju}])$, $([\underline{b}_{il}, \underline{b}_{iu}], [\overline{b}_{il}, \overline{b}_{iu}])$, $([\underline{d}_{kjl}, \underline{d}_{kju}], [\overline{d}_{kjl}, \overline{d}_{kju}])$ and $([\underline{e}_{kl}, \underline{e}_{ku}], [\overline{e}_{kl}, \overline{e}_{ku}])$ $(i = 1,2,...,m;\ j = 1,2,...,n;\ k = 1,2,...,p)$ are rough intervals. Let $x = (x_1, x_2, ..., x_n)^t$ be the vector form of decision variables.

Now, according to Hamzehee et al. [36], here we state two theorems which help us to find the rough optimal range of the problem (1).

Let us consider two LPIC problems which are as follows.

Appendix A.3.1. LPIC-1

$$\left.\begin{aligned}
&\text{Maximize } Z = \sum_{j=1}^{N} [\underline{c}_{jl}, \underline{c}_{ju}]x_j \\
&\text{s.t } \sum_{j=1}^{N} ([\underline{a}_{ijl}, \underline{a}_{iju}])x_j \leq ([\underline{b}_{il}, \underline{b}_{iu}]) \ \forall \ i = 1,2,...,m, \\
&\sum_{j=1}^{N} ([\underline{d}_{kjl}, \underline{d}_{kju}])x_j \geq ([\underline{e}_{kl}, \underline{e}_{ku}]) \ \forall \ k = 1,2,...,p, \\
&\text{and } x_j \geq 0 \quad \forall \ j = 1,2,...,n.
\end{aligned}\right\} \quad \text{(A2)}$$

Appendix A.3.2. LPIC-2

$$\left.\begin{array}{l}\text{Maximize } Z = \sum\limits_{j=1}^{N}([\bar{c}_{jl}, \bar{c}_{ju}])x_j \\ \text{s.t } \sum\limits_{j=1}^{N}([\bar{a}_{ijl}, \bar{a}_{iju}])x_j \leq ([\bar{b}_{il}, \bar{b}_{iu}]) \quad \forall\ i = 1,2,\ldots,m, \\ \sum\limits_{j=1}^{N}([\bar{d}_{kjl}, \bar{d}_{kju}])x_j \geq ([\bar{e}_{kl}, \bar{e}_{ku}]) \quad \forall\ k = 1,2,\ldots,p, \\ \text{and } x_j \geq 0 \quad \forall\ j = 1,2,\ldots,n. \end{array}\right\} \tag{A3}$$

The best optimal solution of the problem (1) according to Hamzehee et al. [36] and Chinneck and Ramadan [39] is i.e., LPIC-1 is found by solving:

LP1:

$$\left.\begin{array}{l}\text{Maximize } Z = \sum\limits_{j=1}^{N} \underline{c}_{ju}x_j \\ \text{s.t } \sum\limits_{j=1}^{N} \underline{a}_{ijl}x_j \leq \underline{b}_{iu} \quad \forall\ i = 1,2,\ldots,m, \\ \sum\limits_{j=1}^{N} \underline{d}_{kju}x_j \geq \underline{e}_{kl}, \quad \forall\ k = 1,2,\ldots,p, \\ \text{and } x_j \geq 0 \quad \forall\ j = 1,2,\ldots,n. \end{array}\right\} \tag{A4}$$

The best optimal solution of the problem (1) i.e., LPIC-1 is found by solving:

LP2:

$$\left.\begin{array}{l}\text{Maximize } Z = \sum\limits_{j=1}^{N} \underline{c}_{jl}x_j \\ \text{s.t } \sum\limits_{j=1}^{N} \underline{a}_{iju})x_j \leq \underline{b}_{il} \quad \forall\ i - 1,2,\ldots,m, \\ \sum\limits_{j-1}^{N} \underline{d}_{kjl}x_j \geq \underline{e}_{ku}) \quad \forall\ k = 1,2,\ldots,p, \\ \text{and } x_j \geq 0 \quad \forall\ j = 1,2,\ldots,n. \end{array}\right\} \tag{A5}$$

The best optimal solution of the problem (1) i.e., LPIC-2 is found by solving:

LP3:

$$\left.\begin{array}{l}\text{Maximize } Z = \sum\limits_{j=1}^{N} \bar{c}_{ju}x_j \\ \text{s.t } \sum\limits_{j=1}^{N} \bar{a}_{ijl}x_j \leq \bar{b}_{iu} \quad \forall\ i = 1,2,\ldots,m, \\ \sum\limits_{j=1}^{N} \bar{d}_{kju}x_j \geq \bar{e}_{kl} \quad \forall\ k = 1,2,\ldots,p, \\ \text{and } x_j \geq 0 \quad \forall\ j = 1,2,\ldots,n. \end{array}\right\} \tag{A6}$$

The best optimal solution of the problem (1) i.e., LPIC-2 is found by solving:

LP4:

$$\left.\begin{aligned} &\text{Maximize } Z = \sum_{j=1}^{N} \overline{c}_{jl} x_j \\ &\quad \text{s.t } \sum_{j=1}^{N} \overline{a}_{iju})x_j \leq \overline{b}_{il} \quad \forall\ i = 1,2,\ldots,m, \\ &\quad \sum_{j=1}^{N} \overline{d}_{kjl} x_j \geq \overline{e}_{ku} \quad \forall\ k = 1,2,\ldots,p, \\ &\quad \text{and } x_j \geq 0 \quad \forall\ j = 1,2,\ldots,n. \end{aligned}\right\} \tag{A7}$$

Appendix A.3.3. Types of Solutions

There are three possible types of solutions of the problem LPIC-1 and LPIC-2 which are listed below:

- If LP-1 and LP-2 (LP-3 and LP-4) have optimal solutions, then the problem LPIC-1 (LPIC-2) has a finite bounded surely optimal (possibly optimal) range. If the maximizing value of LP-1 and LP-2 (LP-3 and LP-4) are respectively $\underline{z}_l, \underline{z}_u (\overline{z}_l, \overline{z}_u)$, then the surely optimal range (possibly optimal range) of LPIC-1 (LPIC-2) is $[\underline{z}_l, \underline{z}_u]([\overline{z}_l, \overline{z}_u])$
- If LP-2 (LP-4) is unbounded, then LPIC-1 (LPIC-2) is unbounded.
- If LP-1 (LP-3) is infeasible, then LPIC-1 (LPIC-2) is infeasible.

Appendix A.4. Algorithm for Conversion from a Rough LPP to a Crisp LPP

The steps for converting a Rough LPP to a Crisp LPP are

Step 1: Decompose the problem (1) into two linear programming part where interval coefficients are considered i.e., LPIC-1 and LPIC-2 which are given by (2) and (3), respectively.

Step 2: In this step, LPIC-1 again transformed into two linear programming problem where crisp coefficients are considered i.e., LP-1 and LP-2 given by respectively the problems (5) and (6) and similarly from LPIC-2 as LP-3 and LP-4 given by (7) and (8), respectively.

Step 3: In this step, LP-1, LP-2, LP-3 and LP-4 to find $\underline{z}_l$, $\underline{z}_u$, $\overline{z}_l$ and $\overline{z}_u$ respectively. $[\underline{z}_l, \underline{z}_u]$ and $[\overline{z}_l, \overline{z}_u]$ are the solutions of LPIC-1 and LPIC-2 respectively, which are called surely optimal range and possibly optimal range.

Step 4: Here three possible consequences are given below:

- If LPIC-1 and LPIC-2 have an optimal range, then the LPRIC problem (1) has an optimal range $([\underline{z}_l, \underline{z}_u], [(\overline{z}_l, \overline{z}_u)])$ which is a rough interval, compared to the surely optimal range
- If LPIC-1 has boundless range, then the LPRIC problem (1) has boundless range.
- If LPIC-2 is infeasible, then the LPRIC problem (1) has an infeasible solution space.

Appendix A.5. Convergence of the GRG Method

Let us consider the general nonlinear programming problem

$$\begin{aligned} &\mathit{minimize}\ f(x) \\ x \in R^n :\ &g(x) = 0, \\ &x \geq 0, \end{aligned} \tag{A8}$$

where $f : \mathbb{R}^n \to \mathbb{R}$ and $g : \mathbb{R}^n \to \mathbb{R}^m$ are continuously differentiable functions. Any nonlinear programming theory is based on the LAGRANGE function of (19), i.e., on

$$L(x, u, \bar{u}) = f(x) - g(x)^T u - x^T \bar{u}, \tag{A9}$$

where $x \in \mathbb{R}^n$, $u \in \mathbb{R}^m$ and $\bar{u} \in \mathbb{R}^n$. Using this LAGRANGE function, the necessary optimality or Kuhn–Tucker conditions are stated in the form

$$\begin{aligned} \nabla_N L(x,u,\bar{u}) &= 0, \\ g(x) &= 0, \\ x &\geq 0, \\ \bar{u} &\geq 0, \\ x^T\bar{u} &= 0, \end{aligned} \tag{A10}$$

which are to be approximated by the GRG algorithm. Note that the first condition can be used to eliminate $\bar{u}$, since we get $\bar{u} = \nabla f(x) - \nabla g(x)^T u$. A further function which will play an important role for proving a global convergence theorem is the augmented LAGRANGE function

$$\Psi_r(x,v,\bar{v}) := f(x) - g(x)^T v + \frac{1}{2} r g(x)^T g(x) - \bar{v}_J^T x_J + \frac{1}{2} r x_J^T x_J - \frac{1}{2r}\bar{v}_K^T \bar{v}_K, \tag{A11}$$

where $r \in \mathbb{R}$, $x \in \mathbb{R}^n$ $v \in \mathbb{R}^m$ and $\bar{v} \in \mathbb{R}^n$.

Assume that the GRG Algorithm produces a sequence of iterates $x^k, v^k, \bar{v}^k, \bar{u}^k, r^k$.

First, we note again that the boundedness of $\{x^k\}$ implies the boundedness of $\{u^k\}$, $\{v^k\}$ and $\{\bar{v}^k\}$.

Again, we denote this set of iteration indices by $\mathbb{K}$. Since only finitely many different index sets R_k and F_k are allowed, there is an infinite subset $\mathbb{K}'$ of $\mathbb{K}$ and index sets R, F with $R_k = R \subset N$, $F_k = F \subset N$, for all $k \in \mathbb{K}'$. For the same reason, we find a constant index set $D \subset N$, i.e., we can also assume that $D_k = D$ and $I_{\bar{k}} = I := N\ D$ for all $k \in \mathbb{K}'$.

Thus, there are accumulation points x^*, u^* and $\bar{u}^*$ of $\{x^k\}, \{u^k\}, and \{\bar{u}^k\}$, respectively, so that

$$\begin{aligned} \lim_{k\in\mathbb{K}''} x^k &= x^*, \\ \lim_{k\in\mathbb{K}''} u^k &= u^*, \\ \lim_{k\in\mathbb{K}''} \bar{u}^k &= \bar{u}^* \end{aligned}$$

Then, for an infinite subset $\mathbb{K}$

$$\begin{aligned} \nabla_D L(x^k,u^k,\bar{u}^k) &= \nabla_D f(x^k) - \nabla_D g(x^k)^T u^k = 0, \\ \nabla_F L(x^k,u^k,\bar{u}^k) &= \nabla_F f(x^k) - \nabla_F g(x^k)^T u^k = 0, \qquad \text{for all } k \in \mathbb{K}''. \end{aligned} \tag{A12}$$

We conclude that

$$\nabla_N L(x^*,u^*,\bar{u}^*) = 0.$$

In addition, we have $x^* \geq 0$, $x^{*T}\bar{u}^* = 0$, $g(x^*) = 0$ and $\bar{u}^* \geq 0$
$x^*, u^*$and $\bar{u}^* = 0$ define a Kuhn–Tucker point for problem (19).

References

1. Hitchcok, F.L. The distribution of a product form several sources to numerous localities. *J. Math. Phys.* **1941**, *20*, 224–230. [CrossRef]
2. Schell, E.D. Distribution of a product by several properties. In Proceedings of the 2nd Symposium in Linear Programming, DCS/Comptroller, HQ US Air Force, Washington, DC, USA, 24–25 October 1955; pp. 615–642.
3. Yang, L.; Liu, P.; Li, S.; Gao, Y.; Ralescu, D.A. Reduction methods of type-2 uncertain variables and their applications to solid transportation problem. *Inf. Sci.* **2015**, *291*, 204–237. [CrossRef]

4. Kocken, H.G.; Sivri, M. A simple parametric method to generate all optimal solutions of fuzzy solid transportation problem. *Appl. Math. Model.* **2016**, *40*, 4612–4624. [CrossRef]
5. Halder, S.; Das, B.; Panigrahi, G.; Maiti, M. Some special fixed charge solid transportation problems of substitute and breakable items in crisp and fuzzy environments. *Comput. Ind. Eng.* **2017**, *111*, 272–281. [CrossRef]
6. Das, A.; Bera, U.K.; Maiti, M. A Profit Maximizing Solid Transportation Model Under a Rough Interval Approach. *IEEE Trans. Fuzzy Syst.* **2017**, *25*, 485–498. [CrossRef]
7. Bera, S.; Giri, P.K.; Jana, D.K.; Basu, K.; Maiti, M. Multi-item 4D-TPs under budget constraint using rough interval. *Appl. Soft Comput.* **2018**, *71*, 364–385. [CrossRef]
8. Sarkar, M.; Lee, Y.H. Optimum pricing strategy for complementary products with reservation price in a supply chain model. *J. Ind. Manag. Optim.* **2017**, *13*, 1553–1586. [CrossRef]
9. Sarkar, M.; Hur, S.; Sarkar, B. Effects of Variable Production Rate and Time-Dependent Holding Cost for Complementary Products in Supply Chain Model. *Math. Probl. Eng.* **2017**, *2017*, 2825103. [CrossRef]
10. Khanna, A.; Kishore, A.; Sarkar, B.; Jaggi, C. Supply Chain with Customer-Based Two-Level Credit Policies under an Imperfect Quality Environment. *Mathematics* **2018**, *6*, 299. [CrossRef]
11. Hirsch, W.M.; Dantzig, G.B. The fixed charge problem. *Naval Res. Logist. NRL* **1968**, *15*, 413–424. [CrossRef]
12. Kowalski, K.; Lev, B. On step fixed-charge transportation problem. *Omega* **2008**, *36*, 913–917. [CrossRef]
13. Gen, M.; Li, Y.Z. Spanning tree-based genetic algorithm for bicriteria transportation problem. *Comput. Ind. Eng.* **1998**, *35*, 531–534. [CrossRef]
14. Verma, R.; Biswal, M.P.; Biswas, A. Fuzzy programming technique to solve multi-objective transportation problems with some non-linear membership functions. *Fuzzy Sets Syst.* **1997**, *91*, 37–43. [CrossRef]
15. Shafaat, A.; Goyal, S.K. Resolution of degeneracy in transportation problems. *J. Oper. Res. Soc.* **1988**, *39*, 411–413. [CrossRef]
16. Saad, O.M.; Abbas, S.A. A parametric study on transportation problem under fuzzy environment. *J. Fuzzy Math.* **2003**, *11*, 115–124.
17. Iqbal, M.W.; Sarkar, B. Recycling of lifetime dependent deteriorated products through different supply chains. *RAIRO-Oper. Res.* **2019**, *53*, 129–156. [CrossRef]
18. Sarkar, B.; Mandal, B.; Sarkar, S. Preservation of deteriorating seasonal products with stock-dependent consumption rate and shortages. *J. Ind. Manag. Optim.* **2017**, *13*, 187–206. [CrossRef]
19. Pramanik, S.; Maity, K.; Jana, D.K. A multi-objective solid transportation problem with reliability for damageable items in random fuzzy environment. *Int. J. Oper. Res.* **2018**, *31*, 1–23. [CrossRef]
20. Sarkar, B.; Shaw, B.K.; Kim, T.; Mitali, S.; Shin, D. An integrated inventory model with variable transportation cost, two-stage inspection, and defective items. *J. Ind. Manag. Optim.* **2017**, *13*, 1975–1990. [CrossRef]
21. Ojha, A.; Mondal, S.K.; Maiti, M. A solid transportation problem with partial nonlinear transportation cost. *J. Appl. Comput. Math.* **2014**, *3*, 1–6.
22. Ishfaq, N.; Sayed, S.; Akram, M.; Smarandache, F. Notions of Rough Neutrosophic Digraphs. *Mathematics* **2018**, *6*, 18. [CrossRef]
23. Bit, A.K.; Biswal, M.P.; Alam, S.S. Fuzzy programming approach to multiobjective solid transportation problem. *Fuzzy Sets Syst.* **1993**, *57*, 183–194. [CrossRef]
24. Jiménez, F.; Verdegay, J.L. Solving fuzzy solid transportation problems by an evolutionary algorithm based parametric approach. *Eur. J. Oper. Res.* **1999**, *117*, 485–510. [CrossRef]
25. Li, T.H.S.; Su, Y.T.; Lai, S.W.; Hu, J.J. Walking motion generation, synthesis, and control for biped robot by using PGRL, LPI, and fuzzy logic. *IEEE Trans. Syst. Man Cybern. Part B Cybern.* **2011**, *41*, 736–748. [CrossRef]
26. Dey, A.; Pal, A.; Pal, T. Interval type 2 fuzzy set in fuzzy shortest path problem. *Mathematics* **2016**, *4*, 62. [CrossRef]
27. Kundu, P.; Kar, S.; Maiti, M. Fixed charge transportation problem with type-2 fuzzy variables. *Inf. Sci.* **2014**, *255*, 170–186. [CrossRef]
28. Tao, Z.; Xu, J. A class of rough multiple objective programming and its application to solid transportation problem. *Inf. Sci.* **2012**, *188*, 215–235. [CrossRef]
29. Haley, K.B. New methods in mathematical programming—The solid transportation problem. *Oper. Res.* **1962**, *10*, 448–463. [CrossRef]
30. Ojha, A.; Das, B.; Mondal, S.K.; Maiti, M. A multi-item transportation problem with fuzzy tolerance. *Appl. Soft Comput.* **2013**, *13*, 3703–3712. [CrossRef]

31. Liu, P.; Yang, L.; Wang, L.; Li, S. A solid transportation problem with type-2 fuzzy variables. *Appl. Soft Comput.* **2014**, *24*, 543–558. [CrossRef]
32. Giri, P.K.; Maiti, M.K.; Maiti, M. Fully fuzzy fixed charge multi-item solid transportation problem. *Appl. Soft Comput.* **2015**, *27*, 77–91. [CrossRef]
33. Schittkiowski, K. On the convergence of a generalized reduced gradient algorithm for nonlinear programming problems. Optimization **1986**, *17*, 731–755. [CrossRef]
34. Khanra, A.; Maiti, M.K.; Maiti, M. Profit maximization of TSP through a hybrid algorithm. *Comput. Ind. Eng.* **2015**, *88*, 229–236. [CrossRef]
35. Manerba, D.; Mansini, R.; Riera-Ledesma, J. The traveling purchaser problem and its variants. *Eur. J. Oper. Res.* **2017**, *259*, 1–18. [CrossRef]
36. Hamzehee, A.; Yaghoobi, M.A.; Mashinchi, M. Linear programming with rough interval coefficients. *J. Intell. Fuzzy Syst.* **2014**, *26*, 1179–1189.
37. Rebolledo, M. Rough intervals—Enhancing intervals for qualitative modeling of technical systems. *Artif. Intell.* **2006**, *170*, 667–685. [CrossRef]
38. Xiao, S.; Lai, E.M.K. A rough programming approach to power-balanced instruction scheduling for VLIW digital signal processors. *IEEE Trans. Signal Process.* **2008**, *56*, 1698–1709. [CrossRef]
39. Chinneck, J.W.; Ramadan, K. Linear programming with interval coefficients. *J. Oper. Res. Soc.* **2000**, *51*, 209–220. [CrossRef]

6

Imperfect Multi-Stage Lean Manufacturing System with Rework under Fuzzy Demand

Muhammad Tayyab [1], Biswajit Sarkar [1,*] and Bernardo Nugroho Yahya [2]

1 Department of Industrial & Management Engineering, Hanyang University, Ansan Gyeonggi-do 155 88, Korea; mtayyabntu@yahoo.com

2 Industrial and Management Engineering Department, Hankuk University of Foreign Studies, 81, Oedae-ro, Mohyeon-myeon, Cheoin-gu, Yongin-si, Gyeonggi-do 17035, Korea; bernardo@hufs.ac.kr

* Correspondence: bsbiswajitsarkar@gmail.com.

Abstract: Market conditions fluctuate abruptly in today's competitive environment and leads to imprecise demand information. In particular, market demand data for freshly launched products is highly uncertain. Further, most of the products are generally manufactured through complex multi-stage production systems that may produce defective items once they enter the out-of-control state. Production management of a multi-stage production system in these circumstances requires robust production model to reduce system costs. In this context, this paper introduces an imperfect multi-stage production model with the consideration of defective proportion in the production process and uncertain product demand. Fuzzy theory is applied to handle the uncertainty in demand information and the center of gravity approach is utilized to defuzzify the objective function. This defuzzified cost objective is solved through the analytical optimization technique and closed form solution of optimal lot size and minimum cost function are obtained. Model analysis verifies that it has successfully achieved global optimal results. Numerical experiment comprising of three examples is conducted and optimal results are analyzed through sensitivity analysis. Results demonstrate that larger lot sizes are profitable as the system moves towards a higher number of stages. Sensitivity analysis indicates that the processing cost is the most influencing factor on the system cost function.

Keywords: Lean manufacturing; multi-stage production system; fuzzy demand; imperfect products; reworking

---

1. Introduction

Lean manufacturing concepts are becoming rapidly popular in global production industries subjected to growing market competitions and increased manufacturing costs. Initiated by Toyota's engineer Taiichi Ohno, credited with developing the principles of lean production after World War II; eliminating unwanted activities, empowering workforce, cut down inventories, and improving productivity are the focused themes of lean philosophy. Unlike Henry Ford, who maintained resources in expectation of what might be required for future manufacturing, the management team at Toyota built partnership with suppliers and consequently became made-to-order. Thus, they can easily handle quick changes, and opposite to what their competitors could, they became able to respond quicker to market demands.

Highly inflexible and having high volume/low variety products of automatic machinery are the constraints in the textile industries. Due to this nature of textile industry, implementation of lean manufacturing is a challenge in this industry. Therefore, value stream mapping, kaizen, 5S, kanban, poka-yoke, and visual controls are practiced in combination with each other to improve the textile manufacturing process [1]. Most of the textile manufacturing processes include more than one production stages, and among all other concerns, optimum work in process inventory levels and scrap control are the premier ones. Thus, managers are interested in working out ways to minimize total system costs through optimal lot sizing in the production system. This builds importance of economic production quantity model in this sector to escape high inventory levels and cut down production costs. Figure 1 shows a general multi-stage production flow of an apparel manufacturing industry.

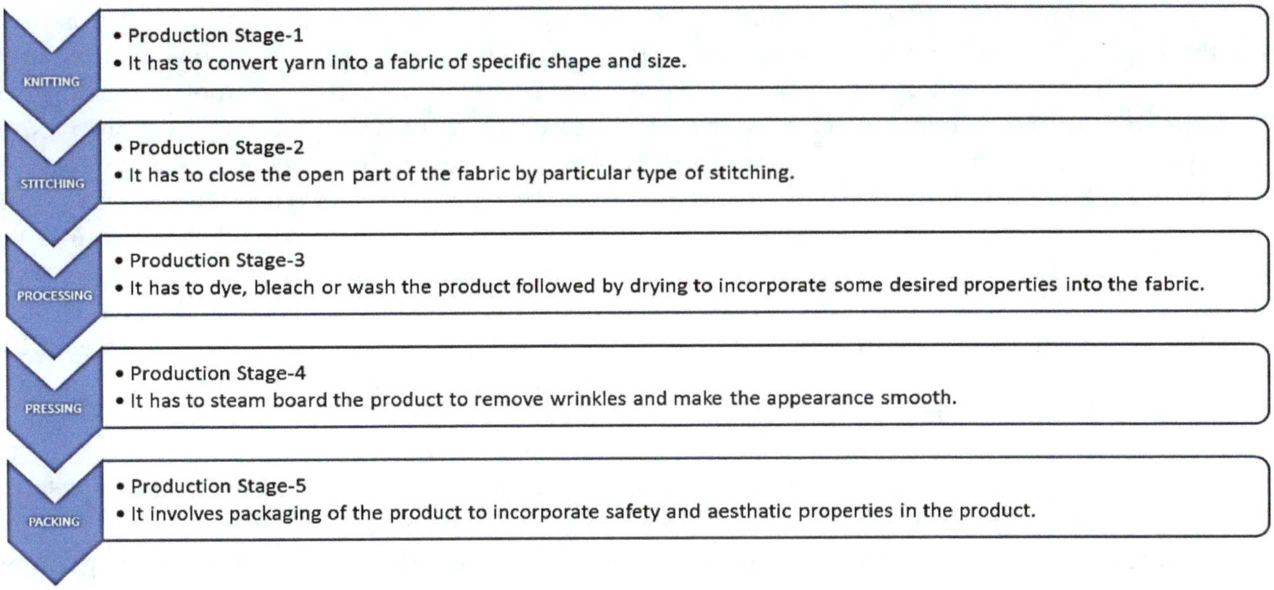

Figure 1. General production flow of an apparel manufacturing industry.

Since Harris [2] and Taft [3] provided Economic Order Quantity (EOQ) and Economic Production Quantity (EPQ) formulas, respectively, economic lot size models are broadly developed by the researchers. Due to their simplicity, these inventory models are used in many applied situations. In the inventory management research, Lee [4], Gupta and Chakraborty [5], and Tayi and Ballou [6] studied reprocessing of imperfect items, in which shortages due to reworking were not taken into account by the last two. Lee et al. [7] and Glock and Jaber [8] provided production models for imperfect production systems without considering rework opportunity. C.S Lin [9] studied the production process for its out-of-control state along with raw material resource constraints. Salameh and Jaber [10] provided an inventory model for imperfect quality items where these items are withdrawn from stock resulting in lower holding cost per unit time. Their model was extended by Khan et al. [11], and Jaber et al. [12] considers strategies for handling these imperfect quality items. Ouyang et al. [13] and Cárdenas-Barrón [14] formulated inventory models in which backordering is allowed and Eroglu and Ozdemir [15] considered shortages in their studies. Jaber and Guiffrida [16] studied Wright's learning curve along with preventive maintenance, and Daya [17] likewise extended the inventory model with planned maintenance schedules.

Most of the researchers developed inventory models for single-stage imperfect production systems, where numerous of them focused on analysis of the defective rates. Reworking in single-stage inventory models with constant defective rates has been established by Jamal et al. [18], Biswas and Sarker [19], Chung [20], and Taleizadeh et al. [21], in which the latter two included backordering in their models. Cárdenas-Barrón [22] extended the inventory model provided by Jamal et al. [18] with the addition of the algebraic solution approach. Ma et al. [23] analyzed imperfect production system without providing reworking alternatives in their model, and Barzoki et al. [24] presented discounted sales of non-reworkable items. Other researchers modified inventory model with random-defective rate in single-stage manufacturing system, where the reworking option was assumed by Chiu et al. [25], Chiu et al. [26], Noorollahi et al. [27], Haider et al. [28], and Chiu et al. [29] using the renewal reward theorem. Taleizadeh et al. [30] allowed a backordering alternative along with the rework prospect in their inventory model, whereas Chiu et al. [31] and Taleizadeh et al. [32] developed inventory models, ignoring the reworking facility. Random defect-rate contemplation was further established by Sana [33], Sarkar and Moon [34], Yoo et al. [35], Chakraborty and Giri [36], Wee et al. [37], Chiu et al. [38], and Sarkar et al. [39] using a number of other shop-floor conditions.

In real situations, determining the precise value of product demand is very difficult. To deal with such conditions, managers need to gather demand data from the experts. If, according to the experts, demand about some randomly chosen quantity is fixed; their opinion is imprecise and then the expected annual demand is vaguely expressed. This explanation of demand is through a fuzzy number. As defined by Dang and Hong [40], fuzziness is some parametric property of the demand captured when its sharp boundaries cannot be determined by the decision makers. Zadeh [41] introduced fuzzy terminologies in 1978. After him, several researches including Petrovic and Sweeney [42], Roy and Maiti [43], Yao and Su [44], Chang and Ouyang [45], Mahata and Goswami [46], and De and Goswami [47] formulated inventory models considering parameters like demand, total average cost, storage space, and holding cost as fuzzy numbers. Numerous other researches solved inventory models considering fuzzy demand. Chang and Yeh [48], Sadeghu and Niaki [49], Lin et al. [50], Cosgun et al. [51], and Sadeghi et al. [52] treated demand as a trapezoidal fuzzy number in their inventory models. Some other researchers including Rong and Maiti [53], Zhang et al. [54], Moghaddam [55], and Huang et al. [56] considered demand as a Triangular Fuzzy Number (TFN). In addition, Taleizadeh et al. [57] presented a multi-constraint EOQ model with incremental discounts and uncertain item cost under fuzzy environment, and Jana et al. [58] considered storage space and available budget as fuzzy variables. Das et al. [59] assumed credit period in a supply chain as a fuzzy number, and Kumar and Goswami [60] developed a production inventory model under fuzzy shifting time to out-of-control state and fuzzy defective proportion.

In fact, in real situations, production processes are imperfect, thus producing defective items at a certain rate. That is why many researchers and practitioners have formulated inventory models expressing different real world circumstances [21–24]. In addition, parametric information of the real world production systems is not always known with precision [48]. This imprecise data can be random or uncertain in nature. Probability theory can handle the situations with random information, whereas the computational methods provided by the fuzzy theory are required to tackle the uncertain conditions. Several approaches including the signed distance method, min-max approximation, weighted average method, the center of gravity approach (centroid method), and center of the largest area are available in the uncertainty control literature to model the uncertain environment. This research work has used the center of gravity approach to grasp the uncertainty in demand information because of its simplicity and ease of use.

Vast research has been done on single-stage production system, in which machine breakdowns and defective proportion are considered as constant, random, and occasionally fuzzy in nature [22,34,60]. Regarding multistage production systems, only a few economic batch quantity models are available with limited real world production scenarios to provide succor in decision making for the managers dealing with multi-stage production systems [61,62]. This paper attempts to add in the literature by formulating an economic lot size model for multi-stage production facility with setup time requirement, uncertain annual demand rate, and imperfection constraints. The aim of this research work is to initiate a multi-stage production model by considering a real world shop-floor situation of defective production and uncertain market conditions regarding demand information.

Literature analysis indicates that no such model exists in the current research stream of the inventory and production management. This fact provides motivation for the development of multi-stage production model with the aforementioned attributes. Thus, the novelty of this study is the introduction of imperfect production proportion at each stage of the multi-stage production system and imprecise product demand. The objective of the research is to minimize the total cost of the imperfect multi-stage production system by determining optimal lot size under uncertain environments. Fuzzy theory is applied to grasp the uncertainty in demand information and analytical optimization technique is applied to obtain global optimum results of the model. The model is analyzed through sensitivity analysis and significant conclusions are obtained that verifies the substantial contribution of the proposed model in the inventory research literature.

Framework of the paper is as follows: The mathematical model is developed in Section 2. Section 3 provides experimental study and optimal results. Section 4 presents discussion and analysis of the proposed production model. Finally, in Section 5, concluding remarks and future research directions of this paper are discussed.

2. Materials and Methods

This research work studies an n-stage imperfect production system where defective items are produced along with the perfect quality items. These defective items are reworked at an additional processing cost, and product demand is fulfilled at the nth production stage. Figures 2 and 3 show inventory behavior of production stage-k and production stage-n, respectively. The objective of the model is to minimize the total system cost by determining optimal lot size.

2.1. Assumptions

Following assumptions are considered in the development of multi-stage production model.

1. A single type of item is produced in an n-stage production system.
2. Production runs are fairly consistent, i.e., system is assumed to be a lean manufacturing system where inventory behavior follows a similar trend in repetitive production cycles.
3. Uncertain product demand is measured as a Triangular Fuzzy Number (TFN).
4. Inline inspections provides effective results, determines the defective items immediately, and helps the managers to make immediate decisions over it. This model assumes that defective items are detected during the hundred percent inline inspection.
5. In textile industries, defective items are generally reworked. This model assumes reworking of defective items at each production stage. The reworking process is considered as perfect and no item is scrapped.
6. Defective rate is constant at each production stage, but it may vary from stage to stage.
7. Setup time of each production stage is ρ percent of the total time of that production stage.
8. Transportation time and cost among the production stages is assumed to be negligible.

2.2. Model Formulation

Keeping in view the above-mentioned assumptions, a mathematical model is formulated.

Figure 2 shows the inventory behavior of kth production stage. Maximum inventory level (I_{3k}) of the kth production stage ($k = 1, 2, 3, \ldots, n-1$) is computed as $I_{3k} = I_{1k} + I_{2k}$, where I_{1k} is the inventory level after production phase and I_{2k} is the inventory level after reworking phase.

Production time of stage-k is $T_{1k} = \frac{Q}{P_k}$, and its reworking time is $T_{2k} = \frac{Q\alpha_k}{P_k}$. It is obvious from triangle abc in Figure 2, that $\tan\omega_1 = \frac{I_{1k}}{T_{1k}}$, which gives $I_{1k} = Q(1-\alpha_k)$. Similarly, from triangle bef, one can observe that $\tan\omega_2 = \frac{I_{2k}}{T_{2k}}$, which provides $I_{2k} = Q\alpha_k$. From above the maximum inventory level of kth production stage is obtained as $I_{3k} = Q$.

Total inventory of kth production stage is the sum of the triangular areas abc, bcd, and bdf (Figure 2). One can simply compute these areas and the total inventory of kth production stage is now formulated as:

$$I_k = \frac{Q^2}{2P_k}(1 + \alpha_k - {\alpha_k}^2) \tag{1}$$

Total production time of stage-k can be obtained by the sum $T_k = T_{sk} + T_{1k} + T_{2k}$, where the setup time of stage-k, $T_{sk} = \rho(T_{1k} + T_{2k})$. Total time of stage-$k$ (T_k) is obtained as:

$$T_k = \frac{Q}{P_k}(1+\rho)(1+\alpha_k) \tag{2}$$

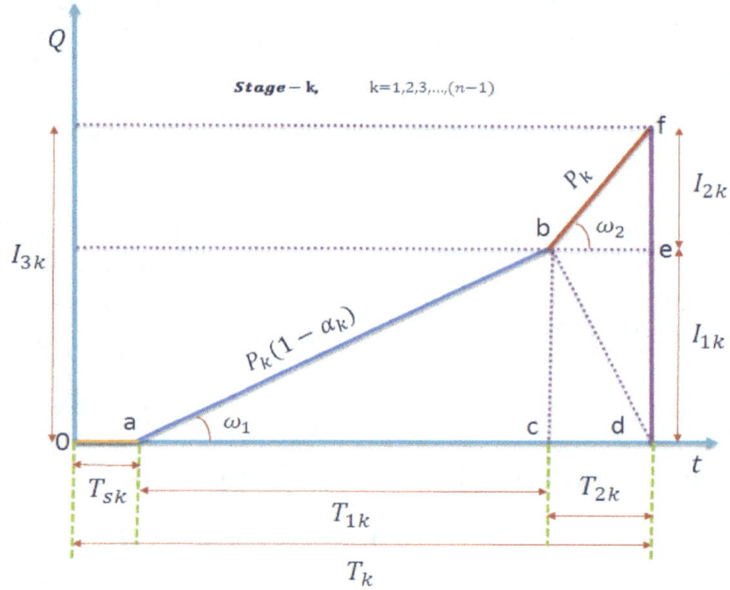

Figure 2. Inventory diagram of kth production stage.

Applying similar method for production stage-n (Figure 3), one can simply obtain production time $T_{1n} = \frac{Q}{P_n}$, reworking time $T_{2n} = \frac{Q\alpha_n}{P_n}$, pure consumption time $T_D = \frac{Q}{D}\left(1 - \frac{D}{P_n}(1+\alpha_n)\right)$, inventory level after production phase $I_{1n} = Q\left(1 - \alpha_n - \frac{D}{P_n}\right)$, and inventory level after reworking phase as $I_{2n} = Q\alpha_n\left(1 - \frac{D}{P_n}\right)$ for production stage-n. Finally, maximum inventory level of stage-n is computed as $I_{\max(n)} = Q\left(1 - \frac{D}{P_n}(1+\alpha_n)\right)$.

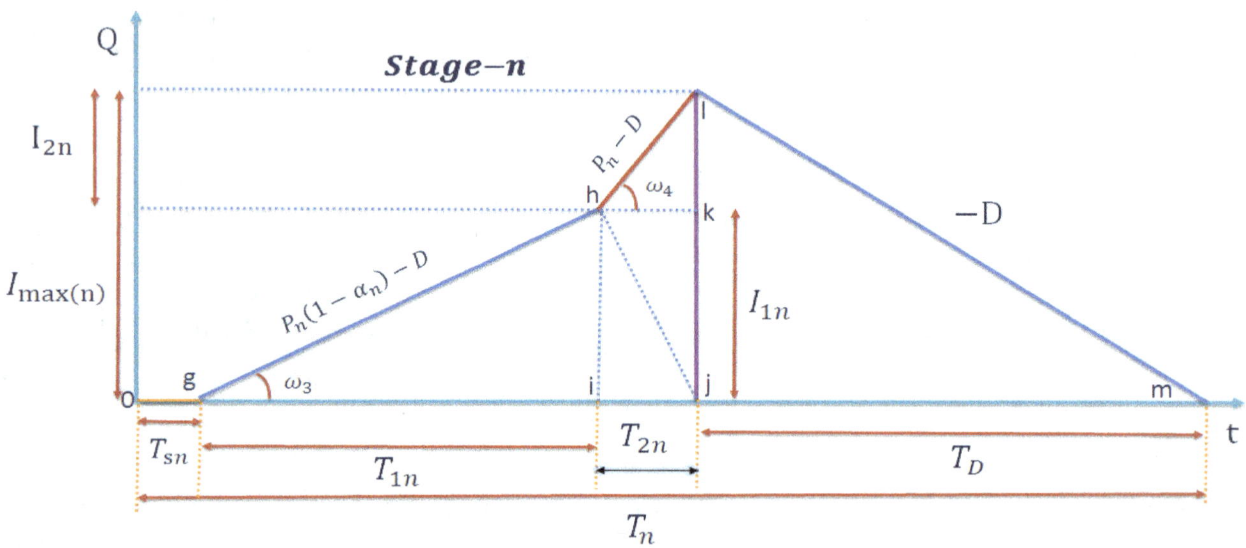

Figure 3. Inventory diagram of nth production stage.

Proceeding towards determination of total inventory level of finished items (I), we have

$$I = \frac{Q^2}{2D}\left(1 - \frac{D}{P_n}(1+\alpha_n+{\alpha_n}^2)\right) \tag{3}$$

Total time of production stage-n is obtained as

$$T_n = (1+\rho)\frac{Q}{D} \tag{4}$$

Total time of the complete production system can be determined by the sum $T = \sum_{k=1}^{n-1} T_k + T_n$, and is given by

$$T = \frac{Q(1+\rho)}{D}\left(1 + D\sum_{k=1}^{n-1}\left(\frac{1+\alpha_k}{P_k}\right)\right) \tag{5}$$

Average inventory level of the finished items $(\bar{I})$ in the system is calculated as $\bar{I} = \frac{I}{T}$, which provides

$$\bar{I} = \frac{Q(P_n - D(1+\alpha_n+\alpha_n{}^2))}{2P_n(1+\rho)\left(1+D\sum_{k=1}^{n-1}\left(\frac{1+\alpha_k}{P_k}\right)\right)} \tag{6}$$

Total cost of the system by considering setup cost, order processing cost, inspection cost, and inventory holding cost is formulated as follows:

Setup cost (K_S) *($K/lot/stage)*

Setup cost is a major part of the total cost in a production system. It is required to be incurred prior to the production of every lot. For an n-stage production system, the total setup cost is

$$K_s = \sum_{i=1}^{n} K_i \tag{7}$$

Order processing cost (K_P) *($C/item/stage)*

Order processing cost and reworking cost are considered as the same in this model. The total order processing cost is the sum of processing cost of good items as well as the reworking cost of defective items:

$$K_P = Q\sum_{i=1}^{n} C_i(1+\alpha_i) \tag{8}$$

Inspection cost (K_I) *($J/item/stage)*

Inspection cost is incurred on both good quality items as well as on reworked items and is assumed to be the same. Total inspection cost:

$$K_I = Q\sum_{i=1}^{n} J_i(1+\alpha_i) \tag{9}$$

Inventory holding cost (K_C) *($H/item/year)*

Under linear assumptions, total inventory holding cost is proportional to the holding cost of average inventory of finished items in the cycle as

$$K_C = H\bar{I} = H\frac{Q(P_n - D(1+\alpha_n+\alpha_n{}^2))}{2P_n(1+\rho)\left(1+D\sum_{k=1}^{n-1}\left(\frac{1+\alpha_k}{P_k}\right)\right)} \tag{10}$$

Total system cost function by considering aforementioned cost components is given by

$$TC(Q) = \frac{\sum_{i=1}^{n} K_i + Q\sum_{i=1}^{n} C_i(1+\alpha_i) + Q\sum_{i=1}^{n} J_i(1+\alpha_i)}{\frac{Q(1+\rho)}{D}\left(1+D\sum_{k=1}^{n-1}\left(\frac{1+\alpha_k}{P_k}\right)\right)} + H\frac{Q(P_n - D(1+\alpha_n+\alpha_n{}^2))}{2P_n(1+\rho)\left(1+D\sum_{k=1}^{n-1}\left(\frac{1+\alpha_k}{P_k}\right)\right)} \tag{11}$$

and after some simplifications

$$TC(Q) = \frac{\left(Q^2(HP_n - DH(1+\alpha_n+\alpha_n{}^2)) + 2DP_n\sum_{i=1}^{n}K_i + 2DQP_n\left(\sum_{i=1}^{n}C_i(1+\alpha_i) + \sum_{i=1}^{n}J_i(1+\alpha_i)\right)\right)}{2QP_n(1+\rho)\left(1+D\sum_{k=1}^{n-1}\left(\frac{1+\alpha_k}{P_k}\right)\right)} \tag{12}$$

Cárdenas-Barrón [14] viewed annual demand as a constant parameter. However, in several real world scenarios, some uncertainties may alter annual demand slightly. Thus, annual demand may be considered as a fuzzy number. To replace the annual demand D by a fuzzy number $\widetilde{D}$, consider the fuzzy number $\widetilde{D}$ as a TFN $\widetilde{D} = (D - \Delta_1, D, D + \Delta_2)$, where $0 < \Delta_1 < D$ and $0 < \Delta_2$ as shown in Figure 4. The Fuzzy Membership Function (FMF) of $\widetilde{D}$ is formulated as

$$\mu_{\widetilde{D}}(x) = \begin{cases} \frac{x-D+\Delta_1}{\Delta_1} & if \quad D-\Delta_1 \le x \le D \\ \frac{D+\Delta_2-x}{\Delta_2} & if \quad D \le x \le D+\Delta_2 \\ 0 & otherwise\ . \end{cases}$$

The centroid of $\mu_{\widetilde{D}}(x)$ is given by $D^* = D + \frac{1}{3}(\Delta_2 - \Delta_1)$. This result is considered to obtain the annual demand in fuzzy sense. For any $T > 0$, let $TC(Q)\ (x) = y(> 0)$. Using the extension principle of Kaufmann and Gupta [63] and Zimmermann [64], the membership function of the fuzzy cost $TC(Q)(\widetilde{D})$ is formulated as

$$\mu_{TC(Q)\widetilde{D}}(y) = \begin{cases} sup_{x\epsilon TC(Q)^{-1}(y)}\mu_{(\widetilde{D})}(x) & if \quad TC(Q)^{-1}(y) \neq \varnothing \\ 0 & if \quad TC(Q)^{-1}(y) = \varnothing \end{cases} .$$

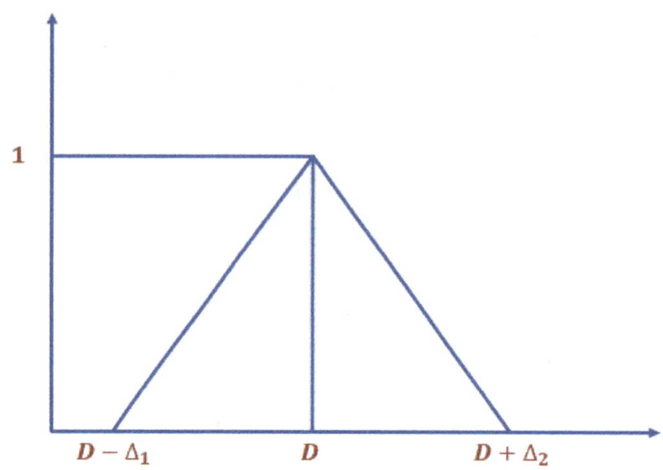

Figure 4. Triangular fuzzy number $\widetilde{D}$.

From $TC(Q)\ (x) = y$ and the system cost function, we obtain

$$y = \frac{\left(Q^2(HP_n - DH(1+\alpha_n+\alpha_n{}^2)) + 2DP_n\sum_{i=1}^{n}K_i + 2DQP_n\left(\sum_{i=1}^{n}C_i(1+\alpha_i) + \sum_{i=1}^{n}J_i(1+\alpha_i)\right)\right)}{2QP_n(1+\rho)\left(1+D\sum_{k=1}^{n-1}\left(\frac{1+\alpha_k}{P_k}\right)\right)} \tag{13}$$

For simplification purpose, let us assume
$B = HP_n$,
$F = H(1+\alpha_n+\alpha_n{}^2)$,
$L = 2P_n\sum_{i=1}^{n}K_i$,

$G = 2P_n\left(\sum_{i=1}^{n} C_i(1+\alpha_i) + \sum_{i=1}^{n} J_i(1+\alpha_i)\right),$
$M = 2P_n(1+\rho),$

and

$R = 2P_n(1+\rho)\sum_{k=1}^{n-1}\left(\frac{1+\alpha_k}{P_k}\right).$

This provides

$$y = \frac{BQ^2 + D(L+Q(G-FQ))}{Q(M+DR)} \tag{14}$$

and

$$x = \frac{Q(BQ-My)}{FQ^2+QRy-L-GQ} \tag{15}$$

From the standard triangular FMF and values of x and y, the FMF of $TC(Q)(\widetilde{D})$ is given by

$$\mu_{TC(Q)\widetilde{D}}(y) = \begin{cases} \frac{Q(BQ-My)}{(FQ^2+QRy-L-GQ)\Delta_1} - \frac{D-\Delta_1}{\Delta_1} & if\ \ y_1 \le y \le y_2 \\ \frac{D+\Delta_2}{\Delta_2} - \frac{Q(BQ-My)}{(FQ^2+QRy-L-GQ)\Delta_2} & if\ \ y_2 \le y \le y_3 \end{cases}.$$

Figure 5 represents the FMF of $TC(Q)(\widetilde{D})$.

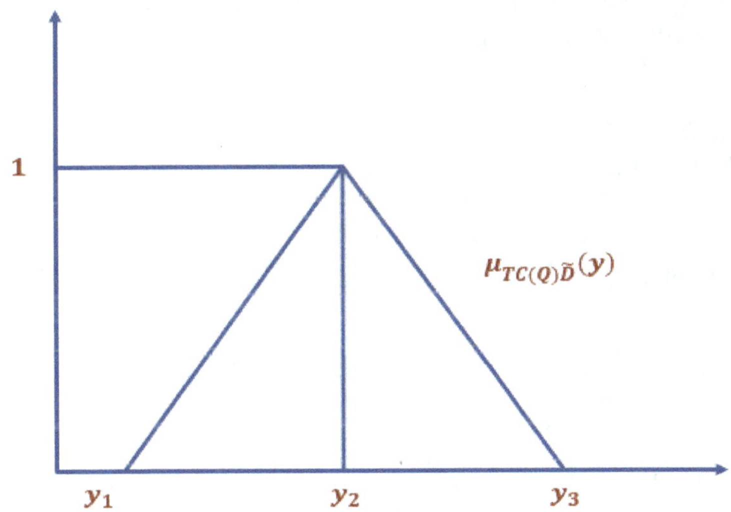

Figure 5. Triangular fuzzy number $TC(Q)(\widetilde{D})$.

Now centroid of $\mu_{TC(Q)\widetilde{D}}(y)$ is formulated as

$$TC(Q) = \frac{\int_{-\infty}^{\infty}\left[y\mu_{TC(Q)\widetilde{D}}(y)\right]dy}{\int_{-\infty}^{\infty}\left[\mu_{TC(Q)\widetilde{D}}(y)\right]dy} \tag{16}$$

After defuzzification, crisp total cost function becomes

$$TC(Q) = \frac{\left(\frac{BQ^2+D(L+Q(G-FQ))}{M+DR} + \frac{BQ^2+(L+Q(G-FQ))(D-\Delta_1)}{M+R(D-\Delta_1)} + \frac{BQ^2+(L+Q(G-FQ))(D+\Delta_2)}{M+R(D+\Delta_2)}\right)}{3Q} \tag{17}$$

As lot size Q is a decision variable, we differentiate the cost function (17) with respect to Q. Thus, from the necessary conditions, we have

$$\frac{d(TC)}{dQ}=\frac{\left(\Delta_1\left(\begin{array}{c}(M+DR)\left(L(M+3DR)+Q^2\left(\begin{array}{c}FM-2BR\\+3DFR\end{array}\right)\right)\\+R\left(L(2M+3DR)+Q^2\left(\begin{array}{c}2FM-BR\\+3DFR\end{array}\right)\right)\Delta_2\end{array}\right)-(M+DR)\left(\begin{array}{c}3(-BQ^2+D(L+FQ^2))(M+DR)\\+\left(\begin{array}{c}L(M+3DR)\\+Q^2(FM-2BR+3DFR)\end{array}\right)\Delta_2\end{array}\right)\right)}{3Q^2(M+DR)(M+DR-R\Delta_1)(M+DR+R\Delta_2)}=0 \quad (18)$$

After some simplification, Equation (18) can be written in standard quadratic form as

$$\frac{d(TC)}{dQ}=\frac{\left(\begin{array}{l}L(-(M+dR)(3d(M+dR)+(M+3dR)\Delta_2)+\Delta_1((M+dR)(M+3dR)+R(2M+3dR)\Delta_2))\\+Q^2(\Delta_1((M+dR)(FM-2BR+3dFR)+R(2FM-BR+3dFR)\Delta_2)+(M+dR)\ (3(B-dF)(M+dR)\\+(2BR-F(M+3dR))\Delta_2))\end{array}\right)}{3Q^2(M+dR)(M+dR-R\Delta_1)(M+dR+R\Delta_2)}=0 \quad (19)$$

Quadratic equation (19) provide two roots, one is negative and the other is positive. It is a well-known fact that quadratic equation has only one positive root. Thus, the two solutions for the optimal lot size are obtained as below

$$Q^*=\pm\sqrt{\left(\frac{L((M+DR)(3D(M+DR)+(M+3DR)\Delta_2)-\Delta_1((M+DR)(M+3DR)+R\Delta_2(2M+3DR)))}{\left(\Delta_1\left(\begin{array}{c}(M+DR)(FM-2BR+3DFR)+\\R\Delta_2(2FM-BR+3DFR)\end{array}\right)+(M+DR)\left(\begin{array}{c}3(B-DF)(M+DR)\\+(2BR-F(M+3DR))\Delta_2\end{array}\right)\right)}\right)}$$

Negative root seems quite impossible candidate solution for optimal lot size problem, and is thus discarded. The positive root is taken as the optimal lot size Q^*,

$$Q^*=\sqrt{\left(\frac{L((M+DR)(3D(M+DR)+(M+3DR)\Delta_2)-\Delta_1((M+DR)(M+3DR)+R\Delta_2(2M+3DR)))}{\left(\Delta_1\left(\begin{array}{c}(M+DR)(FM-2BR+3DFR)+\\R\Delta_2(2FM-BR+3DFR)\end{array}\right)+(M+DR)\left(\begin{array}{c}3(B-DF)(M+DR)\\+(2BR-F(M+3DR))\Delta_2\end{array}\right)\right)}\right)} \quad (20)$$

To satisfy sufficient conditions,

$$\frac{d^2(TC)}{(dQ)^2}=\frac{2L\left(3+M\left(-\frac{1}{M+DR}-\frac{1}{M+DR-R\Delta_1}-\frac{1}{M+DR+R\Delta_2}\right)\right)}{3Q^3R}>0 \quad (21)$$

As the solution to the above expression is positive, the sufficient condition is also satisfied. Thus the system cost attains minimum value in the interval (0, T) for Q^*.

3. Numerical Experiment and Results

This section analyzes the impact of various numerical values of parameters on the optimal cost of the system by taking appropriate units. Three numerical experiments are carried out for a five-stage production system ($n = 5$).

3.1. Example 1

Parameters K_i = \$100/stage, D = 50,000 items/year, P_n = 200,000 items/year, H = \$5/item/year, and ρ = 2.00% per stage are taken from Wee et al. [37] and Biswas and Sarker [19]. Parameters α_i = 1.00% for each stage, C_i = \$3/item for each stage, J_i = \$0.02/item for each stage, P_{n-1} = 210,000 items/year, P_{n-2} = 220,500 items/year, P_{n-3} = 231,525 items/year, P_{n-4} = 243,102 items/year, and Δ_1 = 8000, and Δ_2 = 12,000 are considered randomly.

3.2. Example 2

Numerical values of the parameters in this example are taken from Tayyab and Sarkar [61] as $K_i = \$400/\text{stage}$, $D = 15{,}000$ items/year, $P_n = 70{,}000$ items/year, $H = \$4/\text{item/year}$, $\rho = 2.00\%$ per stage, $C_i = \$35/\text{item}$ for each stage, $J_i = \$1/\text{item}$ for each stage, $P_{n-1} = 73{,}500$ items/year, $P_{n-2} = 77{,}175$ items/year, $P_{n-3} = 81{,}033$ items/year, $P_{n-4} = 85{,}085$ items/year, and $\alpha_i = 1.00\%$ for each production stage is the same as in Example 1. Parameters $\Delta_1 = 3000$ and $\Delta_2 = 4000$ are considered randomly.

3.3. Example 3

Numerical values of the parameters in this example are taken from Tayyab and Sarkar [61] as $K_i = \$200/\text{stage}$, $D = 12{,}000$ items/year, $P_n = 60{,}000$ items/year, $H = \$20/\text{item/year}$, $\rho = 2.00\%$ per stage, $C_i = \$100/\text{item}$ for each stage, $J_i = \$0.5/\text{item}$ for each stage, $P_{n-1} = 63{,}000$ items/year, $P_{n-2} = 66{,}150$ items/year, $P_{n-3} = 69{,}457$ items/year, $P_{n-4} = 72{,}930$ items/year, and $\alpha_i = 1.00\%$ for each production stage is same as in Example 1 and 2. Parameters $\Delta_1 = 2000$ and $\Delta_2 = 3000$ are considered randomly.

Optimal results of the above numerical examples are provided in Table 1.

Table 1. Optimal results of the numerical experiments.

Production Stages	Example 1		Example 2		Example 3	
	Q* (items)	TC* ($)	Q* (items)	TC* ($)	Q* (items)	TC* ($)
Single-stage	1664.93	159,552	1984.44	552,648	557.94	1,236,016
Two-stage	2346.56	252,080	2795.5	905,477	786.79	2,051,893
Three-stage	2867.7	316,070	3414.82	1,160,906	961.72	2,652,656
Four-stage	3306.35	364,808	3935.67	1,360,773	1108.92	3,127,405
Five-stage	3692.53	404,441	4393.98	1,525,950	1238.47	3,522,058

4. Discussion

Decision makers of the multi-stage production systems can get the advantage of the proposed production model to reduce system cost in the presence of defective production and imprecise information. The proposed production model has provided global optimal results for lot size decision to achieve the minimum cost of the multi-stage production system. The detailed procedure of fuzzy theory application provided for uncertainty control can enable the managers and decision-makers to conveniently address the uncertain conditions of respective shop-floors. Each numerical example is solved for five-stage production systems and results are summarized in Table 1. Figures 6–8 illustrate the convexity of minimum cost objective in each case, which verifies the robustness of the model. Optimal results indicate that the larger lot sizes provide more benefit to the production systems comprising of higher number of stages. For instance, in Example 3, optimal lot size for single-stage production system is 557.94 items and 1238.47 items for five-stage production system. Consistency in the obtained results assures that this research work has achieved the desired goal of analyzing product imperfection and imprecise information in multi-stage production modeling.

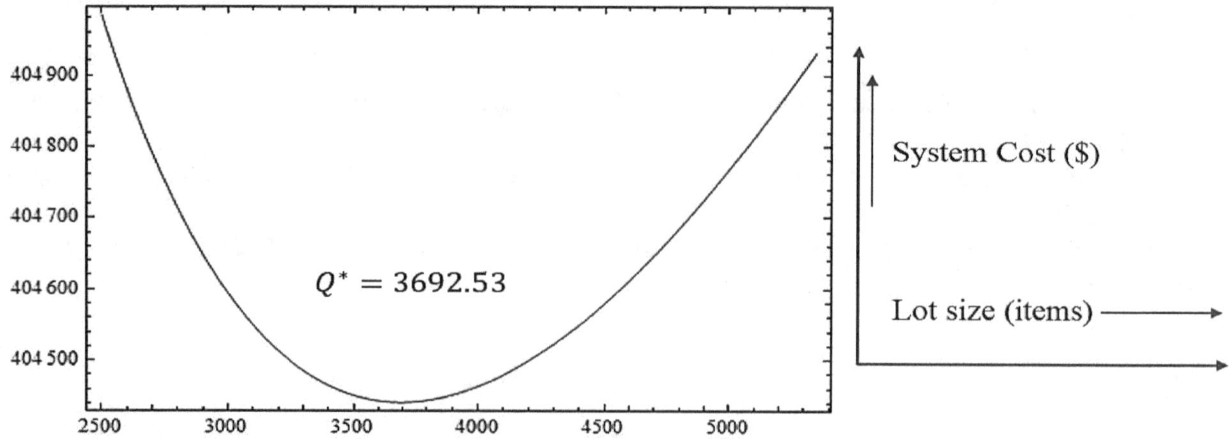

Figure 6. Objective function convexity diagram for Example 1 (Five-stage production system).

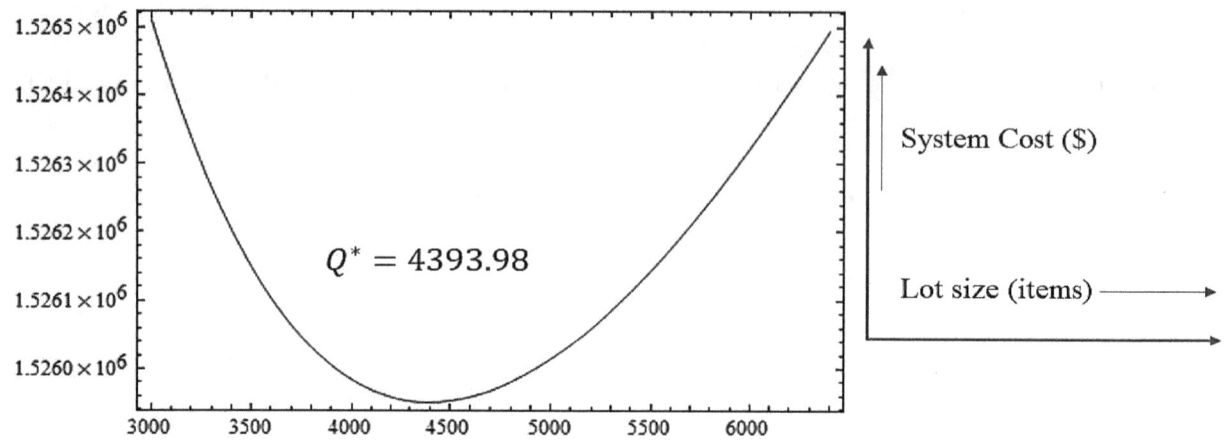

Figure 7. Objective function convexity diagram for Example 2 (Five-stage production system).

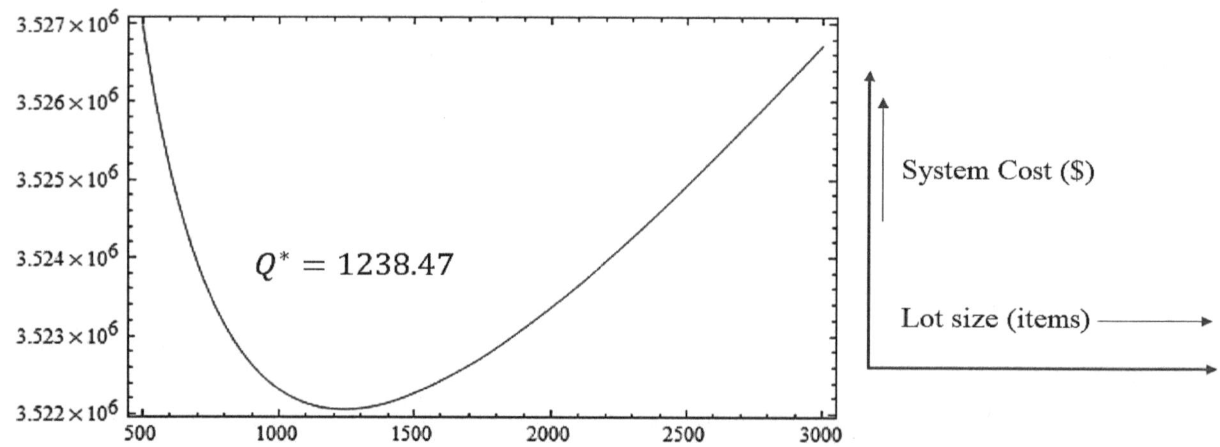

Figure 8. Objective function convexity diagram for Example 3 (Five-stage production system).

Optimal results are further evaluated for the possible variations in key parameters of the given examples. The effect of key parameters on the optimal cost function for up-to five-stage production system is summarized in Tables 2–4 for Example 1 to Example 3, respectively. The values of the parameters are varied from −50% to +50%.

1. It is evident from the change in optimal cost function that the order processing cost (C_i) is the most sensitive factor than other parameters. One can observe that with each percentage increase in C_i there exists almost similar amount of percentage increase in the system cost. For instance, regarding four-stage production system in Example 2, system cost increases by 24.17% by the

increment of 25% in order of processing cost, and 48.34% increase in system cost is observed by a 50% rise in the order of processing cost.

2. Next to the effect of order processing cost is the impact of production rate (P_i) on the system cost. It can be noted that the effect of production rate on the production system with higher number of stages is more than the system with a lower number of production stages. This illustrates the necessity of wisely adjusting production rate at each production stage to keep the system cost at its minimum value.

3. An interesting observation from the sensitivity analysis of all the numerical examples reveal that the impact of setup cost (K_i) and the inventory holding cost (H) is same on the system cost for all the production systems with any number of stages. Thus, the decision makers have to make intelligent decision in putting efforts to reduce one of these costs first. Here arises the importance of lean philosophy, as Single Minute Exchange of Dyes (SMED) is the best possible solution for setup cost reduction. Therefore, managers should make efforts to apply SMED for setup cost reduction first, as it is much more beneficial than cutting down stock management equipment and inventory holding staff.

4. Inspection cost (J_i) and defective proportion (α_i) bear trivial impact on the system cost that gradually varies as the system moves towards a higher number of stages.

The sensitivity analysis provides important inferences for the managers to develop optimal policies that can achieve the best outcomes at their shop floors.

Table 2. Sensitivity analysis of key parameters for Example 1.

Parameters	Change (%)	Percentage Change in $TC(Q)$*				
		Single-Stage Process	**Two-Stage Process**	**Three-Stage Process**	**Four-Stage Process**	**Five-Stage Process**
P_i	−50	−0.73	−17.09	−24.92	−29.53	−32.57
	−25	−0.23	−6.41	−9.94	−12.24	−13.85
	+25	+0.13	+4.27	+7.09	+9.13	+10.67
	+50	+0.21	+7.33	+12.40	+16.20	+19.15
K_i	−50	−1.11	−0.80	−0.65	−0.57	−0.51
	−25	−0.51	−0.36	−0.30	−0.26	−0.23
	+25	+0.45	+0.32	+0.26	+0.23	+0.21
	+50	+0.85	+0.61	+0.50	+0.44	+0.39
C_i	−50	−47.79	−48.32	−48.56	−48.70	−48.80
	−25	−23.89	−24.16	−24.28	−24.35	−24.40
	+25	+23.89	+24.16	+24.28	+24.35	+24.40
	+50	+47.79	+48.32	+48.56	+48.70	+48.80
J_i	−50	−0.32	−0.32	−0.32	−0.32	−0.33
	−25	−0.16	−0.16	−0.16	−0.16	−0.16
	+25	+0.16	+0.16	+0.16	+0.16	+0.16
	+50	+0.32	+0.32	+0.32	+0.32	+0.33
H	−50	−1.11	−0.80	−0.65	−0.57	−0.51
	−25	−0.51	−0.36	−0.30	−0.26	−0.23
	+25	+0.45	+0.32	+0.26	+0.23	+0.21
	+50	+0.85	+0.61	+0.50	+0.44	+0.39
α_i	−50	−0.47	−0.38	−0.32	−0.28	−0.25
	−25	−0.24	−0.19	−0.16	−0.14	−0.12
	+25	+0.24	+0.19	+0.16	+0.14	+0.12
	+50	+0.47	+0.38	+0.32	+0.28	+0.25

Table 3. Sensitivity analysis of key parameters for Example 2.

Parameters	Change (%)	Percentage Change in $TC(Q)$*				
		Single-Stage Process	Two-Stage Process	Three-Stage Process	Four-Stage Process	Five-Stage Process
P_i	−50	−0.17	−15.12	−22.75	−27.42	−30.57
	−25	−0.05	−5.61	−8.95	−11.20	−12.81
	+25	+0.03	+3.70	+6.28	+8.20	+9.69
	+50	+0.05	+6.33	+10.94	+14.48	+17.28
K_i	−50	−0.32	−0.23	−0.19	−0.16	−0.15
	−25	−0.15	−0.10	−0.09	−0.07	−0.07
	+25	+0.13	+0.09	+0.08	+0.07	+0.06
	+50	+0.25	+0.18	+0.14	+0.12	+0.11
C_i	−50	−48.08	−48.23	−48.30	−48.34	−48.37
	−25	−24.04	−24.12	−24.15	−24.17	−24.18
	+25	+24.04	+24.12	+24.15	+24.17	+24.18
	+50	+48.08	+48.23	+48.30	+48.34	+48.37
J_i	−50	−1.37	−1.38	−1.38	−1.38	−1.38
	−25	−0.69	−0.69	−0.69	−0.69	−0.69
	+25	+0.69	+0.69	+0.69	+0.69	+0.69
	+50	+1.37	+1.38	+1.38	+1.38	+1.38
H	−50	−0.32	−0.23	−0.19	−0.16	−0.15
	−25	−0.15	−0.10	−0.09	−0.07	−0.07
	+25	+0.13	+0.09	+0.08	+0.07	+0.06
	+50	+0.25	+0.18	+0.14	+0.12	+0.11
α_i	−50	−0.49	−0.40	−0.35	−0.31	−0.27
	−25	−0.24	−0.20	−0.17	−0.15	−0.14
	+25	+0.24	+0.20	+0.17	+0.15	+0.14
	+50	+0.49	+0.40	+0.35	+0.30	+0.27

Table 4. Sensitivity analysis of key parameters for Example 3.

Parameters	Change (%)	Percentage Change in $TC(Q)$*				
		Single-Stage Process	Two-Stage Process	Three-Stage Process	Four-Stage Process	Five-Stage Process
P_i	−50	−0.10	−14.38	−21.92	−26.61	−29.80
	−25	−0.03	−5.31	−8.57	−10.79	−12.41
	+25	+0.02	+3.48	+5.97	+7.84	+9.31
	+50	+0.03	+5.95	+10.37	+13.80	+16.55
K_i	−50	−0.21	−0.15	−0.12	−0.10	−0.09
	−25	−0.09	−0.07	−0.05	−0.05	−0.04
	+25	+0.08	+0.06	+0.05	+0.04	+0.04
	+50	+0.16	+0.11	+0.09	+0.08	+0.07
C_i	−50	−49.40	−49.50	−49.55	−49.58	−49.59
	−25	−24.70	−24.75	−24.77	−24.79	−24.80
	+25	+24.70	+24.75	+24.77	+24.79	+24.80
	+50	+49.40	+49.50	+49.55	+49.58	+49.59
J_i	−50	−0.25	−0.25	−0.25	−0.25	−0.25
	−25	−0.12	−0.12	−0.12	−0.12	−0.12
	+25	+0.12	+0.12	+0.12	+0.12	+0.12
	+50	+0.25	+0.25	+0.25	+0.25	+0.25
H	−50	−0.21	−0.15	−0.12	−0.10	−0.09
	−25	−0.09	−0.07	−0.05	−0.05	−0.04
	+25	+0.08	+0.06	+0.05	+0.04	+0.04
	+50	+0.16	+0.11	+0.09	+0.08	+0.07

Table 4. *Cont.*

Parameters	Change (%)	Percentage Change in $TC(Q)$*				
		Single-Stage Process	Two-Stage Process	Three-Stage Process	Four-Stage Process	Five-Stage Process
α_i	−50	−0.49	−0.41	−0.35	−0.31	−0.28
	−25	−0.25	−0.20	−0.18	−0.16	−0.14
	+25	+0.25	+0.20	+0.18	+0.16	+0.14
	+50	+0.49	+0.41	+0.35	+0.31	+0.28

5. Conclusions

Agreeing to lean philosophy, high inventory levels are considered as one of the major non-value added aspect of the production system. For true implementation of lean culture, inventory as a form of non-value added activity should be reduced to its minimum. Therefore, there must be a defined optimal lot size in accordance with the input requirements of each production stage at minimum system cost. Cárdenas-Barrón [14] provided interpretation of considering defective rate in his inventory model to achieve optimal cost at the single-stage production facility. This inference is carried forward in this paper for its implication on a multi-stage production system.

This study is made-up to support production and planning managers, especially in the textile production sector where defective production and uncertain product demand are faced. It introduces an imperfect multi-stage production model with the consideration of defective proportion in the production process and uncertain product demand. Thus, the novelty of this study is the introduction of imperfect production proportion at each stage of the multi-stage production system and imprecise product demand. Fuzzy theory is applied to handle the uncertainty in demand information and center of gravity approach is utilized to defuzzify the objective function. This defuzzified cost objective is solved through the analytical optimization technique and closed form solution of the optimal lot size is obtained.

Numerical experiment comprising of three examples is conducted and optimal results are analyzed through sensitivity analysis. Optimal results indicate that as the number of stages increase, lots of larger sizes are cost effective. Sensitivity analysis indicates that the order processing cost bears the highest effect on the system cost and its effect increases slightly as the system moves towards higher number of stages. Creditably, this study can work as an elementary support in implementing lean culture in the production system. Analysis of the model illustrates the major advantage of this study by highlighting the importance of lean activities to reduce the system cost, as the results suggest application of SMED for setup cost reduction in the multi-stage production system prior to making efforts for inventory holding cost reduction.

Despite various benefits, this study has some limitations. It has considered constant defective proportion in the production system, whereas defective rate becomes imprecise in long run production processes. Secondly, the proposed model has incorporated a perfect reworking opportunity that is very rare in the real life production systems, as defective products are also produced during reworking. The analysis of the results state that the effect of order processing cost on the total system cost is highest among all other system parameters. However, this study has not provided order processing cost reduction policy to minimize system cost for future runs. Further, the model can be extended with several ideas including consideration of shortages backordered [39,65,66], variable production rate [67–69], trade credit policies [70,71], allocation problem [72,73], and different rework options [19]. The immediate possible extension to this paper can be the incorporation of uncertain defective rate [61,62].

Author Contributions: Conceptualization, M.T. and B.S.; Methodology, M.T.; Software, M.T.; Validation, B.S.; Formal analysis, M.T. and B.S.; Investigation, M.T.; Resources, M.T. B.S. and B.N.Y.; Data curation, M.T. and B.S.; Writing—original draft preparation, M.T.; Writing—review and editing, B.S. and B.N.Y.; Visualization, M.T. B.S. and B.N.Y.; Supervision, B.S.

Nomenclature

Indices

k	production stages ($k = 1, 2, 3 \ldots (n-1)$)
i	production stages ($i = 1, 2, 3 \ldots n$)

Decision variable

Q	lot size (items)

Parameters

n	number of production stages
P_k	production rate for k^{th} production stage (items per unit time)
P_n	production rate for n^{th} production stage (items per unit time, where $P_k > P_n > D$)
$\widetilde{D}$	demand rate (fuzzy, and fulfilled at n^{th} production stage)
α_k	defective proportion at k^{th} production stage
α_n	defective proportion at n^{th} production stage
T_k	cycle time of kth production stage (years)
T_n	cycle time of nth production stage (years)
T	time between production runs (total cycle time of n-stages, years)
K_i	setup cost of ith production stage ($/setup)
H	inventory holding cost of finished items ($/item/year)
C_i	order processing cost at i^{th} production stage ($/item/stage)
J_i	inspection cost at i^{th} production stage ($/item/stage)
ρ	percentage setup time per stage
$\bar{I}$	average inventory of finished items in the system
$TC(Q)$	total cost of the production system ($ per unit time)

Abbreviations

Below enlisted abbreviations are used for the multi-stage production model development in this research work.

EOQ	Economic order quantity
EPQ	Economic production quantity
TFN	Triangular fuzzy number
FMF	Fuzzy membership function
SMED	Single minute exchange of dyes

References

1. Hodge, G.L.; Goforth Ross, K.; Joines, J.A.; Thoney, K. Adapting lean manufacturing principles to the textile industry. *Prod. Plan. Control* **2011**, *22*, 237–247. [CrossRef]
2. Harris, F.W. How many parts to make at once. *Fact. Mag. Manag.* **1913**, *10*, 135–136. [CrossRef]
3. Taft, E.W. The most economical production lot. *Iron Age* **1918**, *101*, 1410–1412.
4. Lee, H.L. Lot sizing to reduce capacity utilization in a production process with defective items, process corrections, and rework. *Manag. Sci.* **1992**, *38*, 1314–1328. [CrossRef]
5. Gupta, T.; Chakraborty, S. Looping in a multistage production system. *Int. J. Prod. Res.* **1984**, *22*, 299–311. [CrossRef]
6. Tayi, G.K.; Ballou, D.P. An integrated production-inventory model with reprocessing and inspection. *Int. J. Prod. Res.* **1988**, *26*, 1299–1315. [CrossRef]
7. Lee, H.H.; Chandra, M.J.; Deleveaux, V.J. Optimal batch size and investment in multistage production systems with scrap. *Prod. Plan. Control* **1997**, *8*, 586–596. [CrossRef]
8. Glock, C.H.; Jaber, M.Y. Learning effects and the phenomenon of moving bottlenecks in a two-stage production system. *Appl. Math. Model.* **2013**, *37*, 8617–8628. [CrossRef]
9. Lin, C.S. Integrated production-inventory models with imperfect production processes and a limited capacity for raw materials. *Math. Comput. Model.* **1999**, *29*, 81–89. [CrossRef]

10. Salameh, M.K.; Jaber, M.Y. Economic production quantity model for items with imperfect quality. *Int. J. Prod. Econ.* **2000**, *64*, 59–64. [CrossRef]
11. Khan, M.; Jaber, M.Y.; Guiffrida, A.L.; Zolfaghari, S. A review of the extensions of a modified EOQ model for imperfect quality items. *Int. J. Prod. Econ.* **2011**, *132*, 1–2. [CrossRef]
12. Jaber, M.Y.; Zanoni, S.; Zavanella, L.E. Economic order quantity models for imperfect items with buy and repair options. *Int. J. Prod. Econ.* **2014**, *155*, 126–131. [CrossRef]
13. Ouyang, L.Y.; Chen, C.K.; Chang, H.C. Quality improvement, setup cost and lead-time reductions in lot size reorder point models with an imperfect production process. *Comput. Oper. Res.* **2002**, *29*, 1701–1717. [CrossRef]
14. Cárdenas-Barrón, L.E. Economic production quantity with rework process at a single-stage manufacturing system with planned backorders. *Comput. Ind. Eng.* **2009**, *57*, 1105–1113. [CrossRef]
15. Eroglu, A.; Ozdemir, G. An economic order quantity model with defective items and shortages. *Int. J. Prod. Econ.* **2007**, *106*, 544–549. [CrossRef]
16. Jaber, M.Y.; Guiffrida, A.L. Learning curves for imperfect production processes with reworks and process restoration interruptions. *Eur. J. Oper. Res.* **2008**, *189*, 93–104. [CrossRef]
17. Ben-Daya, M. The economic production lot-sizing problem with imperfect production processes and imperfect maintenance. *Int. J. Prod. Econ.* **2002**, *76*, 257–264. [CrossRef]
18. Jamal, A.M.; Sarker, B.R.; Mondal, S. Optimal manufacturing batch size with rework process at a single-stage production system. *Comput. Ind. Eng.* **2004**, *47*, 77–89. [CrossRef]
19. Biswas, P.; Sarker, B.R. Optimal batch quantity models for a lean production system with in-cycle rework and scrap. *Int. J. Prod. Res.* **2008**, *46*, 6585–6610. [CrossRef]
20. Chung, K.J. The economic production quantity with rework process in supply chain management. *Comput. Math. Appl.* **2011**, *62*, 2547–2550. [CrossRef]
21. Taleizadeh, A.A.; Jalali-Naini, S.G.; Wee, H.M.; Kuo, T.C. An imperfect multi-product production system with rework. *Sci. Iran* **2013**, *20*, 811–823.
22. Cárdenas-Barrón, L.E. Optimal manufacturing batch size with rework in a single-stage production system—A simple derivation. *Comput. Ind. Eng.* **2008**, *55*, 758–765. [CrossRef]
23. Ma, W.N.; Gong, D.C.; Lin, G.C. An optimal common production cycle time for imperfect production processes with scrap. *Math. Comput. Model.* **2010**, *52*, 724–737. [CrossRef]
24. Barzoki, M.R.; Jahanbazi, M.; Bijari, M. Effects of imperfect products on lot sizing with work in process inventory. *Appl. Math. Comput.* **2011**, *217*, 8328–8336. [CrossRef]
25. Chiu, S.W.; Tseng, C.T.; Wu, M.F.; Sung, P.C. Multi-item EPQ model with scrap, rework and multi-delivery using common cycle policy. *J. Appl. Res. Technol.* **2014**, *12*, 615–622. [CrossRef]
26. Chiu, S.W.; Chiu, Y.S.; Yang, J.C. Combining an alternative multi-delivery policy into economic production lot size problem with partial rework. *Expert Syst. Appl.* **2012**, *39*, 2578–2583. [CrossRef]
27. Noorollahi, E.; Karimi-Nasab, M.; Aryanezhad, M.B. An economic production quantity model with random yield subject to process compressibility. *Math. Comput. Model.* **2012**, *56*, 80–96. [CrossRef]
28. Moussawi-Haidar, L.; Salameh, M.; Nasr, W. Production lot sizing with quality screening and rework. *Appl. Math. Model.* **2016**, *40*, 3242–3256. [CrossRef]
29. Chiu, S.W.; Gong, D.C.; Wee, H.M. Effects of random defective rate and imperfect rework process on economic production quantity model. *Jpn. J. Ind. Appl. Math.* **2004**, *21*, 375. [CrossRef]
30. Taleizadeh, A.A.; Cárdenas-Barrón, L.E.; Mohammadi, B. A deterministic multi product single machine EPQ model with backordering, scraped products, rework and interruption in manufacturing process. *Int. J. Prod. Econ.* **2014**, *150*, 9–27. [CrossRef]
31. Chiu, S.W.; Ting, C.K.; Chiu, Y.S. Optimal production lot sizing with rework, scrap rate, and service level constraint. *Math. Comput. Model.* **2007**, *46*, 535–549. [CrossRef]
32. Taleizadeh, A.A.; Wee, H.M.; Jalali-Naini, S.G. Economic production quantity model with repair failure and limited capacity. *Appl. Math. Model.* **2013**, *37*, 2765–2774. [CrossRef]
33. Sana, S.S. A production–inventory model in an imperfect production process. *Eur. J. Oper. Res.* **2010**, *200*, 451–464. [CrossRef]
34. Sarkar, B.; Moon, I. An EPQ model with inflation in an imperfect production system. *Appl. Math. Comput.* **2011**, *217*, 6159–6167. [CrossRef]

35. Yoo, S.H.; Kim, D.; Park, M.S. Lot sizing and quality investment with quality cost analyses for imperfect production and inspection processes with commercial return. *Int. J. Prod. Econ.* **2012**, *140*, 922–933. [CrossRef]
36. Chakraborty, T.; Giri, B.C. Joint determination of optimal safety stocks and production policy for an imperfect production system. *Appl. Math. Model.* **2012**, *36*, 712–722. [CrossRef]
37. Wee, H.M.; Wang, W.T.; Yang, P.C. A production quantity model for imperfect quality items with shortage and screening constraint. *Int. J. Prod. Res.* **2013**, *51*, 1869–1884. [CrossRef]
38. Chiu, S.W.; Chou, C.L.; Wu, W.K. Optimizing replenishment policy in an EPQ-based inventory model with nonconforming items and breakdown. *Econ. Model.* **2013**, *35*, 330–337. [CrossRef]
39. Sarkar, B.; Cárdenas-Barrón, L.E.; Sarkar, M.; Singgih, M.L. An economic production quantity model with random defective rate, rework process and backorders for a single stage production system. *J. Manuf. Syst.* **2014**, *33*, 423–435. [CrossRef]
40. Dang, J.F.; Hong, I.H. The Cournot game under a fuzzy decision environment. *Comput. Math. Appl.* **2010**, *59*, 3099–3109. [CrossRef]
41. Zadeh, L.A. Fuzzy sets as a basis for a theory of possibility. *Fuzzy Set Syst.* **1999**, *100*, 9–34. [CrossRef]
42. Petrovic, D.; Sweeney, E. Fuzzy knowledge-based approach to treating uncertainty in inventory control. *Comput. Integr. Manuf. Syst.* **1994**, *7*, 147–152. [CrossRef]
43. Roy, T.K.; Maiti, M. Multi-objective inventory models of deteriorating items with some constraints in a fuzzy environment. *Comput. Oper. Res.* **1998**, *25*, 1085–1095. [CrossRef]
44. Yao, J.S.; Chang, S.C.; Su, J.S. Fuzzy inventory without backorder for fuzzy order quantity and fuzzy total demand quantity. *Comput. Oper. Res.* **2000**, *27*, 935–962. [CrossRef]
45. Chang, H.C.; Yao, J.S.; Ouyang, L.Y. Fuzzy mixture inventory model involving fuzzy random variable lead time demand and fuzzy total demand. *Eur. J. Oper. Res.* **2006**, *169*, 65–80. [CrossRef]
46. Mahata, G.C.; Goswami, A. Fuzzy inventory models for items with imperfect quality and shortage backordering under crisp and fuzzy decision variables. *Comput. Ind. Eng.* **2013**, *64*, 190–199. [CrossRef]
47. De, S.K.; Goswami, A.; Sana, S.S. An interpolating by pass to Pareto optimality in intuitionistic fuzzy technique for a EOQ model with time sensitive backlogging. *Appl. Math. Comput.* **2014**, *230*, 664–674. [CrossRef]
48. Chang, S.Y.; Yeh, T.Y. A two-echelon supply chain of a returnable product with fuzzy demand. *Appl. Math. Model.* **2013**, *37*, 4305–4315. [CrossRef]
49. Sadeghi, J.; Niaki, S.T. Two parameter tuned multi-objective evolutionary algorithms for a bi-objective vendor managed inventory model with trapezoidal fuzzy demand. *Appl. Soft Comput.* **2015**, *30*, 567–576. [CrossRef]
50. Lin, J.; Hung, K.C.; Julian, P. Technical note on inventory model with trapezoidal type demand. *Appl. Math. Model.* **2014**, *38*, 4941–4948. [CrossRef]
51. Coşgun, Ö.; Ekinci, Y.; Yanık, S. Fuzzy rule-based demand forecasting for dynamic pricing of a time company. *Knowl. Based Syst.* **2014**, *70*, 88–96. [CrossRef]
52. Sadeghi, J.; Mousavi, S.M.; Niaki, S.T. Optimizing an inventory model with fuzzy demand, backordering, and discount using a hybrid imperialist competitive algorithm. *Appl. Math. Model.* **2016**, *40*, 7318–7335. [CrossRef]
53. Rong, M.; Maiti, M. On an EOQ model with service level constraint under fuzzy-stochastic demand and variable lead-time. *Appl. Math. Model.* **2015**, *39*, 5230–5240. [CrossRef]
54. Zhang, B.; Lu, S.; Zhang, D.; Wen, K. Supply chain coordination based on a buyback contract under fuzzy random variable demand. *Fuzzy Set Syst.* **2014**, *255*, 1–6. [CrossRef]
55. Moghaddam, K.S. Fuzzy multi-objective model for supplier selection and order allocation in reverse logistics systems under supply and demand uncertainty. *Expert Syst. Appl.* **2015**, *42*, 6237–6254. [CrossRef]
56. Huang, M.; Song, M.; Leon, V.J.; Wang, X. Dentralized capacity allocation of a single-facility with fuzzy demand. *J. Manuf. Syst.* **2014**, *33*, 7–15. [CrossRef]
57. Taleizadeh, A.A.; Niaki, S.T.; Nikousokhan, R. Constraint multiproduct joint-replenishment inventory control problem using uncertain programming. *Appl. Soft Comput.* **2011**, *11*, 5143–5154. [CrossRef]
58. Jana, D.K.; Das, B.; Maiti, M. Multi-item partial backlogging inventory models over random planninghorizon in random fuzzy environment. *Appl. Soft Comput.* **2014**, *21*, 12–27. [CrossRef]
59. Das, B.C.; Das, B.; Mondal, S.K. An integrated production inventory model under interactive fuzzy credit period for deteriorating item with several kets. *Appl. Soft Comput.* **2015**, *28*, 453–465. [CrossRef]

60. Ku, R.S.; Goswami, A. A fuzzy random EPQ model for imperfect quality items with possibility and necessity constraints. *Appl. Soft Comput.* **2015**, *34*, 838–850.
61. Tayyab, M.; Sarkar, B. Optimal batch quantity in a cleaner multi-stage lean production system with random defective rate. *J. Clean. Prod.* **2016**, *139*, 922–934. [CrossRef]
62. Kim, M.S.; Sarkar, B. Multi-stage cleaner production process with quality improvement and lead time dependent ordering cost. *J. Clean. Prod.* **2017**, *144*, 572–590. [CrossRef]
63. Kaufman, A.; Gupta, M.M. *Introduction to Fuzzy Arithmetic*; Van Nostrand Reinhold Company: New York, NY, USA, 1991.
64. Zimmermann, H.J. Fuzzy control. In *Fuzzy Set Theory—And Its Applications*; Springer: Dordrecht, The Netherlands, 1996; pp. 203–240.
65. Sarkar, B.; Moon, I. Improved quality, setup cost reduction, and variable backorder costs in an imperfect production process. *Int. J. Prod. Econ.* **2014**, *155*, 204–213. [CrossRef]
66. Sarkar, B. A production-inventory model with probabilistic deterioration in two-echelon supply chain management. *Appl. Math. Model.* **2013**, *37*, 3138–3151. [CrossRef]
67. Sarkar, B.; Sana, S.S.; Chaudhuri, K. Optimal reliability, production lotsize and safety stock: An economic manufacturing quantity model. *Int. J. Manag. Sci. Eng. Manag.* **2010**, *5*, 192–202. [CrossRef]
68. Moon, I.; Shin, E.; Sarkar, B. Min–max distribution free continuous-review model with a service level constraint and variable lead time. *Appl. Math. Comput.* **2014**, *229*, 310–315. [CrossRef]
69. Shin, D.; Guchhait, R.; Sarkar, B.; Mittal, M. Controllable lead time, service level constraint, and transportation discounts in a continuous review inventory model. *Rairo-Oper. Res.* **2016**, *50*, 921–934. [CrossRef]
70. Sarkar, B.; Ahmed, W.; Kim, N. Joint effects of variable carbon emission cost and multi-delay-in-payments under single-setup-multiple-delivery policy in a global sustainable supply chain. *J. Clean. Prod.* **2018**, *185*, 421–445. [CrossRef]
71. Sarkar, B.; Sana, S.S.; Chaudhuri, K. An inventory model with finite replenishment rate, trade credit policy and price-discount offer. *J. Ind. Eng.* **2013**, 1–18. [CrossRef]
72. Ahmed, W.; Sarkar, B. Impact of carbon emissions in a sustainable supply chain management for a second generation biofuel. *J. Clean. Prod.* **2018**, *186*, 807–820. [CrossRef]
73. Cárdenas-barrón, L.E.; Sarkar, B.; Treviño-garza, G. Easy and improved algorithms to joint determination of the replenishment lot size and number of shipments for an EPQ model with rework. *Math. Comput. Appl.* **2013**, *18*, 132–138. [CrossRef]

7

A Generalized Process Targeting Model and an Application Involving a Production Process with Multiple Products

Mohammad A. M. Abdel-Aal * and Shokri Z. Selim

Systems Engineering Department, King Fahd University of Petroleum and Minerals, P.O. Box 5063, Dhahran 31261, Saudi Arabia; selim@kfupm.edu.sa

* Correspondence: mabdelaal@kfupm.edu.sa or m.abdelaal82@gmail.com.

Abstract: This paper presents a generalized targeting model that subsumes most known targeting problems. In this paper, a recurrent state is defined as a condition that requires reprocessing or rework. The generalized model can accommodate one or two specifications limits and can be used for the following quality characteristics: The nominal-the-better, the larger-the-better, and the smaller-the-better. This model can be used to find the optimal mean of a quality characteristic, as well as the optimal specification limits. In addition, the paper studies the conditions under which the solution to the proposed model can provide a global solution. The paper shows that, for some of the special cases and under very general conditions, the optimal lower limit should be zero and the optimal upper limit should be infinity. This paper proves that the expected profits improve for the case where only a lower limit on the quality characteristic is used, if a recurrent state is included by adding an optimized upper limit. A special case of the model is used to study the problem of determining a common mean for multiple products, as well as the optimal upper specification limits for each product. A solution procedure for maximizing the expected profits and obtaining the optimal solution is introduced. A numerical example is presented.

Keywords: quality control; targeting model; process mean; strict quasi-concavity; recurrent states; optimized limits

---

1. Introduction

Targeting models are used to find the optimal process parameters that will result in the least amount of losses, due to reworked or scrapped items, while satisfying the specification limit(s). Achieving minimum costs and high quality are prerequisites for attaining higher market shares and profit. Taguchi introduced three types of loss functions; nominal-the-better quality characteristic, N-type, smaller-the-better quality characteristic, S-type, and larger-the-better quality characteristic, L-type [1]. This paper has three main goals. The first goal is to introduce a general model for determining the optimal process parameters for targeting models, where the manufacturing process involves rework and/or scrapping of non-conforming items. The second goal is to show which of the model special cases can yield a global optimal solution. To this end, the paper investigates the strict quasi-concavity of the expected profit. The third goal is to show that adding a fictitious upper limit to a model that involves only a lower limit will result in a higher expected profits.

Over the last four decades, several papers have presented models, developed for determining the optimal process mean under various manufacturing settings. Table 1 summarizes some of the studies in the literature. The third column in Table 1 specifies which specification limits are considered in the paper, where L and U stand for the lower, and upper specification limits, respectively. The fourth column in the table indicates whether the mentioned paper discusses rework when non-conforming items are produced. The last column in Table 1 specifies the specification limit(s), whose violation will result in rework. In the following paragraphs, a discussion of some of the most important models in the literature is provided.

Table 1. Summary of some of the targeting models.

Reference	Spec. Limits		Rework Included	Rework Conditions	
	L	U		$x < L$	$x > U$
Springer (1951) [2]	✓	✓			
Bettes (1962) [3]	✓	✓	✓	✓	✓
Hunter and Kartha (1977) [4]	✓				
Arcelus and. Banerjee (1985) [5]	✓				
Golhar and Pollack (1988) [6]	✓	✓	✓	✓	✓
Pugh (1988) [7]	✓	✓			
Rahim and Banerjee (1988) [8]		✓			
Bai and Lee (1993) [9]	✓		✓	✓	
Dodson (1993) [10]	✓	✓			
Lee and Kim (1994) [11]	✓	✓	✓	✓	✓
Chen and Chung (1996) [12]	✓				
Pulak and Al-Sultan (1996) [13]	✓				
Al-Sultan and Al-Fawzan (1997) [14]	✓	✓			
Lee and Jang (1997) [15]	✓	✓			
Liu and Raghavachari (1997) [16]	✓	✓	✓	✓	✓
Roan et al. (1997) [17]	✓				
Wen and Mergen (1999) [18]	✓				
		✓			
Kim et al. (2000) [19]	✓		✓	✓	
Lee et al. (2000) [20]	✓	✓	✓	✓	
Rahim and Al-Sultan (2000) [21]	✓	✓			
Roan et al. (2000) [22]	✓				
Williams et al. (2000) [23]	✓				
Chen and Chou (2002) [24]	✓				
		✓			
Lee and Elsayed (2002) [25]	✓	✓	✓	✓	
Duffuaa and Siddiqui (2003) [26]	✓				
Bowling et al. (2004) [27]	✓	✓	✓	✓	✓
Chen (2004) [28]	✓				
		✓			
Rahim and Tuffaha (2004) [29]	✓	✓			
Chen and Chou (2005) [30]	✓				
		✓			
Chen (2006) [31]	✓	✓	✓		✓
Hong et al. (2006) [32]	✓		✓	✓	
Lee et al. (2007) [33]	✓				
Khasawneh et al. (2008) [34]	✓	✓	✓		✓
Chen and Kao (2009) [35]	✓	✓	✓		✓
Darwish (2009) [36]	✓		✓	✓	
Duffuaa et al. (2009) [37]	✓		✓	✓	
Chen and Khoo (2010) [38]	✓	✓	✓		✓
	✓	✓			

Table 1. *Cont.*

Reference	Spec. Limits L	Spec. Limits U	Rework Included	Rework Conditions $x < L$	Rework Conditions $x > U$
Park et al. (2011) [39]	✓				
Selim and Al-Zu'bi (2011) [40]	✓	✓	✓		✓
Darwish et al. (2013) [41]	✓				
Duffuaa and El-Ga'aly (2013a) [42]	✓	✓	✓	✓	
Duffuaa and El-Ga'aly (2013b) [43]	✓		✓	✓	
Peng and Khasawneh (2014) [44]	✓	✓	✓		✓
Chen et al. (2015a) [45]	✓				
Chen et al. (2015b) [46]	✓	✓	✓		✓
Chen and Kan (2015) [47]	✓	✓	✓		✓
Dodd et al. (2015) [48]	✓	✓	✓		✓
Duffuaa and El-Ga'aly (2015) [49]	✓		✓	✓	
Raza and Turiac (2016) [50]	✓	✓	✓	✓	
Raza et al. (2016) [51]	✓	✓	✓	✓	
This paper (General model)	✓				
	✓		✓	✓	
		✓			
		✓	✓		✓
	✓	✓			
	✓	✓	✓	✓	
	✓	✓	✓		✓
	✓	✓	✓	✓	✓

To the best of the authors' knowledge, Springer [2] was the first study that developed a targeting problem with two-sided specification limits. In that paper, a product is accepted if its quality characteristic falls within pre-defined specification limits; otherwise it is rejected. Bettes [3] presented a study for setting the optimum values for the process mean and upper specification limit. Golhar and Pollack [6] presented a model for determining the optimal process mean and upper specification limit for a process where the item is accepted if the considered quality characteristic falls between the specification limits, and is reworked if it does not. Rahim and Al-Sultan [21] studied the problem of jointly determining the optimum mean and variance of a process. Rahim and Tuffaha [29] considered a targeting problem, where the product is sold in a primary market if its quality characteristic falls within the specification limits; otherwise, it is sold in a secondary market.

In order to determine the probability distribution of the quality characteristic of the production process, a sample of data on the quality characteristic should be collected from the production line. The observations are to be tested for the best probability distribution fit. The adequacy of the fit should be assessed by goodness-of-fit tests, i.e., chi-square test. Interested readers are referred to [37]. Another technique is to apply Burr's density function to fit the set of collected data of the quality characteristic, given the first four moments of the data of the process characteristic; i.e., the mean, standard deviation, skewness coefficient, and kurtosis coefficient, can be reasonably accurately estimated. Then, Burr distribution can be easily transferred to any normal/non-normal distribution. Interested readers are referred to [52,53].

In some cases, it would be impractical or even impossible to measure the quality characteristic. In these cases, a variable that is correlated with the quality characteristic is used to estimate the quality of the product. This correlated variable is known as a surrogate variable, and it is known to be relatively easier to measure. Many studies have investigated models with surrogate variables [9,11,15,25,35,39].

Bai and Lee [9] presented the problem of setting the optimum target value of the process mean and the lower specification limit of a cement bag-filling process. Measuring the weight of a bag is difficult, due to the high-speed packing; therefore, the milliampere reading of a load cell is used as a surrogate.

Lee and Kim [11] extended the work of Bai and Lee [9], and considered a filling process with a pre-defined lower specification limit on the quality characteristic of interest. The material in a container is inspected based on a surrogate variable. Under-filled and over-filled containers are emptied and refilled. The optimal process mean, upper and lower specification limits are determined simultaneously by maximizing the profit function.

Lee and Jang [15] considered a production process with three-class screening. They constructed two profit models; one model is based on the performance variable, while the other is based on a surrogate variable.

The preceding studies addressed the case where a single product type is considered. However, situations arise where multiple product types are produced through a common production process. Lee et al. [33] developed a targeting model for a production process with multiple products. They determined the optimal common process mean; where a product is accepted if its quality characteristic is above a predefined lower limit, otherwise it is scrapped. Park et al. [39] extended the work of Lee et al. [33] and discussed a model that contains surrogate variables. The model enabled the estimation of the common process mean and the optimum lower limit for the performance variables.

In Section 2 of this paper, a generalized targeting model, that subsumes most of the known N-type targeting models, is developed. Special cases of the model cover the S-type and the L-type. This model can be used to find the optimal mean of a quality characteristic and the optimal specification limits. Section 3 provides a discussion on the quasi-concavity of the model and its special cases. This section identifies cases where an optimal lower limit is zero and an optimal upper limit is infinity. The section also identifies those cases where the upper limit is a finite value that can be easily obtained using line search methods. In Section 4, a case for introducing an upper limit on the quality characteristic for a process targeting model, that aims to improve expected profits, is presented. Section 5 provides an application for the proposed model by considering the model presented in [33], where several products have a common quality characteristic, while holding different cost parameters. The objective is to find the optimal common mean of the quality characteristic. Each product has its own specification limits. The solution procedure of the multiple products model is presented in Section 6. An illustrative numerical example and discussion of the results are given in Section 7. Finally, Section 8 concludes the paper.

The main contributions and novel merits of the manuscript can be summarized as:

1. Developing a generalized targeting model for process quality control. This generalized model subsumes most of the known N-type targeting models. Special cases of the model cover the S-type and the L-type.
2. Providing a solution procedure to find the optimal mean of a quality characteristic and optimal specification limits for the generalized targeting model.
3. Very useful managerial insights have been extracted from the numerical experiments.

The following notation is used throughout the paper:

X	the random variable of the quality characteristic under consideration
x	the realized value of X
L	the lower specification limit of X
U	the upper specification limit of X
$g_0(x)$	the production cost function
$g_a(x)$	the net profit function for an accepted product
$g_L(x)$	the cost incurred when $x < L$
$g_U(x)$	the cost incurred when $x > U$
TP	the total profit function
$E(TP)$	the expected total profit
$f(\cdot)$	the probability density function of X with mean μ and variance σ_i^2
$F(\cdot)$	the cumulative distribution function of X

In addition to the above notation, the following notation is used in the multiple product model:

X_i	the random value of the quality characteristic of the ith product type, $i = 1, 2, \ldots, n$
x_i	the realized value of X_i
μ^*	the optimal common process mean
L_i	the lower specification limit of X_i
U_i	the upper specification limit of X_i
A_i	the selling price of one unit of the ith product
B	the fixed production cost per unit
C	the variable production cost proportional to X_i
I_c	the inspection cost per unit
S	the scrapping cost per unit
R	the rework cost per unit
α_i	the proportion of the ith product type produced, $\sum_{i=1}^{n} \alpha_i = 1$

2. Generalized Targeting Model

In this section, a general targeting model that incorporates most of the known targeting models is developed. This model may be used for finding the optimal mean and/or variance of a quality characteristic. It may also be used to determine the optimal "specification" limits for the quality characteristic. It is assumed that there is a single quality characteristic, x, with a probability density function $f(x)$. Without loss of generality, it is assumed that $\infty > x \geq 0$.

The cost structure is described as follows: The production cost, $g_0(x)$ is a function of the realized value of the quality characteristic. This includes raw material cost, manufacturing cost, and quality control and/or inspection costs.

It is assumed that there is a two-sided specification limit on the quality characteristic, a lower and an upper specification limit, L, and U, respectively. If the quality characteristic is less than the threshold L, i.e., $x < L$, an extra cost $g_L(x)$ is incurred. This could be the cost of scrapping the item, or reworking it. It is assumed that this cost is dependent on the value of the quality characteristic. This cost includes all expenses associated with scrapping or reworking the product. The expected cost in this case is defined as $G_L(L) = \int_{x=0}^{L} g_L(x) f(x) dx$. If the quality characteristic is between L and U, then the item is accepted. In this case, revenue may be collected, but some additional costs may be incurred, such as finishing, shipping, or packaging. The notation, $g_a(x)$, is used to denote the difference between the revenue and those additional costs. The expected value of this function is given by $G_a(L, U) = \int_{x=L}^{U} g_a(x) f(x) dx$. If the quality characteristic exceeds U, i.e., $x > U$, a cost is incurred. This could be the cost of scrapping or reworking the item. As in the case where $x < L$, this cost may be dependent on x. This cost is denoted by $g_U(x)$. The expected cost in this case is given by $G_U(U) = \int_{x=U}^{\infty} g_U(x) f(x) dx$. The functions g_0, g_a, g_L, and g_U may also reflect any of Taguchi's loss functions and in this case the model may have one or two limits dependent on the type of the quality characteristic.

In addition, the following assumptions are considered:

(a) The quality characteristic assumes non-negative values,
(b) The cost structure and parameters remain the same after rework,
(c) The probability density function of the quality characteristic does not change after rework is performed, and
(d) There is no limit on the number of rework attempts.

Next, the concepts of terminal and recurrent states of an item are introduced. An item is in a terminal state if it does not require any further processing, i.e., the item is scrapped or accepted. An item is in a recurrent state if it requires rework.

The literature of process targeting models discuss the case where a product is in a terminal state once it is produced. These models appear in Table 1 with "No" in the rework column. For some other models, the item is in a terminal state only if the quality characteristic is acceptable, i.e., rework is performed if $x < L$, or $x > U$, models in [3,6,11,16,27]. Other models assume that the terminal state is $x \leq U$, i.e., rework is needed if $x > U$, models [31,34,35,38,40,44,46–48]. Finally, there are models where a recurrent state is limited to the case where $x < L$, models [9,19,20,25,32,36,37,42,43,49–51].

Before introducing the new model, let's turn to the "specification" limits on the quality characteristic. Usually, specification limits are set by outside agents, such as downstream manufacturing processes, regulatory agencies, or clients/consumers. However, one may envision situations where this may not be the case. Consider a filling operation that has a lower specification limit on the fill volume. The lower limit is set in the interest of the consumer. However, the producer will incur a loss for overfilling. In this case, the producer may wish to introduce an upper limit for the fill volume to have a cap on lost material. A similar situation applies to coating processes, where a lower limit on the thickness of the coating is specified by standards. However, the manufacturer may want to set an upper limit on the thickness to avoid the cost of additional coating material. On the other hand, thick coating may have a significant effect on the product quality. It may result in uneven texture finish and may initiate cracking and delamination of paints, due to external forces, which will definitely affect the product life and performance. In these scenarios, the decision maker has to set a "suitable" upper limit.

Consider a quality characteristic that has to satisfy both the lower and upper limits. If the quality characteristic is outside the limits further processing is performed, i.e., a recurrent state is reached. In this case, the total profit, TP is given by:

$$TP = \begin{cases} E(TP) - g_L(x) & L > x \\ -g_0(x) + g_a(x) & L \leq x \leq U. \\ E(TP) - g_U(x) & x > U \end{cases} \tag{1}$$

The expected value of the total profit is given by:

$$E(TP) = \frac{G_a(L,U) - G_0(L,U) - G_L(L) - G_U(U)}{F(U) - F(L)}, \tag{2}$$

where $G_0(L,U) = \int_{x=L}^{U} g_0(x)f(x)dx$ is the expected cost of producing a unit within the limits. One can interpret the expression in Equation (2) as the expected profit, given that the quality characteristic is within the limits. The denominator is the probability of not being in a recurrent state. Now, the following three special cases of the model shown in Equation (2) are considered:

I. Consider the case where the recurrent state is reached when $x > U$. In this case, the random variable, TP, will be the same as in Equation (1) above with the exception that if $x < L$, then $TP = -g_0(x) - g_L(x)$, thus:

$$TP = \begin{cases} -g_0(x) - g_L(x) & L > x \\ -g_0(x) + g_a(x) & L \leq x \leq U. \\ E(TP) - g_U(x) & x > U \end{cases}$$

The probability of not being in a recurrent state is $F(U)$. Hence Equation (2) becomes,

$$E(TP) = \frac{G_a(L,U) - G_{0U}(U) - G_L(L) - G_U(U)}{F(U)}, \tag{3}$$

where $G_{0U}(U) = \int_{x=0}^{U} g_0(x)f(x)dx$. If there is no lower limit; i.e., $L = 0$, then the model shown in Equation (1) becomes,

$$TP = \begin{cases} -g_0(x) + g_a(x) & x \leq U \\ E(TP) - g_U(x) & x > U \end{cases}.$$

Therefore, the model shown in Equation (2) reduces to,

$$E(TP) = \frac{G_a(U) - G_{OU}(U) - G_U(U)}{F(U)},$$

where $G_a(U) = \int_{x=0}^{U} g_a(x)f(x)dx$. This is typically the known model for S-type quality characteristic with recurrent state when $x > U$.

II. Similarly, if the recurrent state is reached when $x < L$, then the model shown in Equation (1) reduces to

$$TP = \begin{cases} E(TP) - g_L(x) & L > x \\ -g_0(x) + g_a(x) & L \leq x \leq U. \\ -g_0(x) - g_U(x) & x > U \end{cases}$$

Hence, the model shown in Equation (2) becomes,

$$E(TP) = \frac{G_a(L,U) - G_{OL}(L) - G_L(L) - G_U(U)}{1 - F(L)}, \quad (4)$$

where $G_{OL}(L) = \int_{x=L}^{\infty} g_0(x)f(x)dx$. For this special case, if there is no upper limit; i.e., $U = \infty$, then the model shown in Equation (1) becomes:

$$TP = \begin{cases} E(TP) - g_L(x) & L > x \\ -g_0(x) + g_a(x) & L \leq x \end{cases}.$$

Hence, the model shown in Equation (2) reduces to,

$$E(TP) = \frac{G_a(L) - G_{OL}(L) - G_L(L)}{1 - F(L)}.$$

This represents the known model for L-type quality characteristic with recurrent state when $L > x$.

III. Finally, if there is no recurrent state; i.e., the item is scrapped if not accepted, then the model shown in Equation (1) becomes,

$$TP = \begin{cases} -g_0(x) - g_L(x) & L > x \\ -g_0(x) + g_a(x) & L \leq x \leq U. \\ -g_0(x) - g_U(x) & x > U \end{cases}$$

Thus, the model shown in Equation (2) reduces to,

$$E(TP) = G_a(L,U) - G_{00} - G_L(L) - G_U(U), \quad (5)$$

where $G_{00} = \int_{x=0}^{\infty} g_0(x)f(x)dx$. Similar to special cases I and II, if there is only a single limit U, and the acceptance condition is $x \leq U$, then one has to substitute 0 for L in any integral involving L, and the model shown in Equation (5) becomes,

$$E(TP) = G_a(U) - G_{00} - G_U(U).$$

This represents the known model for S-type quality characteristic without recurrent state. Similarly, as in special case II, if the acceptance condition is $x \geq L$, then one has to substitute ∞ for U in any integral involving U, hence the model shown in Equation (5) becomes,

$$E(TP) \;= G_a(L) - G_{00} - G_L(L).$$

This represents the known model for L-type quality characteristic without recurrent state.

The conclusion of this section is as follows, the model shown in Equation (2) is a general model and S-type quality characteristic and L-type quality characteristic models are special cases of the model shown in Equation (2). This general model and its special cases subsume most of the known targeting models that appear in the literature. In addition, this generalized model can be used for setting the optimum mean, variance, upper or lower limits, as will be shown in next sections.

It is important to understand whether the solution obtained is a local or global optimum. In the case of the maximizing a function, a global solution is guaranteed if the function involved is concave. However, concavity of a function is a strict condition. A less restrictive one is that the function is strictly quasi-concave. In the next section, the quasi-concavity of the models given by Equations (2)–(5) will be discussed.

3. Quasi-Concavity of the Generalized Targeting Model

Strictly quasi-concave functions have the property that a local maximum point is also global. A function, $h(x)$, is quasi-concave if the set $S_\alpha \;=\; \{x|h(x) \geq \alpha\}$ is convex for any scalar α and $S_\alpha \neq \Phi$. Also, $h(x)$ is quasi-concave if, for any x_1 and x_2 in its domain and $0 \leq \lambda \leq 1$, $h(\lambda x_1 + (1-\lambda)x_2) \;\geq\; \min\{h(x_1), h(x_2)\}$ [54]. There is an extra method to characterize the strict quasi-concavity of a function of a single variable. A function of one variable is strictly quasi-concave if there is a point, p, in its domain where the function is increasing at points before p, and decreasing at points after p. This implies that the slope before p is positive, while it becomes negative at points after p. This property will be used in this paper.

The next subsections will attempt to prove the obtained results for general differentiable functions f, g_a, g_0, g_U, and g_L. In some cases no result can be obtained for general functions, for those cases, this paper adopts a cost structure that is widely used in most papers in the literature, where the cost of violating the limits, and that of accepting the quality characteristic, are constants, while that of production is a linear function, i.e., $g_L(x) \;=\; S$, $g_U(x) \;=\; R$, $g_a(x) \;=\; A$, and $g_0(x) \;=\; B + Cx$.

3.1. *Quasi-Concavity of Equation (2) with Respect to* μ, L, *and* U

In this subsection, the quasi-concavity of the function in Equation (2) with respect to the mean, and the limits, L and U is studied.

Assume that the limits L and U are known and the mean, μ, is to be determined. No result can be obtained about the quasi-convexity of Equation (2) as a function of μ. If the above cost structure is adopted, the expected profit function given by Equation (6) becomes:

$$E(TP) \;=\; A - B - \frac{C\int_{x=L}^{U} xf(x)dx + SF(L) + R[1-F(U)]}{F(U) - F(L)}. \tag{6}$$

Counter-examples can be constructed to show that the function given by Equation (6) is not quasi-concave if $f(x)$ is a normal density function.

Next, consider the case where μ and U are known and the lower limit, L, is to be determined. The partial derivative of the profit function in Equation (6) with respect to L is given by:

$$\frac{\partial E(TP)}{\partial L} \;=\; f(L)\frac{CL[F(U)-F(L)] - C\int_{x=L}^{U} xf(x)dx - SF(U) - R[1-F(U)]}{[F(U)-F(L)]^2}.$$

Note that the second term in the numerator exceeds the first term, hence, the partial derivative is negative and the function achieves its maximum value at $L = 0$. This result is valid for any differentiable density function, $f(x)$.

Next, consider the case where μ and L are known and the upper limit, U, is to be determined. Assume that $f(x)$ is differentiable. Using the same cost structure presented earlier, then the partial derivative of $E(TP)$ shown in Equation (6) with respect to U is given by:

$$\frac{\partial E(TP)}{\partial U} = f(U)\frac{-CU[F(U)-F(L)] + C\int_{x=L}^{U} xf(x)dx + SF(L) + R(1-F(L))}{[F(U)-F(L)]^2}.$$

The sum of the first two terms in the numerator is negative and is a strictly decreasing function of U. Eventually, this sum goes to $-\infty$ as U goes to ∞.

Since this part of the discussion is interested in $U \geq \mu$, then there are two cases:

a. If the numerator at $U = \mu$ is negative then it remains negative for $U > \mu$ and the maximum of $E(TP)$ is achieved at $U = \mu$.
b. If the numerator at $U = \mu$ is positive, then as U increasers, the numerator decreases and eventually becomes negative. Hence $E(TP)$ changes sign from positive to negative, and is strictly quasi-concave.

3.2. *Quasi-Concavity of Equation (3) with Respect to μ, L, and U*

Counter examples can be constructed to show that Equation (3) is not quasi-concave with respect to μ for the general functions and even for the special cost structure introduced at the beginning of this section.

The partial derivative of Equation (3) with respect to L is given by:

$$\frac{\partial E(TP)}{\partial L} = -f(L)\,\frac{g_a(L) + g_L(L)}{F(U)} < 0.$$

Hence, the maximum of Equation (3) is achieved if L is set to 0.

The partial derivative of Equation (3) with respect to U, adopting the above cost-structure, is given by:

$$\frac{\partial E(TP)}{\partial U} = f(U)\,\frac{-CUF(U) + C\int_{x=0}^{U} xf(x)dx + (A+S)F(L) + R}{F(U)^2}.$$

The proof in this case is similar to that in Section 3.1. If $-C\mu.F(\mu) + C\int_{x=0}^{\mu} xf(x)dx + (A+S)F(L) + R \leq 0$, then the maximum profit is attained when $U = \mu$. Otherwise, the function $E(TP)$ is strictly quasi-concave function in U.

3.3. *Quasi-Concavity of Equation (4) with Respect to μ, L and U*

Counter examples can be constructed to show that Equation (4) is not quasi-concave with respect to μ for the general functions and even for the special cost structure introduced at the beginning of this section. Using the same cost structure presented earlier, then Equation (4) simplifies to:

$$E(TP) = \frac{(A+R)F(U) - (A-B+S)F(L) - C\int_{x=L}^{\infty} xf(x)dx - R - B}{1-F(L)}. \quad (7)$$

The partial derivative of Equation (7) with respect to L is given by:

$$\frac{\partial E(TP)}{\partial L} = f(L)\frac{CL[1-F(L)] - C\int_{x=L}^{\infty} xf(x)dx - (A+R)(1-F(U)) - S}{[1-F(L)]^2}.$$

Note that the second term in the numerator exceeds the first term, hence the numerator is negative and the maximum of the function is achieved at $L = 0$.

Finally, the partial derivative of Equation (4) with respect to U is given by:

$$\frac{\partial E(TP)}{\partial U} = f(U)\frac{g_a(U) + g_U(U)}{1 - F(L)} > 0.$$

Hence, the maximum is achieved by setting U at its largest possible value.

3.4. *Quasi-Concavity of Equation (5) with Respect to μ, L, and U*

If the cost structure described above is used, then Equation (5) simplifies to:

$$E(TP) = (A + R)F(U) - (A + S)F(L) - C\mu - B - R. \tag{8}$$

If L and U are known and $f(x)$ is the density function of the normal distribution, then $F(U)$ is concave for $U > \mu$ and $F(L)$ is convex for $L < \mu$. Hence, Equation (8) is a concave function in μ.

For the general function shown in Equation (5), it is straightforward to show that:

$$\frac{\partial E(TP)}{\partial L} = -f(L)[g_a(L) + g_L(L)] < 0,$$

$$\frac{\partial E(TP)}{\partial U} = f(U)[g_a(U) + g_U(U)] > 0.$$

Hence, the maximum of the function is achieved if $L = 0$ and if U is set at the largest possible value.

Table 2 summarizes the above results. In order to show that two parameters are fixed and the third one is a decision variable, that third parameter is written as an argument of the function TP. For example, $TP(\mu)$ means that L and U are fixed and μ is a decision variable.

4. Optimizing the Limits of a Quality Characteristic

Table 2 shows that the optimal lower limit is zero or the least acceptable value. Only the cases modeled by Equations (2) and (3) provide a chance to optimize the upper limit. Usually, specification limits are determined based on technical or contractual conditions. Hence, optimizing the upper limit does not seem natural in this case. However, there is room for improving the expected profit by including an upper limit on the quality characteristic to reduce the cost of raw material usage, as shown in the next sections. Consider a case modeled by Equation (4) with $U = \infty$. Here, the product is accepted as long as the quality characteristic exceeds the lower specification limit. If an upper limit is introduced along with an associated recurrent state, then the model described by Equation (2) is on hand. A similar case is obtained when the case is modeled by Equation (5) with $U = \infty$. Then, introducing an upper limit and a recurrent state will result in a model given by Equation (3). In Section 5, this approach is implemented. This implementation shows that the expected profit has improved.

Table 2. Quasi-concavity of *TP* as a function of one parameter.

Model	*TP*(μ)	*TP*(*U*)	*TP*(*L*)
Equation (2)	Not strictly quasi-concave [2]	Strictly quasi-concave [2]	Optimal $L = 0$ [2]
Equation (3)		Optimal $U = \infty$ or the function is strictly quasi-concave [2]	Optimal $L = 0$ [1]
Equation (4)		Optimal $U = \infty$ [1]	Optimal $L = 0$ [2]
Equation (5)	Concave [2]		Optimal $L = 0$ [1]

[1] The is true for general functions. [2] This is true for the special cost structure.

5. Multiple Products Model

In this section, the generalized model that is presented in Section 2 is applied and an application, where a model given by Equation (5), is modified as described in the previous section, resulting in an improvement in the objective function. Lee et al. [33] considered a facility that manufactures multiple products (these products can be considered as different grades of the same product). The products are sold at different prices. Consider a quality characteristic that is common to all of these products, where the mean of the characteristic is the same among all the products. A product is scrapped if $x < L$, otherwise it is accepted. The current paper extends their work in different directions. To limit the losses of the manufacturer, an upper limit is added for the quality characteristic of each product. The upper limit assigned to each product is to be determined. If the upper limit is exceeded, a penalty is incurred. This penalty can be viewed as a rework cost. The resulting model involves the products' selling prices, production costs, inspection costs, scrap costs, and rework costs.

Consider a case of n products, and assume that the quality characteristic of product i is normally distributed and given by X_i with variance σ_i^2. $X_1, X_2, \ldots, X_n$ are assumed to be independent normal random variables with a common mean μ. Every item is inspected for conformance to the specification limits L_i and U_i. The inspection operation incurs a fixed inspection cost I_c, independent of product type. If $U_i \geq x_i \geq L_i$, the item is accepted and sold at a price A_i, i.e., $g_a(x) = A_i - I_c$. On the other hand, if $x_i < L_i$, the item fails to meet the lower specification limit; thus it is scrapped at a cost S, which is a fixed scrapping cost for all product types, i.e., $g_L(x) = S + I_c$. Moreover, if $x_i > U_i$, the item is reworked where a penalty or rework cost, R, is incurred, and hence, $g_U(x) = R + I_c$. The reworked item is inspected again and may be accepted, scrapped or reworked again. The per unit production cost is assumed to be a linear function of x_i, hence, $g_0(x) = B + Cx_i$, where B and C are fixed, and variable costs per unit, respectively. The scenario described above fits the model given by Equation (3), therefore, the expected profit is given by:

$$E(TP) = \sum_{i=1}^{n} \alpha_i \frac{G_a(U_i) - G_{OU}(U_i) - G_L(L_i) - G_U(U_i)}{F(U_i)}, \tag{9}$$

where:

α_i is the proportion of product i of the total production,

$$G_a(U_i) = \int_{x_i = L_i}^{U_i} (A_i - I_c) f(x_i) dx_i\,,$$

$$G_{OU}(U_i) = \int_{x_i = 0}^{U_i} (B + Cx_i) f(x_i) dx_i,$$

$$G_L(L_i) = \int_{x_i = 0}^{L_i} (S + I_c) f(x_i) dx_i,$$

$$G_U(U_i) = \int_{x_i = U_i}^{\infty} (R + I_c) f(x_i) dx_i.$$

By substitution for the above equations into Equation (9), the final form of the total expected profit is given by:

$$E(TP) = H(\mu, \mathrm{U}) = R - B - C\mu + \sum_{i=1}^{n} \alpha_i H_i(\mu, U_i), \tag{10}$$

where $\mathrm{U} = (U_1, U_2, \ldots, U_n)$ and

$$H_i(\mu, U_i) = \frac{C\sigma_i f(U_i) - (R + I_c) - (A_i + S)F(L_i)}{F(U_i)} + A_i. \tag{11}$$

The problem is to maximize $H(\mu, U)$ by determining the optimum common mean, μ, and the upper limits, U_i for $i = 1, 2, \ldots, n$. Note that if $U_i = \infty$, and $R = 0$, the model of [33] is retrieved.

6. Solution Procedure

The function $H(\mu, U)$ is neither concave nor quasi-concave in μ. Hence, it may have local optima. Figure 1 shows the function behavior versus changes in μ.

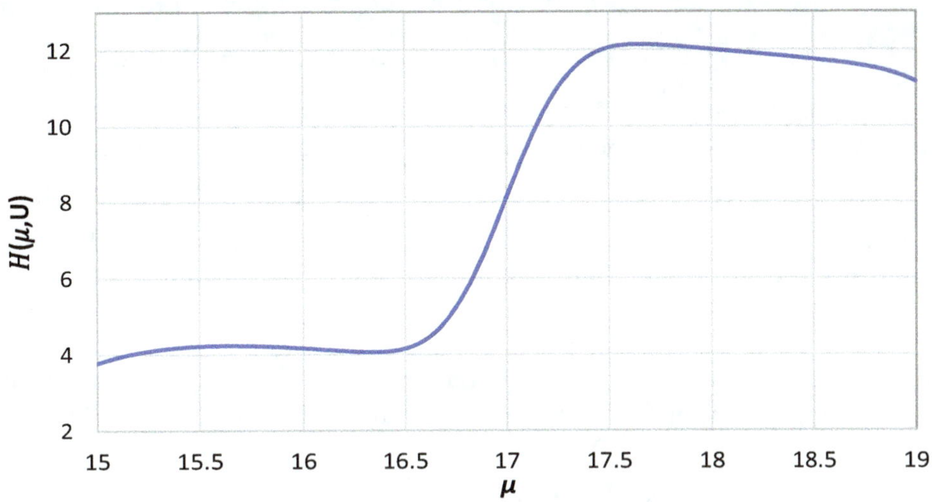

Figure 1. An illustrative example of $H(\mu, U)$ as a function of μ.

The proposed solution procedure includes an exhaustive search of μ. μ is searched in the range $[\mu_{\min}, \mu_{\max}]$. For each value of μ, the optimum values of $U_1, U_2, \ldots, U_n$ are obtained. Section 3 proved that the single variable function $H(U_i)$ is strictly quasi-concave in U_i. Therefore, the following solution procedure is applied:

1. Set $\mu_{\min} = \max_{i=1,2,\ldots,n}\{L_i\}$, $\mu_{\max} = \mu_{\min} + \lambda \max_{i=1,2,\ldots,n}\{\sigma_i\}$, where λ is a positive scalar. For normally distributed quality characteristics, it is recommended to take $\lambda \in [4, 7]$, this will guarantee covering more than 99.99% of the expected output of the production process. Also, set $H^* = 0$, and $\mu^* = \mu_1 = \mu_{\min}$.
2. Solve the problems $\max_{U_i > \mu_k} H_i(\mu_k, U_i)$ for $i = 1, \ldots, n$ using any line search method. Let U_{ik} be the solution for $i = 1, \ldots, n$.
3. Calculate $H(\mu_k, U_k)$ where $U_k = (U_{1k}, U_{2k}, \ldots, U_{nk})$. If $H(\mu_k, U_k) > H^*$, set $H^* = H(\mu_k, U_k)$, $U^* = U_k$ and $\mu^* = \mu_k$. Go to Step 4.
4. Update the value of the process mean; set $\mu_{k+1} = \mu_k + \delta$, where δ is a small positive scalar. If $\mu_{k+1} > \mu_{\max}$, stop; otherwise go to Step 2.

In the next section, the solution approach discussed above is implemented and a discussion of the effect of including an upper bound on the quality characteristic is presented.

7. Results and Discussion

This section presents a numerical example to find the optimum common mean for three products using the solution procedure presented in Section 6. The same data of the example presented in [33] is utilized. Subsequently, the impact on the total expected profit as a result of setting an upper limit to the quality characteristic of each product is investigated.

7.1. An Illustrative Example

This example considers a process where three types of electronic devices are processed through a common process of copper plating. The input data of the parameters are identical to the data in [33] and are shown below:

$$B = \$0.6,\ C = \$0.1,\ S = \$0.25,\ I_c = \$0.008,$$

$$A_1 = \$3.05,\ \alpha_1 = 0.4,\ \sigma_1^2 = 1.11\ \mu\text{m}^2,\ L_1 = 13\ \mu\text{m},$$

$$A_2 = \$3.25,\ \alpha_2 = 0.3,\ \sigma_2^2 = 1.22\ \mu\text{m}^2,\ L_2 = 14\ \mu\text{m},$$

$$A_3 = \$3.45,\ \alpha_3 = 0.3,\ \sigma_3^2 = 1.25\ \mu\text{m}^2,\ L_3 = 15\ \mu\text{m}.$$

In addition to the above parameters, a rework cost is introduced; $R = \$0.2$.
In the exhaustive search of μ, an increment $\delta = 0.001$, and $\lambda = 3$ are used.

The optimal solution is given by: $\mu^* = 17.1$, $U_1^* = 19.4052$, $U_2^* = 19.8687$, $U_3^* = 21.8496$, and $E(TP) = 0.8549$. The optimum common process mean μ^* is the same as that of [33]. However, the expected total profit is greater than that of [33] with a 9.25% enhancement. This is due to the upper limits being set.

The following section discusses the impact of the rework cost value, R, on the profit improvement associated with including an upper limit on the value of the quality characteristic.

7.2. The Impact of Setting Upper Specification Limits

The following section discusses the impact of the rework cost value, R, on the profit improvement associated with including an upper limit on the value of the quality characteristic.

Specification limits are considered in targeting models because quality control issues require them. However, this paper proposes including them as a *stop loss* measure. In the model presented in [33], a product is accepted once the quality characteristic exceeds the lower limit, regardless of how large it may be. This means that material loss could be high. It is proposed to introduce an upper limit on the quality characteristic. The introduction of the upper limit is associated with a penalty or rework cost. This subsection investigates the impact of the value of the rework cost on the improvement in the expected profit, given by Equation (10) and that of [33]. The improvement is computed using Equation (12) below, where $E(TP)^*$ is the maximum expected profit as obtained by the current model in Equation (10) and $E(TP)^{\text{Lee}}$ is the maximum profit obtained by Lee et al. [33]:

$$\%\ \text{Improvement in}\ E(TP) = \frac{E(TP)^* - E(TP)^{\text{Lee}}}{E(TP)^{\text{Lee}}} * 100(\%). \tag{12}$$

Figure 2 shows that as the rework cost, R, increases, the percentage of improvement decreases. The probability of rework becomes almost nil for $R \geq 0.4$.

It is clear from Figure 2 that setting upper specification limits has a major impact on $E(TP)$, and there is a significant improvement in its value. It is also worth noting that, regardless of the value of the rework cost, R, there will be an improvement in $E(TP)$. As R increases, the percentage of improvement approaches 9.12%. If the model does not include rework, then it will consider the upper limits at infinity; i.e., the model described in [33], and this will result in losses from the excessive use of raw material. Therefore, rework in the analysis of the model is needed if one is interested in reducing the loss from the excess raw material.

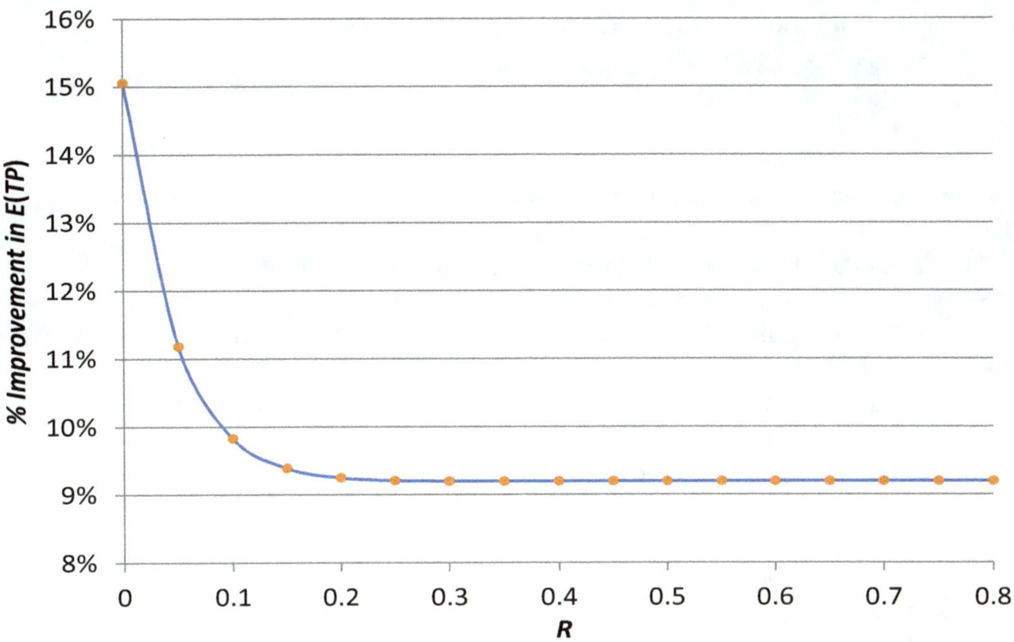

Figure 2. The impact on $E(TP)$ by setting upper limits with different values of R.

8. Conclusions

This paper proposed a generalized targeting model that subsumes most of the known targeting models of that category. Firstly, a generalized model for the N-type targeting models is introduced. The concept of a terminal and recurrent state of an item is introduced. An item reaches a terminal state if no further processing is required; otherwise, it is in a recurrent state. A recurrent state implies further work will be performed on the item. This differentiation of states gave rise to several special cases of the general model. These special cases can be used to model L-type and S-type quality characteristics. An important issue, discussed in this paper, is the conditions under which the obtained solution can be a global solution. To accomplish this task, the strict quasi-concavity of the developed models is examined. Strict quasi-concavity implies that a local maximum is also global. It is shown that the profit function is strictly quasi-concave if one or both specification limits are to be found, assuming that the mean and the other limit are fixed. If one is interested in finding the optimal mean of the quality characteristic, while the specification limits are fixed, one will face the disappointing fact that the profit function is not quasi-concave for all cases, except for the model where there are no recurrent states. This paper used a special case of the proposed model to obtain the optimal common mean of a quality characteristic and the upper specification limits for multiple products. The quality characteristics of the products were assumed to follow independent normal distributions with known variances. A procedure for determining the optimal solution is introduced and implemented on an example having three products. The results show that, incorporating upper specification limits on the quality characteristic has a significant improvement in the value of the optimal expected profit.

A practical case that needs further research is when there is a limit on the number of times an item is allowed to be in a recurrent state. This could be due to excessive loss of material or deterioration of some physical properties. In this paper, it is assumed that the cost of rework is independent of the number of rework cycles the item goes through. This assumption may not be applicable in all cases. Finally, the assumption that the quality characteristic keeps the same probability distribution regardless of number of rework cycles may be relaxed and further research will be needed in this case.

Author Contributions: Modeling, M.A.M.A. and S.Z.S.; formal analysis, M.A.M.A. and S.Z.S.; methodology, M.A.M.A. and S.Z.S.; supervision, S.Z.S.; writing—original draft, M.A.M.A.

Acknowledgments: The authors wish to acknowledge the support of King Fahd University of Petroleum and Minerals.

References

1. Taguchi, G. *Introduction to Quality Engineering: Designing Quality into Products and Processes*; Quality Resources: Tokyo, Japan, 1986.
2. Springer, H.C. A method for determining the most economic position of a process mean. *Ind. Qual. Control* **1951**, *8*, 36–39.
3. Bettes, D.C. Finding an Optimum Target Value in Relation to a Fixed Lower Limit and an Arbitrary Upper Limit. *J. R. Stat. Soc. Ser. C Appl. Stat.* **1962**, *11*, 202–210. [CrossRef]
4. Hunter, W.G.; Kartha, C.P. Determining the Most Profitable Target Value for a Production Process. *J. Qual. Technol.* **1977**, *9*, 176–181. [CrossRef]
5. Arcelus, F.J.; Banerjee, P.K. Selection of the Most Economical Production Plan in a Tool-Wear Process. *Technometrics* **1985**, *27*, 433–437. [CrossRef]
6. Golhar, D.Y.; Pollock, S.M. Determination of the Optimal Process Mean and the Upper Limit for a Canning Problem. *J. Qual. Technol.* **1988**, *20*, 188–192. [CrossRef]
7. Pugh, A.G. An algorithm for economically setting a uniformly-shifting process. *Comput. Ind. Eng.* **1988**, *14*, 237–240. [CrossRef]
8. Rahim, M.A.; Banerjee, P.K. Optimal production run for a process with random linear drift. *Omega* **1988**, *16*, 347–351. [CrossRef]
9. Bai, D.S.; Lee, M.K. Optimal target values for a filling process when inspection is based on a correlated variable. *Int. J. Prod. Econ.* **1993**, *32*, 327–334. [CrossRef]
10. Dodson, B.L. Determining the Optimal Target Value for a Process with Upper and Lower Specifications. *Qual. Eng.* **1993**, *5*, 393–402. [CrossRef]
11. Lee, M.K.; Kim, G.S. Determination of the optimal target values for a filling process when inspection is based on a correlated variable. *Int. J. Prod. Econ.* **1994**, *37*, 205–213. [CrossRef]
12. Chen, S.L.; Chung, K.J. Determination of the optimal production run and the most profitable process mean for a production process. *Int. J. Prod. Res.* **1996**, *34*, 2051–2058. [CrossRef]
13. Pulak, M.F.S.; Al-Sultan, K.S. The optimum targeting for a single filling operation with rectifying inspection. *Omega* **1996**, *24*, 727–733. [CrossRef]
14. Al-Sultan, K.S.; Al-Fawzan, M.A. An extension of Rahim and Banerjee's model for a process with upper and lower specification limits. *Int. J. Prod. Econ.* **1997**, *53*, 265–280. [CrossRef]
15. Lee, K.; Jang, J.S. The optimum target values for a production process with three-class screening. *Int. J. Prod. Econ.* **1997**, *49*, 91–99. [CrossRef]
16. Liu, W.; Raghavachari, M. The target mean problem for an arbitrary quality characteristic distribution. *Int. J. Prod. Res.* **1997**, *35*, 1713–1728. [CrossRef]
17. Roan, J.; Gong, L.; Tang, K. Process mean determination under constant raw material supply. *Eur. J. Oper. Res.* **1997**, *99*, 353–365. [CrossRef]
18. Wen, D.; Mergen, A.E. Running a Process with Poor Capability. *Qual. Eng.* **1999**, *11*, 505–509. [CrossRef]
19. Kim, Y.J.; Cho, B.R.; Phillips, M.D. Determination of the Optimal Process Mean with the Consideration of Variance Reduction and Process Capability. *Qual. Eng.* **2000**, *13*, 251–260. [CrossRef]
20. Lee, M.K.; Hong, S.H.; Kwon, H.M.; Kim, S.B. Optimum process mean and screening limits for a production process with three-class screening. *Int. J. Reliab. Qual. Saf. Eng.* **2000**, *7*, 179–190. [CrossRef]
21. Rahim, M.A.; Al-Sultan, K.S. Joint determination of the optimum target mean and variance of a process. *J. Qual. Maint. Eng.* **2000**, *6*, 192–199. [CrossRef]
22. Roan, J.; Gong, L.; Tang, K. Joint determination of process mean, production run size and material order quantity for a container-filling process. *Int. J. Prod. Econ.* **2000**, *63*, 303–317. [CrossRef]
23. Williams, W.W.; Tang, K.; Gong, L. Process improvement for a container-filling process with random shifts. *Int. J. Prod. Econ.* **2000**, *66*, 23–31. [CrossRef]
24. Chen, C.H.; Chou, C.Y. Determining the Optimum Process Mean of a One-Sided Specification Limit. *Int. J. Adv. Manuf. Technol.* **2002**, *20*, 439–441. [CrossRef]

25. Lee, M.K.; Elsayed, E.A. Process mean and screening limits for filling processes under two-stage screening procedure. *Eur. J. Oper. Res.* **2002**, *138*, 118–126. [CrossRef]
26. Duffuaa, S.O.; Siddiqui, A.W. Process targeting with multi-class screening and measurement error. *Int. J. Prod. Res.* **2003**, *41*, 1373–1391. [CrossRef]
27. Bowling, S.R.; Khasawneh, M.T.; Kaewkuekool, S.; Cho, B.R. A Markovian approach to determining optimum process target levels for a multi-stage serial production system. *Eur. J. Oper. Res.* **2004**, *159*, 636–650. [CrossRef]
28. Chen, C.H. Determining the Optimum Process Mean of a One-sided Specification Limit with the Linear Quality Loss Function of Product. *J. Appl. Stat.* **2004**, *31*, 693–703. [CrossRef]
29. Rahim, M.A.; Tuffaha, F. Integrated model for determining the optimal initial settings of the process mean and the optimal production run assuming quadratic loss functions. *Int. J. Prod. Res.* **2004**, *42*, 3281–3300. [CrossRef]
30. Chen, C.H.; Chou, C.Y. Determining the Optimum Process Mean under a Log-normal Distribution. *Qual. Quant.* **2005**, *39*, 119–124. [CrossRef]
31. Chen, C.H. The Optimum Selection of Imperfect Quality Economic Manufacturing Quantity and Process Mean by Considering Quadratic Quality Loss Function. *J. Chin. Inst. Ind. Eng.* **2006**, *23*, 12–19. [CrossRef]
32. Hong, S.H.; Kwon, H.-M.; Lee, M.K.; Cho, B.R. Joint optimization in process target and tolerance limit for L-type quality characteristics. *Int. J. Prod. Res.* **2006**, *44*, 3051–3060. [CrossRef]
33. Lee, M.K.; Kwon, H.M.; Hong, S.H.; Kim, Y.J. Determination of the optimum target value for a production process with multiple products. *Int. J. Prod. Econ.* **2007**, *107*, 173–178. [CrossRef]
34. Khasawneh, M.T.; Bowling, S.R.; Cho, B.R. A Markovian approach to determining process means with dual quality characteristics. *J. Syst. Sci. Syst. Eng.* **2008**, *17*, 66–85. [CrossRef]
35. Chen, C.H.; Kao, H.S. The determination of optimum process mean and screening limits based on quality loss function. *Expert Syst. Appl.* **2009**, *36*, 7332–7335. [CrossRef]
36. Darwish, M.A. Economic selection of process mean for single-vendor single-buyer supply chain. *Eur. J. Oper. Res.* **2009**, *199*, 162–169. [CrossRef]
37. Duffuaa, S.O.; Al-Turki, U.M.; Kolus, A.A. A process targeting model for a product with two dependent quality characteristics using 100% inspection. *Int. J. Prod. Res.* **2009**, *47*, 1039–1053. [CrossRef]
38. Chen, C.-H.; Khoo, M.B.C. Optimum Process Mean Setting for Product with Rework Process. *Tamkang J. Sci. Eng.* **2010**, *13*, 375–384.
39. Park, T.; Kwon, H.M.; Hong, S.-H.; Lee, M.K. The optimum common process mean and screening limits for a production process with multiple products. *Comput. Ind. Eng.* **2011**, *60*, 158–163. [CrossRef]
40. Selim, S.Z.; Al-Zu'bi, W.K. Optimal means for continuous processes in series. *Eur. J. Oper. Res.* **2011**, *210*, 618–623. [CrossRef]
41. Darwish, M.A.; Abdulmalek, F.; Alkhedher, M. Optimal selection of process mean for a stochastic inventory model. *Eur. J. Oper. Res.* **2013**, *226*, 481–490. [CrossRef]
42. Duffuaa, S.O.; El-Ga'aly, A. A multi-objective mathematical optimization model for process targeting using 100% inspection policy. *Appl. Math. Model.* **2013**, *37*, 1545–1552. [CrossRef]
43. Duffuaa, S.O.; El-Ga'aly, A. A multi-objective optimization model for process targeting using sampling plans. *Comput. Ind. Eng.* **2013**, *64*, 309–317. [CrossRef]
44. Peng, C.-Y.; Khasawneh, M.T. A Markovian approach to determining optimum process means with inspection sampling plan in serial production systems. *Int. J. Adv. Manuf. Technol.* **2014**, *72*, 1299–1323. [CrossRef]
45. Chen, C.H.; Chou, C.Y.; Kan, C.C. Modified economic production and raw material model with quality loss for conforming product. *J. Ind. Prod. Eng.* **2015**, *32*, 196–203. [CrossRef]
46. Chen, C.H.; Khoo, M.B.C.; Chou, C.Y.; Kan, C.C. Joint determination of process quality level and production run time for imperfect production process. *J. Ind. Prod. Eng.* **2015**, *32*, 219–224. [CrossRef]
47. Chen, C.H.; Kan, C.C. Tolerance design based on specified process capability value. *Int. J. Manag. Sci. Eng. Manag.* **2015**, *10*, 210–214. [CrossRef]
48. Dodd, C.; Scanlan, J.; Marsh, R.; Wiseall, S. Improving profitability of optimal mean setting with multiple feature means for dual quality characteristics. *Int. J. Adv. Manuf. Technol.* **2015**, *81*, 1767–1780. [CrossRef]
49. Duffuaa, S.O.; El-Ga'aly, A. Impact of inspection errors on the formulation of a multi-objective optimization process targeting model under inspection sampling plan. *Comput. Ind. Eng.* **2015**, *80*, 254–260. [CrossRef]

50. Raza, S.A.; Turiac, M. Joint optimal determination of process mean, production quantity, pricing, and market segmentation with demand leakage. *Eur. J. Oper. Res.* **2016**, *249*, 312–326. [CrossRef]
51. Raza, S.A.; Abdullakutty, F.C.; Rathinam, S. Joint determination of process mean, price differentiation, and production decisions with demand leakage: A multi-objective approach. *Appl. Math. Model.* **2016**, *40*, 8446–8463. [CrossRef]
52. Lin, Y.C.; Chou, C.Y. On the design of variable sample size and sampling intervals $\overline{X}$ charts under non-normality. *Int. J. Prod. Econ.* **2005**, *96*, 249–261. [CrossRef]
53. Chen, C.H.; Chou, C.Y. Optimum process mean, standard deviation and specification limits settings under the Burr distribution. *Eng. Comput.* **2017**. [CrossRef]
54. Bazaraa, M.S.; Sherali, H.D.; Shetty, C.M. *Nonlinear Programming: Theory and Algorithms*, 3th ed.; Wiley-Interscience: Hoboken, NJ, USA, 2006.

8

Joint Inventory and Pricing Policy for an Online to Offline Closed-Loop Supply Chain Model with Random Defective Rate and Returnable Transport Items

Biswajit Sarkar [1], Mehran Ullah [1] and Seok-Beom Choi [2,*]

1 Department of Industrial & Management Engineering, Hanyang University, Ansan Gyeonggi-do 15588, Korea; bsarkar@hanyang.ac.kr (B.S.); mehrandirvi@gmail.com (M.U.)
2 Department of International Business, Cheju Halla University, Jeju-do 63092, Korea
* Correspondence: sbchoi@cau.ac.kr.

Abstract: Environmental deterioration is one of the current hot topics of the business world. To cope with the negative environmental impacts of corporate activities, researchers introduced the concept of closed-loop supply chain (CLSC) management and remanufacturing. This paper studies joint inventory and pricing decisions in a multi-echelon CLSC model that considers online to offline (O2O) business strategy. An imperfect production process is examined with a random defective rate that follows a probability distribution. The results show that the O2O channel increases the profit of the system. For the defective rate, three different distributions are considered and three examples are solved. The results of the three examples conclude that the highest profit is generated when the defective rate follows a uniform distribution. Furthermore, based on the salvage value of defective items, two cases were studied. Results and sensitivity analysis show that the increase in defective rate does not reduce total profit in every situation, as perceived by the existing literature. Sensitivity analysis and numerical examples are given to show robustness of the model and draw important managerial insights.

Keywords: CLSC management; O2O channel; random defective rate; hybrid manufacturing-remanufacturing strategy

---

1. Introduction

The widespread usage of the internet and competitive business environment have resulted in the development of a dual channel sales strategy called the O2O strategy [1]. Companies are integrating online and offline sales channels to compete in the market by increasing service level, and to attract more demand. The literature termed integrating offline and online channels as a multi-channel context [2]. Traditional enterprises can deliver diverse types of products and services by incorporating online sales channels for their businesses. The number of companies taking advantage of this opportunity is increasing. This enables potential customers to browse the catalogs, price information, availability of the products, and even order the products before visiting the physical stores. Therefore, online sales channels can improve the sales of physical (offline) stores. A case study conducted by Chang et al. [3] provided factual proof that integrating offline and online sales channels increases the sales order. However, O2O sales channel requires the customer acquisition cost to convince the potential consumer or customer to purchase a specific product; literature considers this cost a primary business metric for O2O channeling. It determines the worth of the end customer to the business and, through it, the return on investment from the customers can be obtained. This enables businesses to analyze their investment decisions on a single customer to improve profitability [4]. Organizations make investments for online marketing and offline marketing of their products. Then, to compute the customer acquisition cost, each customer, who buys the product or service, is asked about the channel of information through which he came to know about that product. In this way, organizations collect a handful of data for a specific period and calculate customer acquisition cost for both types of channels.

Empirical evidence shows that the O2O channel plays a vital role for value creation in the business strategy [4]. However, O2O brings many management complexities and costs into supply chain management. These costs include online and offline marketing investments, subsidies, commissions, administration costs, and bonuses. Successful investment in O2O channeling is directly associated with future corporate profits. It increases customer retention and expands market shares by capturing new customers [5]. To incorporate an e-commerce strategy, this study considers a supply chain management model that captures extra market demand through the O2O channel and considers the required investment and maintenance cost for the O2O channel. In this context, this paper studies joint inventory and pricing decisions in a multi-echelon closed-loop supply chain management with remanufacturing and returnable transport items (RTI). Even though many papers had explored CLSC management, they either did not focus on O2O or they did not consider all other important features such as the consideration of transportation packaging management. In fact, from our literature review, we find that none of the existing studies examine the packaging management system and investment decisions in a CLSC system with O2O channel. Notice that the most closely related paper is Sarkar et al. [6], however, they did not consider dual-channel structure. Furthermore, they also neglect defective production and assume that all the produced products are perfect items, which is an unrealistic assumption. Moreover, they only studied inventory policies and considered constant demand, while this paper studies joint inventory and pricing policies with price and investment dependent demand.

2. Literature Review

This section gives an overview of the existing literature relevant to this paper. This paper contributes to research in the field of CLSC management, O2O channel in CLSC management, and RTI management and design, therefore, we focus on the related literature in these three areas.

Because of the increasing appreciation of environmental sustainability, CLSC management becomes a very important topic. More specifically, the remanufacturing process has grown into a big business. According to Bulmus et al. [7], $43 billion valued products were sold from remanufacturing used products in the U.S. in 2011. Many giant well-established enterprises, including Xerox, Kodak, and HP, have extensively incorporated their remanufacturing processes into their regular production lines and operations [8]. Research in this field dates back to the 1960s, with the first work reported from Schrady [9]. This paper studied an inventory model with single manufacturing batch followed by many repair batches, this policy is called (R, I) policy. Later on, Richter [10] and Richter [11] extended Schrady's model and introduced a hybrid manufacturing-remanufacturing system. According to Richter assumptions, market demand is fulfilled by manufacturing new products from raw material, and by repairing used products. Continuing his work, Richter [12] found that a complete recovery or complete disposal is the optimal choice. Many others extended this work with similar results including Richter and Dobos [13], Dobos and Richter [14], and Dobos and Richter [15]. Maiti and Giri [16] considered the used product quality in CLSC management and examined analytical models in five different scenarios including the centralized, vertical Nash, and three decentralized cases. Recently, Moshtagh and Taleizadeh [17] considered return rate with quality and used three distinct distributions to model the return rate. Tian and Zhang [18] considered disassembly scheduling and pricing of returned goods with price based yield. For a more comprehensive review of CLSC management, it is suggested that readers study Govindan and Soleimani [19] and Diallo et al. [20].

Today, with the progress of the Internet, online shopping is gaining popularity; the consumer buys products both through online and offline channels. The online purchasing channel has various advantages such as energy and time savings. Consumers can also compare the product prices among different websites and choose the best product. However, some customers like to observe, feel and examine finished products before buying, which could not be accomplished without the traditional offline channel. Therefore, organizations operate both the channels to get the associated advantages. Many business giants like Lenovo, Sony, IBM, Dell, Nike, and HP use this dual-channel policy to improve their sales [21–23]. In the literature, many articles have examined the impact of introducing the online channel on performances of companies and supply chain management. For example, Tsay and Agrawal [24] examined the channel coordination policies to reduce channel disputes between supply chain members. Yao and Liu [25] used Stackelberg and Bertrand games to investigate channel conflict and competition among suppliers and retailers. Huang and Swaminathan [26] andYan [27] also used game theory to investigate optimal pricing decisions in O2O supply chain management. Dan et al. [28] studied dual-channel supply chain management with retailer manipulation of the retail services; they found that the optimal pricing strategies can be affected by retail service. Huang et al. [22] considered pricing policies for an O2O supply chain management including demand disruptions. Li et al. [29] reviewed the difference between single-channel and dual-channel supply chain management. The O2O business model is characterized by high transportation demand because of last mile supply systems; furthermore, the returns from CLSC model also increase the transportation demand of the system; therefore, energy efficient transportation becomes a primary part of O2O supply chain. Bányai et al. [30] and Bányai [31] studied energy efficiency in first mile and last mile logistics systems. Despite all these excellent research works, papers that deal with product remanufacturing, reuse, and transportation packaging with the imperfect production system in the O2O supply chains are sparse in the literature. This study bridges this gap by studying a dual-channel CLSC model with remanufacturing, and returnable transport items considering random imperfection in the production system.

Solid waste generation is one of the main environmental threats posed by supply chain management [32,33]. RTIs are used by organizations to reduce solid waste of supply chain management [34,35]. In addition, disposable packaging also increases the cost of the product. Livingstone and Sparks [36] stated that packaging is responsible for 10–40% of product prices. Moreover, RTI also increases the protection and security of the finished products and improves working conditions and manual handling [37]. Historically, these RTIs were studied independently from inventory management, for a detailed review, readers may study [38]. Nevertheless, contemporary research recognized packaging as a vital component of supply chain inventory management. In this regard, [39] studied RTI and inventory policies jointly. They investigated a two-level CLSC inventory management and used RTIs to transport finished goods among supply chain members. Glock and Kim [40] studied RTI and inventory management in a single-vendor multi-retailer CLSC structure. They also considered RTI design and considered RTI size and the required number of RTIs as decision variables. Sarkar et al. [6] extended Glock and Kim [40] with RTI design and management policy in hybrid manufacturing-remanufacturing CLSC management. Others, for example, Kelle and Silver [41] proposed several methods to forecast the expected demand and returns of RTI; and Goh and Varaprasad [42] developed a system for evaluating the container return distribution by considering RTIs shrinkage.

Many authors, for example,Lee [43], Gupta and Chakraborty [44], and Tayi and Ballou [45] studied imperfect items in inventory models. Among these, the last two did not consider shortages due to reworking. Lee et al. [46] and Glock and Jaber [47] presented imperfect production models without considering rework of defective products. Salameh and Jaber [48] studied an inventory model for imperfect quality items where these items are withdrawn from stock, resulting in lower holding cost per unit time. Jaber et al. [49] consider policies for handling imperfect quality products. Ouyang et al. [50] and Cárdenas-Barrón [51] developed inventory models and considered back ordering, while Eroglu and Ozdemir [52] considered shortages. Ben-Daya [53] likewise extended the inventory model with planned maintenance schedules. Sarkar et al. [54] considered quality of products in multi-echelon supply chain model. Sarkar [55] provided different approaches for defective product management. An improved way to calculate imperfect items is given by Kim et al. [56], and Sett et al. [57] determined optimal buffer inventory in imperfect production system. Sarkar et al. [58] studied warranty, optimal run time with inspection errors in the imperfect production system, and Kang et al. [59] discussed optimal ordering policies for the imperfect production system. Recently, Khanna et al. [60] considered two-level credit policies with imperfect quality, Tayyab et al. [61] considered process uncertainty, and Tayyab et al. [62] studied the multistage imperfect production system. The majority of the existing literature considered rework or scrap option for defective items, however, this paper considers salvage option for defective items.

3. Problem Definition

This paper considers a three-layer dual-channel hybrid CLSC model with RTI design and management. The manufacturer fulfills market demand through multiple retailers. To increase market share, each retailer opens an online selling channel, for which the retailer pays information cost and transportation cost. Furthermore, the demand rate jointly depends on finished product price and investment in the online channel. The manufacturer produced two types of products, the manufactured products from raw material, and the remanufactured products from used products collected from consumers. An imperfect production system is considered where the perfect items are sold into the primary market, and the defective items are marketed into a secondary market with a lower price. Finished products are transported in reusable secondary packages called RTIs that are owned by a third party logistics (3PL) provider. The 3PL also collects used products from the end customer and delivers them to the manufacturer for remanufacturing. The whole scenario is depicted in Figure 1.

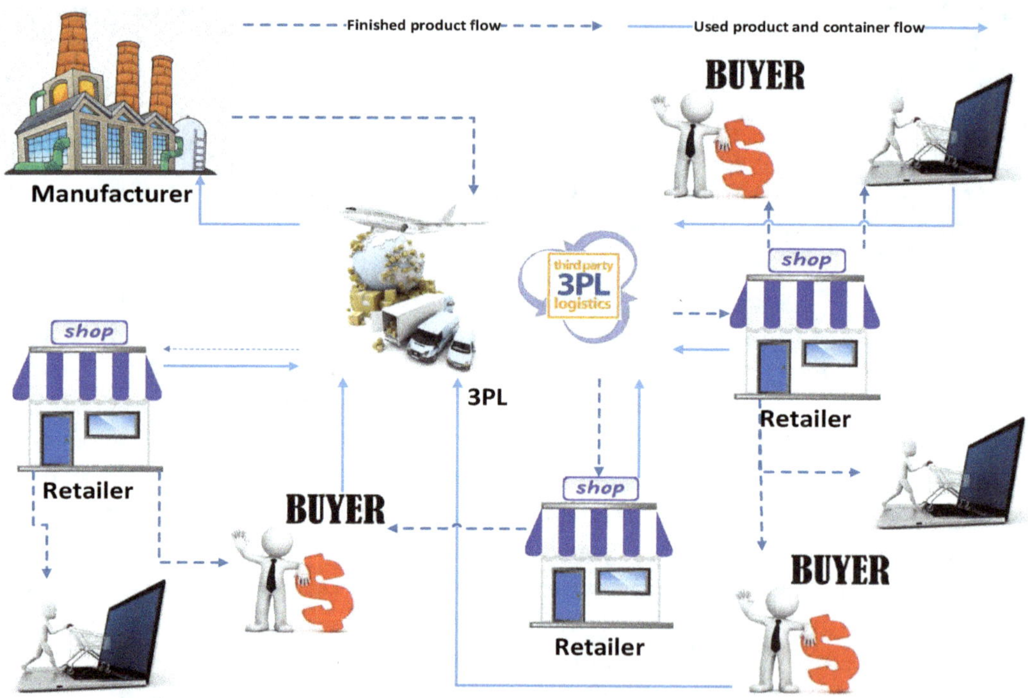

Figure 1. Logistics diagram of the proposed closed-loop supply chain (CLSC).

It is assumed that the total initial deterministic demand (offline market share) for the ith retailer is a_i, which follows demand price sensitivity, b, in a linear relationship. Other possible relationships, in which the demand function can be modeled, are quadratic and exponential. However, linear demand curves are widely used in the literature because of their simplicity and more importantly their sufficiency to catch important managerial implications. Thus, the total price-dependent deterministic demand at the ith retailer is $a_i - bp$. Now we consider the second part of demand, the online channel demand; which for the ith retailer can be given as $k\text{IN}_i$. Where k is the parameter and IN_i is the potential demand. Thus the net realized demand at retailer i is $d_i = a_i - bp + k\mathrm{IN}_i$. It is obvious that the maximum price in the market would be $p = a/b$.

4. Mathematical Model

This paper considered a centralized supply chain system. To model the proposed centralized system, this section first calculates total revenue of the supply chain followed by the cost of individual player and, finally, the net profit is calculated by subtracting the costs of all players from total revenue.

4.1. Total Revenue of the Supply Chain

The proposed system generates revenue by selling the finished products of perfect quality in the dual-channel primary market and imperfect items in the secondary market. Total revenue can be expressed mathematically as,

$$REV = \sum_{i=1}^{n} pd_i + cwr \sum_{i=1}^{n} d_i \tag{1}$$

The first term in (1) shows revenue generated by n retailers, while the second term shows revenue generated from selling imperfect items.

4.2. Retailer's Model

Total cost of the ith retailer consists ordering cost $\frac{A_i}{T}$, holding cost $\frac{d_i h_i}{2}$, which can be calculated from Figure 2. Investment or information cost of O2O channel $\frac{IN_i}{T}$, and transportation cost for the proposed O2O channel $\frac{TR_i}{T}$ are also part of retailer cost function. For n retailers the total cost can be stated as,

$$TC_1 = \sum_{i=1}^{n} \frac{A_i + \mathrm{IN}_i + \mathrm{TR}_i}{T} + \sum_{i=1}^{n} \frac{1}{2} T d_i h_i. \tag{2}$$

4.3. Manufacturer's Model

The manufacturer produces finished products based on the market demand in a single setup per cycle, for which the setup cost per unit time is $\frac{S_m}{T}$. In each cycle, the manufacturer orders used products, for which the ordering cost is $\frac{A_m}{T}$. The manufacturer inventory diagram is shown in Figure 2, the holding cost for the finished product is formulated in Appendix A.1 and can be written as follows,

$$h_f \sum_{i=1}^{n-1} d_{i+1} l_i + \frac{d^2 T}{2(1-r)P_m}.$$

The manufacturer inspects each product for which total inspection cost is $ck \sum_{i=1}^{n} d_i$, and $h_u d\tau \left(1 - \frac{\sum_{i=1}^{n} d_i}{2(1-r)P_m}\right)$ is the holding cost of used products. During production, $r\%$ of the total items produced are defective, to avoid shortages the manufacturer outsources the defective quantity with total cost $ow * r * \sum_{i-1}^{n} d_i$. τd is the remanufactured quantity and $d(1-\tau)$ is the quantity of new products. Therefore, the cost of remanufacturing used products is $bd\tau\,(C_r)$, and the cost of manufacturing is $C_m d(1-\tau)$. The finished products quality is $0 < q \leq 1$, to attain this quality, the manufacturer total quality cost is $d(1-\tau)q^2 C_q$ for manufacturing and $d\tau C_q\,(q^2 - q_r^2)$ for remanufacturing. The manufacturer incurs goodwill lost cost for producing low quality products which is $dg(1-q)$. Finally, the total emission cost from production activities can be expressed as $e_m \sum_{i=1}^{n} C_{ec} d_i$. Hence, the total cost per unit time for the manufacturer (TC_2) can be expressed as follows:

$$\begin{aligned} TC_2 &= \tfrac{A_m+S_m}{T} + h_f\left(\textstyle\sum_{i=1}^{n-1} l_i \sum_{k=i+1}^{n} d_k + \tfrac{d^2T}{2(1-r)P_m}\right) + (\mathrm{ck} + ow * r)\textstyle\sum_{i=1}^{n} d_i + h_u d\tau\left(1 - \tfrac{\sum_{i=1}^{n} d_i}{2(1-r)P_m}\right) \\ &+ d(1-\tau)C_m + d\tau C_r + d\tau C_q\left(q^2 - q_r^2\right) + dq^2(1-\tau)C_q + dg(1-q) + e_m \textstyle\sum_{i=1}^{n} C_{ec} d_i. \end{aligned} \quad (3)$$

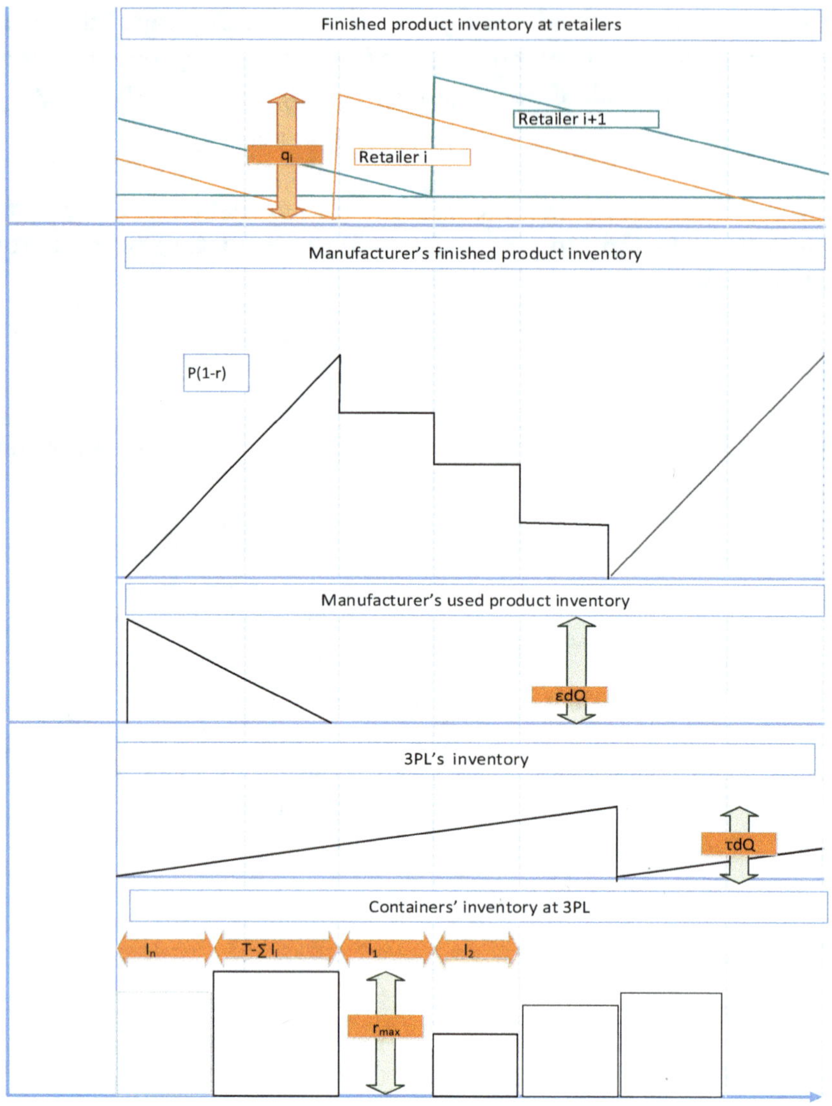

Figure 2. Logistics diagram of the proposed CLSC.

4.4. 3PL's Model

3PL's total cost comprises of collection setup cost per cycle $\frac{S_3}{T}$. The collection cost paid to customers is $\tau A_t \sum_{i=1}^{n} d_i$. The average holding cost of used product at 3PL is $h_u \frac{1}{2} d\tau T$. 3PL also provides transportation services, total transportation cost comprised of three parts: (1) finished products transportation from manufacturer to retailer i, $(\frac{d_i l_{im}}{\lambda})$, (2) empty RTI transportation from retailer i to 3PL $\frac{d_i l_{ij}}{\lambda}$, and (3) used products transportation from 3PL to manufacturer $\frac{dl_{\mathrm{jm}}}{\lambda}$. For n retailers total transportation cost is the sum of forward and reverse transportation costs represented by $C_t\left(\sum_{i=1}^{n} \frac{d_i l_{ij}}{\lambda} + \sum_{i=1}^{n} \frac{d_i l_{im}}{\lambda} + \frac{dl_{\mathrm{jm}}}{\lambda}\right)$. Holding cost of RTIs is given by $h_R\left(\sum_{i=1}^{n} \frac{l_i(d_{\max} - d_i)}{\lambda} + \frac{d_1\left(T - \sum_{i=1}^{n} l_i\right)}{\lambda}\right)$, see Appendix A.2. RTI management cost is $C_\lambda d_{\max} \lambda^{s-1}$. For setting up collection centers and arranging the inspection system for determining the quality of the EOL/EOU products, 3PL makes an initial investment which can be expressed mathematically as $\frac{\gamma\tau^2}{T}$. Finally, 3PL's cost also consists of emissions cost from transportation, which is determined based on the traveled distance, and emissions from fuel required per unit distance for a specific truck is considered to model transportation emissions. Therefore, total emissions due to transportation between retailer i and manufacturer m can be modeled as $\frac{d_i l_{im}}{t_c}(e_r g_m)$, similarly emissions from transportation between retailer i and 3PL j can be written by $\frac{d_i l_{ij}}{t_c}\Big) e_r g_m$, and emissions from 3PL j to manufacturer m is given by $\frac{d\tau l_{jm}}{t_c}$. Now for n retailers, a manufacturer, and a 3PL, total emission cost from transportation can be expressed mathematically as $(C_{ec} e_r g_m)(\sum_{i=1}^{n} \frac{d_i l_{im}}{t_c} + \sum_{i=1}^{n} \frac{d_i l_{ij}}{t_c} + \frac{d\tau l_{jm}}{t_c})$. Total cost per unit of time for the 3PL TC_3 comprises of the above parts and can be expressed as,

$$\begin{aligned} TC_3 &= \frac{S_3}{T} + \tau A_t \sum_{i=1}^{n} d_i + \frac{d\tau Th_u}{2} + C_t \Big(\sum_{i=1}^{n} \frac{d_i l_{ij}}{\lambda} + \sum_{i=1}^{n} \frac{d_i l_{im}}{\lambda} + \frac{dl_{jm}}{\lambda} \Big) + C_\lambda d_{\max} \lambda^{s-1} \\ &+ h_R \Big(\sum_{i=1}^{n} \frac{l_i (d_{\max} - d_i)}{\lambda} + \frac{d_1 (T - \sum_{i=1}^{n} l_i)}{\lambda} \Big) + c_d(\tau - \tau) d + \frac{\gamma \tau^2}{T} + C_{ec} e_r g_m \\ & \sum_{i=1}^{n} \frac{\tau d_i l_{ij}}{t_c} + C_{ec} e_r g_m \sum_{i=1}^{n} \frac{d_i l_{im}}{t_c} + \frac{d\tau C_{ec} e_r g_m l_{jm}}{t_c} \end{aligned} \tag{4}$$

4.5. Net Profit of the Supply Chain

Net profit of the supply chain can be calculated by subtracting cost of retailer (2), cost of manufacturer (3), and cost of 3PL (4) from total generated revenue (1), such that

$$\Pi = Rev - TC_1 - TC_2 - TC_3$$

Using the transformation from demand function, total profit can be expressed as follows,

$$\begin{aligned} \Pi \quad - \quad & cwr \sum_{i=1}^{n} (a_i - bp + k\mathrm{IN}_i) + \sum_{i=1}^{n} p\,(a_i - bp + k\mathrm{IN}_i) - \bigg(\sum_{i=1}^{n} \frac{Th_i}{2} (a_i - bp + k\mathrm{IN}_i) + \sum_{i=1}^{n} \\ & \frac{A_i + \mathrm{IN}_i + \mathrm{TR}_i}{T} \bigg) - \bigg(h_f \bigg(\sum_{i=1}^{n-1} l_i \,(a_{i+1} - bp + k\mathrm{IN}_i) + \frac{T(\sum_{i=1}^{n}(a_i - bp + k\mathrm{IN}_i))^2}{2(1-r)P_m} \bigg) + \frac{A_m + S_m}{T} \\ + \quad & h_u \sum_{i=1}^{n} \tau \,(a_i - bp + k\mathrm{IN}_i) \bigg(1 - \frac{\sum_{i=1}^{n}(a_i - bp + k\mathrm{IN}_i)}{2(1-r)P_m} \bigg) + ((\mathrm{ck} + ow * r)) \sum_{i=1}^{n} a_i - bp + k\mathrm{IN}_i + (C_m * 1 - \tau \\ & \sum_{i=1}^{n} (a_i - bp + k\mathrm{IN}_i)) + \tau C_q \sum_{i=1}^{n} (q^2 - q_r^2)\,(a_i - bp + k\mathrm{IN}_i) + (\tau \sum_{i=1}^{n} C_r \,(a_i - bp + k\mathrm{IN}_i)) + (C_q \\ & q^2 (1-\tau) \sum_{i=1}^{n} (a_i - bp + k\mathrm{IN}_i)) + e_m \sum_{i=1}^{n} C_{ec}\,(a_i - bp + k\mathrm{IN}_i) + \sum_{i=1}^{n} g(1-q)\,(a_i - bp + k\mathrm{IN}_i) \bigg) \\ - \quad & \bigg(\tau A_t \sum_{i=1}^{n} (a_i - bp + k\mathrm{IN}_i) + \frac{T}{2} (\tau h_u \sum_{i=1}^{n} (a_i - bp + k\mathrm{IN}_i)) + \frac{S_3}{T} + C_t \bigg(\sum_{i=1}^{n} \frac{l_{im}(a_i - bp + k\mathrm{IN}_i)}{\lambda} \\ + \quad & \sum_{i=1}^{n} \frac{\tau l_{ij}(a_i - bp + k\mathrm{IN}_i)}{\lambda} + \frac{l_{jm} \sum_{i=1}^{n} \tau (a_i - bp + k\mathrm{IN}_i)}{\lambda} \bigg) + (C_\lambda \lambda^{s-1} (a_{\max} - bp + k\mathrm{IN}_{a_{max}})) \\ + \quad & \frac{h_R((T - \sum_{i=1}^{n} l_i)(a_{\max} - bp + k\mathrm{IN}_{a_{max}}))}{\lambda} + \frac{\gamma \tau^2}{T} + C_{ec} e_r g_m \sum_{i=1}^{n} \frac{\tau l_{ij}(a_i - bp + k\mathrm{IN}_i)}{t_c} + C_{ec} e_r g_m \\ & \sum_{i=1}^{n} \frac{l_{im}(a_i - bp + k\mathrm{IN}_i)}{t_c} + \frac{\sum_{i=1}^{n} \tau C_{ec} e_r g_m l_{jm} (a_i - bp + k\mathrm{IN}_i)}{t_c} \bigg) \end{aligned} \tag{5}$$

Assuming T, λ, τ, and p as any real nonzero numbers, then, $\exists$ a unique T^*, λ^*, τ^* and p^* that maximizes the total profit and satisfies the following first-order conditions (FOC):

$$\frac{\partial\,(\Pi)}{\partial p} = 0, \tag{6}$$

$$\frac{\partial\,(\Pi)}{\partial \tau} = 0, \tag{7}$$

$$\frac{\partial\,(\Pi)}{\partial T} = 0, \tag{8}$$

$$\frac{\partial\,(\Pi)}{\partial \lambda} = 0. \tag{9}$$

Furthermore, (Π) is concave, if and only if at T^*, λ^*, τ^* and p^* the hessian of (Π) is negative semidefinite.

Proof. See Appendix B for the proof of concavity of objective function. □

Until now, the model considers random defective rates of production, but no probability distribution is considered. Now, one can test the model under different distributions: uniform, triangular, and beta. Thus, if r follows a uniform distribution, then the probability density function $f(r)$ for the uniform distribution with parameter x and y is,

$$f(r) = \begin{cases} \frac{1}{x-y} & x \leq r \leq y \\ 0 & otherwise \end{cases}$$

and its expected value is $E[r] = \frac{y+x}{2}$.

Similarly, if r follows a triangular distribution, then the probability density function for the triangular distribution with parameter x, z, and y is

$$f(r) = \begin{cases} 0 & r < x \\ \frac{2(r-x)}{(y-x)(z-x)} & x \leq r < z \\ \frac{2}{(y-x)} & r = z \\ \frac{2(y-r)}{(y-x)(y-z)} & z < r \leq y \\ 0 & y < r \end{cases}$$

and its expected value is $E[r] = \frac{x+y+z}{3}$.

Finally, if r follows a beta distribution, then the probability density function for the beta distribution with parameter λ and β is

$$f(r) = \left\{ \frac{r^{\lambda-1}(1-r)^{\beta-1}}{B(\lambda,\beta)} \quad \lambda > 0, \beta > 0, 0 \leq r < 1 \right.$$

and its expected value is $E[r] = \frac{\lambda}{\lambda+\beta}$.

4.6. Solution Procedure

Our aim is to derive the optimal values of the decision variables and maximize the total net profit under all the three distributions. The solution procedure is as follows:

Step 1.0 Select the distribution function.

- **Step 1.1** Plug the expected value of r in (5).

Step 2.0 Set $\tau = 0$, $\lambda = 1$, and $T = 1$ find p^* from FOC given in (6).

- **Step 2.1** Using $p = p^*$, $\lambda = 1$, and $T = 1$ find τ^* from FOC given in (7).
- **Step 2.2** Using $p = p^*$, $\lambda = 1$, and $\tau = \tau^*$ find T^* from FOC given in (8).
- **Step 2.3** Using $p = p^*$, $T = T^*$, and $\tau = \tau^*$ find λ^* from FOC given in (9).
- **Step 2.4** Using $\tau = \tau^*$, $\lambda = \lambda^*$, and $T = T^*$ find p^* from FOC given in (6).
- **Step 2.5** Using $p = p^*$, $\lambda = \lambda^*$, and $T = T^*$ find τ^* from FOC given in (7).
- **Step 2.6** Using $p = p^*$, $\lambda = \lambda^*$, and $\tau = \tau^*$ find T^* from FOC given in (8).
- **Step 2.7** Using $p = p^*$, $T = T^*$, and $\tau = \tau^*$ find λ^* from FOC given in (9).
- **Step 2.8** Repeat Steps 2.4 - 2.7 until the values of τ^*, λ^*, T^*, and p^* stop changing.
- **Step 2.9** Update the optimal values τ^*, λ^*, T^*, and p^*.

Step 3.0 Set $\lambda = \lfloor \lambda^* \rfloor$, calculate total profit from (5).

Step 4.0 Set $\lambda = \lceil \lambda^* \rceil$, calculate total profit from (5).
Step 5.0 If $\Pi(\tau^*, \lfloor \lambda^* \rfloor, T^*, p^*) < \Pi(\tau^*, \lceil \lambda^* \rceil, T^*, p^*)$, update $\lambda^* = \lfloor \lambda^* \rfloor$, else update $\lambda^* = \lceil \lambda^* \rceil$.

Step 5.1 Repeat Steps 2.4–2.6 until the values of τ^*, T^*, and p^* stop changing.

Step 6.0 Find total profit from (5).
Step 7.0 Repeat the process for all the three probability distributions.

5. Numerical Experiment

The developed model is tested with numerical experiment and sensitivity analysis of the input parameters. The numerical experiment considers a three-layer supply chain network consisting of four retailers, a manufacturer, and a 3PL. The values of input parameters are obtained from [6] and are given in Table 1.

Table 1. General Input parameter values for numerical examples.

A_m	20	S_3	10	n	4	γ	2000	A_t	20	C_λ	0.5	h_R	5	s	2
C_m	110	C_r	84	C_q	5	q_r	0.3	g_m	0.25	e_r	0.01	C_{ec}	18	g	10
C_t	0.004	l_{jm}	50	t_c	80	e_m	0.002066	P_m	10,000	h_f	5.2	h_u	2.5	S_m	60
l_1	0.009	l_2	0.008	l_3	0.007	l_4	0.008	l_m	373	l_{2m}	226	l_{3m}	216	l_{4m}	371
l_j	373	l_{2j}	226	l_{3j}	216	l_{4j}	371	h_1	8.	h_2	7.4	h_3	8.2	h_4	8.1
A_1	63	A_2	51	A_3	39	A_4	63	IN_1	500	IN_2	489	IN_3	623	IN_4	585
TR_1	300	TR_2	283	TR_3	250	TR_4	310	a_1	1200	a_2	720	a_3	520	a_4	600
b	0.9	q	0.9	ck	5	cw	200	k	0.4	x	0.15	y	0.4	z	0.2
λ	0.15	β	0.4	ow	115										

The optimal results for all the three distribution are given in Table 2. We obtain the optimal cycle time as $T = 0.64$ years (for uniform distribution), $T = 0.66$ years (for triangular distribution) and $T = 0.67$ years (for beta distribution). The optimal container capacity is $\lambda = 3$ units (for uniform distribution), $\lambda = 4$ units (for triangular distribution) and $\lambda = 5$ units (for beta distribution). For optimal remanufacturing rate, the results are $\tau = 65$ percent (for uniform distribution), $\tau = 69$ percent (for triangular distribution) and $\tau = 73$ percent (for beta distribution). The optimal prices are $p = \$581.21$ (for uniform distribution), $p = \$582.6$ (for triangular distribution) and $p = \$581.4$ (for beta distribution). From the total profit, it is clear that the highest profit is yielded by uniform distribution \$828,278 followed by the beta distribution \$827,970. The least profit is obtained from triangular distribution, which is \$824,728.

Table 2. Optimal results of three examples.

	Example 1	**Example 2**	**Example 3**
Distribution of Defective Rate	**Uniform**	**Triangular**	**beta**
$\Pi(\$)$	828,278	824,728	827,970
$T^*(days)$	0.634	0.66	0.67
$\lambda^*(units)$	3	4	5
$\tau^*(\%)$	0.65	0.69	0.73
$P^*(\$)$	581.21	582.64	581.43

5.1. Sensitivity Analysis

A sensitivity analysis is performed for all the key parameters and the results are compiled in Tables 3–8. Table 3 provides sensitivity analysis of key parameters related to manufacture and 3PL. From the results, the following insights are obtained:

- The most influential parameter for all the three examples is manufacturing cost. Decreasing the manufacturing cost increases the profit of the system. However, the percentage change in profit to both positive and negative changes in manufacturing cost is asymmetric. Decreasing manufacturing cost by 50 percent increases the total profit by around 11%, on the other hand, an equal increase reduces the profit by less than one percent. Thus, it can be concluded that the profit is more sensitive to negative changes in manufacturing cost compared to the positive changes. Hence, in hybrid systems, supply chain managers must focus on technologies to reduce manufacturing cost to improve overall profitability.

- Almost similar results are shown by remanufacturing cost. Decreasing remanufacturing cost by 50% increases profit by more than 8% for all examples, while increasing remanufacturing cost by 50% reduces profit by less than one percent.
- In the case of investment for the collection of used product, the models showed opposite asymmetry. In this case, the profit is more sensitive towards positive changes in the investment cost. A 50% increase in investment cost reduces total profit by 0.37% and a 50% decrease only increases the profit by 0.16%. Hence it can be concluded that the sensitivity of profit towards positive change is almost double of the negative change.
- An interesting result is obtained from the sensitivity of production rate. The results are different for different examples. For beta distribution, the percent change is almost symmetric in both directions. For uniform distribution, the results are less asymmetric, however, for triangular distribution, the results show a higher degree of asymmetry compared to the other two models. This shows that for supply chains, in which the defective rate follows a triangular distribution, increasing the production rate will not increase the profit. However, in the case of uniform and beta, distribution production rate can be used as a tool to increase the profit of the supply chain.
- Among the three examples, the model with triangular distribution is more sensitive towards manufacturer's setup cost compared to uniform and beta distributions. This means that for supply chain systems, where the defective rate follows a triangular distribution, managers must focus on reduction of the manufacturer's setup cost.

Table 3. Sensitivity analysis for parameters related to manufacturer and 3PL.

Parameter	% Change in Value	% Change in Π (Uniform)	% Change in Π % (Triangular)	% Change in Π (beta)
C_m	−50%	11.19	11.28	11.19
	−25%	5.45	5.472	5.43
	+25%	−0.086	−0.106	−0.08
	+50%	−0.123	−0.138	−0.14
C_r	−50%	8.603	8.625	8.582
	−25%	4.254	4.259	4.256
	+25%	−0.159	−0.17	−0.151
	+50%	−0.163	−0.17	−0.163
γ	−50%	0.169	0.138	0.13
	−25%	0.065	0.0591	0.07
	+25%	−0.033	−0.043	−0.03
	+50%	−0.374	−0.13	−0.37
P_m	−50%	−0.037	-0.035	-0.031
	−25%	0.001	-0.012	0.001
	+25%	0.022	0.001	0.022
	+50%	0.028	0.013	0.028
S_m	−50%	0.019	0.004	0.019
	−25%	0.017	0.0026	0.002
	+25%	0.001	−0.002	0.001
	+50%	0.009	−0.106	−0.008

Sensitivity of all other parameters, related to manufacturer and 3PL, is given in Table A1, Appendix C.

This paper assumed that the demand depends upon price and investment in O2O channel that attracts more customers. Table 4 provides a sensitivity analysis of the demand parameters b and k. The following insights are obtained from the results:

- The profit of the system increases with a decrease in the demand parameter b, and decreases with a decrease in k. However, the percent change is different for both the parameters.
- Considering the demand parameter b, a 50% decrease produces a 114.8% increase for both uniform and beta distributions. The increase for the triangular distribution is, however, slightly higher, 115.3%. Similarly, on the positive side, triangular distribution is also slightly more sensitive compared to the beta distribution and uniform distribution.
- In the case of demand parameter k, the triangular distribution shows a 21.03% decrease to 50% decrease in k, which is slightly higher than beta distribution, 20.92%, and uniform distribution, 20.90%. On the positive side, however, the results are comparable with slight differences. For a 50% increase in k, the total profit increased by 23.3% for the uniform distribution, 23.48% for the triangular distribution, and 23.38% for the beta distribution.

Table 4. Sensitivity analysis for demand parameters.

Parameter	% Change in Value	% Change in Π (Uniform)	% Change in Π % (Triangular)	% Change in Π (beta)
b	−50%	114.8	115.3	114.8
	−25%	38.2	38.59	38.26
	+25%	−22.7	−23.46	−22.87
	+50%	−38.09	−38.46	−38.19
k	−50%	−20.90	−21.03	−20.92
	−25%	−10.7	−10.81	−10.76
	+25%	11.38	11.42	11.40
	+50%	23.3	23.48	23.38

This paper assumes random defective rates. It is a common assumption in the traditional modeling approaches, that higher defective rates generate low profits. However, this is not the case for this paper. Traditional modeling approaches either considered rework or scrap option for defective products. However, our modeling approach considered that defective products are salvaged. Now two situations arise: (1) salvage value of the defective product is more than the outsourcing cost, (2) salvage value of defective products is less than the outsourcing cost. In the first case, the generated revenue is more than cost and hence decreasing the defective rate decreases the total profit. While in the second case, the generated revenue from defective products is less than the total production cost, hence, the total profit increases with the decrease in defective rate.

5.1.1. Case 2

To investigate this phenomenon in detail, we consider a second numerical example. All parametric values are taken from example 1, except that salvage value cw is considered less than the outsourcing cost of the products. In this particular case, we considered $cw = 100 < ow = 115$. The results of case 2 are compiled in Table 5. The results show that the profit of the supply chain is reduced; however, this is obvious as the revenue from defective product salvaging is reduced by 50%. Among the three different distributions, the highest profit is generated by triangular distribution compared to the previous case, in which triangular distribution generated the lowest profit.

Table 5. Optimal results of three examples, Case 2.

	Example 1	**Example 2**	**Example 3**
Distribution of Defective Rate	**Uniform**	**Triangular**	**beta**
$\Pi(\$)$	781,267	781,911	781,326
$T^*(days)$	0.681203	0.681946	0.681272
$\lambda^*(units)$	5	5	5
$\tau^*(\%)$	0.712269	0.710318	0.712085
$P^*(\$)$	594.779	594.59	594.762

In order to clearly understand the relationship of salvage value, defective percentage, and profit of the system, a sensitivity analysis for defective rate parameters is performed for both the examples. The results of sensitivity are summarized in Tables 6 and 7. The results very clearly show that the sensitivity of profit is reversed for the second case. Although the negative and positive changes in the defective rate result in asymmetric changes in the total profit, the trend remains the same. The following important insights can be drawn from the sensitivity analysis:

- For the uniform distribution, increasing the parameter x increases the profit of the system under Case 1 and decreases it under Case 2. Increasing the distribution parameter x increases the expected value of the defective rate, this means, increasing the defective rate increases the profit of the system in Case 1 and decreases the profit in Case 2. The results are also similar for triangular distribution, in which the expected value increases with increasing the parameter x. However, x is more effective in uniform distribution compared to triangular distribution, this is because the net effect in the expected value is more in uniform distribution compared to triangular distribution.

- Increasing the parameter y also increases the expected value of the defective rate. Therefore, increasing the parameter y increases the profit of the system under Case 1 and decreases profit under Case 2, for both uniform and triangular distributions. The impact of y is more than double compared to the impact of x for uniform distribution; for triangular distribution, the impact of x, on profit, is about 50% the effect of y.

- The parameter z only applies to the triangular distribution, therefore, it has no effect on uniform distribution. For triangular distribution, increasing z increases the profit of Case 1 and decreases the profit of Case 2.

- For beta distribution, increasing λ increases the expected value of the defective rate, therefore, it increases the profit in Case 1 and decreases it in Case 2. On the other hand, increasing β decreases the expected value of the defective rate and, therefore, when β is increased, the profit of Case 1 is reduced and the profit of Case 2 is increased.

From the above results it is clear that, for Case 1, total profit increases with an increase in the defective rate, while for Case 2, the total profit decreases with an increase in the defective rate. The cost of outsourcing can also be considered as the cost of shortage and lost sales. Whatever be the situation or assumptions, if the cost is more than salvage value, supply chain managers need not worry about the defective rate. However, when the cost of shortage, outsourcing, or lost sales is more than the salvage value of imperfect items, supply chain managers should focus on reduction of defective rate.

Table 6. Sensitivity analysis for parameter related to defective rate (uniform and triangular distributions).

		Case 1, $cw > ow$		Case 2, $cw < ow$	
Parameter	**% Changes in Value**	**% Change in Π (Uniform)**	**% Change in Π (Triangular)**	**% Change in Π (Uniform)**	**% Change in Π (Triangular)**
x	−50%	−0.65	−0.44	0.13	0.09
	−25%	−0.32	−0.22	0.001	0.05
	+25%	0.33	0.21	−0.06	−0.04
	+50%	0.67	0.44	−0.14	−0.08
y	−50%	−1.76	−0.60	0.34	0.23
	−25%	−0.88	−0.29	0.16	0.12
	+25%	0.88	0.59	−0.16	−0.10
	+50%	1.77	1.15	−0.35	−0.21
z	−50%	0.00	−1.35	0.00	0.12
	−25%	0.00	−0.67	0.00	0.06
	+25%	0.00	0.29	0.00	−0.05
	+50%	0.00	0.59	0.00	−0.10

Table 7. Sensitivity analysis for parameter related to defective rate (beta distribution).

		Case 1, $cw > ow$	Case 2, $cw < ow$
Parameter	**% Changes in Value**	**% Change in Π (beta)**	**% Change in Π (beta)**
λ	−50%	−2.04	0.39
	−25%	−0.95	0.18
	+25%	0.78	−0.15
	+50%	1.40	−0.29
β	−50%	2.78	−0.54
	−25%	1.04	−0.18
	+25%	−0.74	0.15
	+50%	−1.30	0.25

Finally, we considered the impacts of investment cost in the O2O channel on total profit. Sensitivity analysis is performed for information cost and transportation cost. The results are compiled in Table 8. The results clearly show that rise in investment cost increases total profit. The effect is slightly asymmetric; the profit is more sensitive towards positive changes compared to negative changes.

Table 8. Sensitivity analysis of O2O channel parameters.

Parameter	**% Change in Value**	**% Change in Π (Uniform)**	**% Change in Π % (Triangular)**	**% Change in Π (beta)**
IN_i	−50%	−20.76	−20.83	−20.7
	−25%	−10.67	−10.72	−10.6
	+25%	11.293	11.34	11.29
	+50%	23.142	23.27	23.15
TR_i	−50%	0.080	0.091	0.080
	−25%	0.064	−0.016	0.064
	+25%	−0.043	−0.058	−0.04
	+50%	−0.090	−0.096	−0.09

6. Conclusions

This paper studies joint pricing and inventory policies for a multi-echelon CLSC model with an O2O sales channel. In order to reduce solid waste generation in supply chain, RTI is used to transport finished products from the manufacturer to retailers. A 3PL collects used products from the consumer and delivers them to the manufacturer for remanufacturing. Furthermore, carbon emissions from transportation and production activities are also considered to improve the environmental performance of supply chain management. Imperfect production is investigated with a random defective rate, whereas the defective rate follows three different probability distributions. Traditionally, supply chain inventory management considered rework or scrap option for the defective product. However, this model considered that defective products are salvaged, which generated extra revenue. Based on this assumption, two cases were developed, and the results showed that unlike traditional modeling approaches, total profit increases with an increase in defective rate when the salvage value of defective products is more than the cost of outsourcing. When the cost of outsourcing is higher compared to the salvage value, our model converged to traditional modeling approaches, in which profit decreases with an increase in defective rate.

The results further showed that in CLSC models the most important parameters are manufacturing cost and remanufacturing cost, their decrease produces more positive impacts compared to the negative impacts caused by their increments. Moreover, the sensitivity analysis provided that investment in the O2O channel leads to a substantial increase in total profit. This research can be extended in many directions, for example, as we considered deterministic demand, considerations of stochastic demand would be a more realistic example. Furthermore, the modeling approach of defective items needs to be explored further in the presences of shortage cots, lost sales and back ordering. Another important extension of this model is to consider deteriorated products, such as the study done by Ullah et al. [63].

Author Contributions: Conceptualization, B.S. and M.U.; Methodology, B.S. and M.U.; Software, M.U.; Validation, B.S.; Formal analysis, B.S. and M.U.; Investigation, B.S. and M.U.; Resources, B.S. and M.U.; Data curation, B.S. and M.U.; Writing—original draft preparation, M.U.; Writing—review and editing, M.U.; Visualization, B.S. and M.U.; Supervision, B.S. and S.-B.C.

Abbreviations

The following notation are used to develop mathematical model.

Notation

Decision variables

T	replenishment cycle time (year)
τ	remanufacturing rate (percentage of demand)
p	price of the finished product
λ	capacity of a single container (units)

Parameters

d_i	net demand realized by the ith retailer (units/unit time)
IN_i	demand from online channel at i^{th} retailer (units/unit time)
b	demand parameter
k	demand parameter
TR_i	Fixed cost of online channel per unit time ($/unit time)
h_i	holding cost of produced products at i^{th} retailer ($/unit/unit time)
l_i	lead time of the ith retailer
A_i	ordering cost per order of ith retailer ($/order)
n	number of retailers
P_m	production rate of the manufacturer (units/unit time)
S_m	manufacturer setup cost per setup ($/setup)
S_r	3PL setup cost per setup for collecting used products($/setup)
A_m	manufacturer ordering cost per order being placed to 3PL ($/order)
h_f	holding cost of produced products for the manufacturer ($/unit/unit time)
h_u	holding cost of collected products ($/unit/unit time)

C_m	manufacturing cost (\$/product)
C_r	remanufacturing cost (\$/product)
C_q	quality improving cost (\$/product)
g	goodwill lost cost for the manufacturer (\$/product)
A_t	average incentives paid to consumer for collected products (\$/unit)
C_t	transportation cost per container per unit distance (\$/container/unit distance)
l_{im}	distance between the manufacturer and retailer i (kilometers)
l_{ij}	distance between retailer i and 3PL (kilometers)
l_{jm}	distance between 3PL and the manufacturer (kilometers)
C_λ	cost of managing RTIs, including depreciation and repair per unit capacity of RTI (\$/RTI capacity)
s	scaling factor for how the container capacity affects the cost of managing a container
h_R	holding cost of containers at 3PL (\$/unit/unit time)
γ	effective investment by 3PL to collects EOL/EOU products in dollars (\$)
e_m	carbon emissions per product from manufacturing (kg/product)
e_r	carbon emissions per gallon of fuel used by truck (kg/gallon)
g_m	fuel required per mile (gallon/miles)
C_{ec}	carbon tax per unit of emitted carbon (\$/unit of emitted carbon)
C_q	quality upgradation cost (\$/product)
t_c	truck capacity (number of products)
cw	salvage value of defective products (\$/unit)
ow	cost of out sourcing one product (\$/unit), $ow > C_m$
ck	inspection cost per product (\$/unit)
r	defective rate (random variable)
TC_1	cost of multiple retailers (\$/cycle)
TC_2	cost of manufacturer (\$/cycle)
TC_3	cost of 3PL (\$/cycle)

The following abbreviations are used in this manuscript.

O2O	Online to offline
CLSC	Closed-loop supply chain
RTI	Returnable transport item
3PL	Third party logistics

Appendix A

Appendix A.1

From Figure 2, manufacturer inventory can be written as,

$$\frac{Q^2}{2(1-r)P_m} + l_1\left(Q - q_1\right) + l_2\left(Q - q_1 - q_2\right) \ldots\ldots\ldots\ldots + l_{n-1}\left(Q - q_1 - q_2 \ldots\ldots - q_{n-1}\right)$$

$$Q = \left(q_1 + q_2 + q_3 + q_4 \ldots\right).\left(+q_{n-1}\right) + q_n$$

$$\frac{Q^2}{2(1-r)P_m} + T\sum_{i=1}^{n-1} l_i d_i$$

Therefore, the time-weighted average inventory is

$$\frac{d^2 T}{2(1-r)P_m} + \sum_{i=1}^{n-1} l_i d_{i+1}$$

Appendix A.2

The average inventory of RTI at 3PL can be calculated with the help of Figure 2.

$$= r_{max}\left(T-\sum_{i=1}^{n} l_i\right)+\sum_{i=1}^{n} l_i\left(r_{max}-r_i\right)$$

$$= \frac{d_{max}T}{\lambda}\left(T-\sum_{i=1}^{n} l_i\right)+\sum_{i=1}^{n} l_i\left(\frac{d_{max}T}{\lambda}-\frac{T}{\lambda}d_i\right).$$

Therefore, the time weighted average inventory of RTI at 3PL

$$= \frac{d_{max}}{\lambda}\left(T-\sum_{i=1}^{n} l_i\right)+\sum_{i=1}^{n} l_i\left(\frac{d_{max}}{\lambda}-\frac{1}{\lambda}d_i\right).$$

Appendix B

If,

$a_{1,1}=\frac{\partial}{\partial T}\frac{\partial \Pi}{\partial T}$, $a_{1,2}=\frac{\partial}{\partial \tau}\frac{\partial \Pi}{\partial T}$, $a_{1,3}=\frac{\partial}{\partial \lambda}\frac{\partial \Pi}{\partial T}$, $a_{1,4}=\frac{\partial}{\partial p}\frac{\partial \Pi}{\partial T}$, $a_{2,1}=\frac{\partial}{\partial T}\frac{\partial \Pi}{\partial \tau}$, $a_{2,2}=\frac{\partial}{\partial \tau}\frac{\partial \Pi}{\partial \tau}$, $a_{2,3}=\frac{\partial}{\partial \lambda}\frac{\partial \Pi}{\partial \tau}$, $a_{2,4}=\frac{\partial}{\partial p}\frac{\partial \Pi}{\partial \tau}$, $a_{3,1}=\frac{\partial}{\partial T}\frac{\partial \Pi}{\partial \lambda}$, $a_{3,2}=\frac{\partial}{\partial \tau}\frac{\partial \Pi}{\partial \lambda}$, $a_{3,3}=\frac{\partial}{\partial \lambda}\frac{\partial \Pi}{\partial \lambda}$, $a_{3,4}=\frac{\partial}{\partial p}\frac{\partial \Pi}{\partial \lambda}$, $a_{4,1}=\frac{\partial}{\partial T}\frac{\partial \Pi}{\partial p}$, $a_{4,2}=\frac{\partial}{\partial \tau}\frac{\partial \Pi}{\partial p}$, $a_{4,3}=\frac{\partial}{\partial \lambda}\frac{\partial \Pi}{\partial p}$, $a_{4,4}=\frac{\partial}{\partial p}\frac{\partial \Pi}{\partial p}$,

Then the Hessian matrix of (5) can be expressed as

$$H=\begin{pmatrix} a_{1,1} & a_{1,2} & a_{1,3} & a_{1,4} \\ a_{2,1} & a_{2,2} & a_{2,3} & a_{2,4} \\ a_{3,1} & a_{3,2} & a_{3,3} & a_{3,4} \\ a_{4,1} & a_{4,2} & a_{4,3} & a_{4,4} \end{pmatrix}$$

Appendix B.1. When Defective Rate Follows Uniform Distribution

The Hessian matrix for example 1, when defective rate follows uniform distribution, at T = 0.634; λ = 3; τ = 0.65; p = 581.21 is,

$$\begin{pmatrix} -33524.3 & 4001.85 & 487.243 & 23.4258 \\ 4001.85 & -6199.8 & 300.356 & -9.77975 \\ 487.243 & 300.356 & -503.327 & -0.693627 \\ 23.4258 & -9.77975 & -0.693627 & -7.20307 \end{pmatrix}$$

The first four principle minors are -33524.3, $+1.91829\times10^{8}$, -9.08852×10^{10}, and $+6.5227\times10^{11}$. Hence total profit is strictly concave at T = 0.634; λ = 3; τ = 0.65; p = 581.21.

Appendix B.2. When Defective Rate Follows Triangular Distribution

The Hessian matrix for example 2, when defective rate follows triangular distribution, at T = 0.66; λ = 4; τ = 0.69; p = 582.64 is,

$$\begin{pmatrix} -31987.9 & 4059.47 & 273.633 & 23.041 \\ 4059.47 & -6060.61 & 168.46 & -10.0465 \\ 273.633 & 168.46 & -216.887 & -0.208652 \\ 23.041 & -10.0465 & -0.208652 & -7.20295 \end{pmatrix}$$

The first four principle minors are -31987.9, $+1.77387\times10^{8}$, -3.6737×10^{10}, and $+2.63656\times10^{11}$. Hence total profit is strictly concave at T = 0.66; λ = 4; τ = 0.69; p = 582.64.

Appendix B.3. When Defective Rate Follows Beta Distribution

The Hessian matrix for example 3, when defective rate follows beta distribution, at $T = 0.67$; $\lambda = 5$; $\tau = 0.73$; $p = 581.43$ is,

$$\begin{pmatrix} -31332.3 & 4222.72 & 175.343 & 23.1492 \\ 4222.72 & -5970.15 & 108.056 & -10.3247 \\ 175.343 & 108.056 & -113.68 & 0.0186797 \\ 23.1492 & -10.3247 & 0.0186797 & -7.20296 \end{pmatrix}$$

The first four principle minors are -31332.3, $+1.69227 \times 10^{8}$, -1.85284×10^{10}, and $+1.32964 \times 10^{11}$. Hence total profit is strictly concave at $T = 0.67$; $\lambda = 5$; $\tau = 0.73$; $p = 581.43$.

Appendix C.

Table A1. Sensitivity analysis for parameters related to manufacturer and 3PL.

Parameter	% Change in Value	% Change in Π (Uniform)	% Change in Π % (Triangular)	% Change in Π (beta)
h_f	−50%	0.059	0.010	0.059
	−25%	0.037	0.0211	0.003
	+25%	−0.006	−0.027	−0.007
	+50%	−0.029	−0.049	−79.6
h_u	−50%	0.302	0.290	0.302
	−25%	0.143	0.133	0.147
	+25%	−0.09	−0.091	−0.091
	+50%	−0.15	−0.196	−0.152
A_m	−50%	0.001	0.0017	0.001
	−25%	0.008	0.0008	0.0002
	+25%	0.008	−0.0008	0.013
	+50%	0.012	−0.001	0.002
C_t	−50%	0.039	0.064	0.039
	−25%	0.044	0.030	0.028
	+25%	−0.02	−0.034	−0.020
	+50%	−0.04	−0.057	−0.044
A_t	−50%	2.011	1.999	2.01
	−25%	0.999	0.989	1.00
	+25%	−0.1596	−0.17	-0.20
	+50%	0.186	−0.17	−0.16
C_λ	−50%	0.131	0.101	0.115
	−25%	0.06	0.04	0.064
	+25%	−0.03	−0.05	−0.030
	+50%	−0.086	−0.09	−0.075
C_q	−50%	0.381	0.345	0.381
	−25%	0.203	0.179	0.192
	+25%	−0.18	−0.182	−0.19
	+50%	−0.37	−0.40	−0.375
q_r	−50%	−0.03	−0.045	−0.033
	−25%	−0.02	−0.021	−0.028
	+25%	0.044	0.030	0.0446
	+50%	0.10	0.085	0.100
ck	−50%	0.522	0.510	0.507
	−25%	0.253	0.2548	0.253
	+25%	−0.256	−0.254	−0.25
	+50%	−0.506	−0.508	−0.50
h_R	−50%	0.0078	0.04	0.007
	−25%	0.0256	−0.024	0.02
	+25%	−0.049	−0.033	−0.02
	+50%	−0.098	−0.089	−0.02

References

1. Cassab, H.; MacLachlan, D.L. Interaction fluency: A customer performance measure of multichannel service. *Int. J. Prod. Perform. Manag.* **2006**, *55*, 555–568. [CrossRef]
2. Yang, S.; Lu, Y.; Chau, P.Y. Why do consumers adopt online channel? An empirical investigation of two channel extension mechanisms. *Decis. Support Syst.* **2013**, *54*, 858–869. [CrossRef]
3. Chang, Y.W.; Hsu, P.Y.; Yang, Q.M. Integration of online and offline channels: A view of O2O commerce. *Internet Res.* **2018**, *28*, 926–945. [CrossRef]
4. Sarkar, B.; Tayyab, M.; Choi, S.B. Product Channeling in an O2O Supply Chain Management as Power Transmission in Electric Power Distribution Systems. *Mathematics* **2019**, *7*, 4. [CrossRef]
5. Livne, G.; Simpson, A.; Talmor, E. Do customer acquisition cost, retention and usage matter to firm performance and valuation? *J. Bus. Financ. Account.* **2011**, *38*, 334–363. [CrossRef]
6. Sarkar, B.; Ullah, M.; Kim, N. Environmental and economic assessment of closed-loop supply chain with remanufacturing and returnable transport items. *Comput. Ind. Eng.* **2017**, *111*, 148–163. [CrossRef]
7. Bulmus, S.C.; Zhu, S.X.; Teunter, R. Competition for cores in remanufacturing. *Eur. J. Oper. Res.* **2014**, *233*, 105–113. [CrossRef]
8. Taleizadeh, A.A.; Sane-Zerang, E.; Choi, T.M. The effect of marketing effort on dual-channel closed-loop supply chain systems. *IEEE Trans. Syst. Man Cybern. Syst.* **2018**, *48*, 265–276. [CrossRef]
9. Schrady, D.A. A deterministic inventory model for reparable items. *Nav. Res. Logist.* **1967**, *14*, 391–398. [CrossRef]
10. Richter, K. The EOQ repair and waste disposal model with variable setup numbers. *Eur. J. Oper. Res.* **1996**, *95*, 313–324. [CrossRef]
11. Richter, K. The extended EOQ repair and waste disposal model. *Int. J. Prod. Econ.* **1996b**, *45*, 443–447. [CrossRef]
12. Richter, K. Pure and mixed strategies for the EOQ repair and waste disposal problem. *Oper. Res. Spekt.* **1997**, *19*, 123–129. [CrossRef]
13. Richter, K.; Dobos, I. Analysis of the EOQ repair and waste disposal problem with integer setup numbers. *Int. J. Prod. Econ.* **1999**, *59*, 463–467. [CrossRef]
14. Dobos, I.; Richter, K. An extended production/recycling model with stationary demand and return rates. *Int. J. Prod. Econ.* **2004**, *90*, 311–323. [CrossRef]
15. Dobos, I.; Richter, K. A production/recycling model with quality consideration. *Int. J. Prod. Econ.* **2006**, *104*, 571–579. [CrossRef]
16. Maiti, T.; Giri, B. A closed loop supply chain under retail price and product quality dependent demand. *J. Manuf. Syst.* **2015**, *37*, 624–637. [CrossRef]
17. Moshtagh, M.S.; Taleizadeh, A.A. Stochastic integrated manufacturing and remanufacturing model with shortage, rework and quality based return rate in a closed loop supply chain. *J. Clean. Prod.* **2017**, *141*, 1548–1573. [CrossRef]
18. Tian, X.; Zhang, Z.H. Capacitated disassembly scheduling and pricing of returned products with price-dependent yield. *Omega* **2019**, *84*, 160–174. [CrossRef]
19. Govindan, K.; Soleimani, H. A review of reverse logistics and closed-loop supply chains: A Journal of Cleaner Production focus. *J. Clean. Prod.* **2017**, *142*, 371–384. [CrossRef]
20. Diallo, C.; Venkatadri, U.; Khatab, A.; Bhakthavatchalam, S. State of the art review of quality, reliability and maintenance issues in closed-loop supply chains with remanufacturing. *Int. J. Prod. Res.* **2017**, *55*, 1277–1296. [CrossRef]
21. Chiang, W.Y.K.; Chhajed, D.; Hess, J.D. Direct marketing, indirect profits: A strategic analysis of dual-channel supply-chain design. *Manag. Sci.* **2003**, *49*, 1–20. [CrossRef]
22. Huang, S.; Yang, C.; Zhang, X. Pricing and production decisions in dual-channel supply chains with demand disruptions. *Comput. Ind. Eng.* **2012**, *62*, 70–83. [CrossRef]
23. Ahmad, W.; Sarkar, B.; Ullah, M. Impact of Reparation for Imperfect Quality Items having Shortages in the System under Multi-Trade-Credit-Period. *DJ J. Eng. Appl. Math.* **2018**, *5*, 1–16. [CrossRef]
24. Tsay, A.A.; Agrawal, N. Channel conflict and coordination in the e-commerce age. *Prod. Oper. Manag.* **2004**, *13*, 93–110. [CrossRef]

25. Yao, D.Q.; Liu, J.J. Competitive pricing of mixed retail and e-tail distribution channels. *Omega* **2005**, *33*, 235–247. [CrossRef]
26. Huang, W.; Swaminathan, J.M. Introduction of a second channel: Implications for pricing and profits. *Eur. J. Oper. Res.* **2009**, *194*, 258–279. [CrossRef]
27. Yan, R. Pricing strategy for companies with mixed online and traditional retailing distribution markets. *J. Prod. Brand Manag.* **2008**, *17*, 48–56. [CrossRef]
28. Dan, B.; Xu, G.; Liu, C. Pricing policies in a dual-channel supply chain with retail services. *Int. J. Prod. Econ.* **2012**, *139*, 312–320. [CrossRef]
29. Li, B.; Zhu, M.; Jiang, Y.; Li, Z. Pricing policies of a competitive dual-channel green supply chain. *J. Clean. Prod.* **2016**, *112*, 2029–2042. [CrossRef]
30. Bányai, T.; Illés, B.; Bányai, Á. Smart scheduling: An integrated first mile and last mile supply approach. *Complexity* **2018**, *2018*, 5180156 [CrossRef]
31. Bányai, T. Real-time decision making in first mile and last mile logistics: How smart scheduling affects energy efficiency of hyperconnected supply chain solutions. *Energies* **2018**, *11*, 1833. [CrossRef]
32. Habib, M.S.; Sarkar, B.; Tayyab, M.; Saleem, M.W.; Hussain, A.; Ullah, M.; Omair, M.; Iqbal, M.W. Large-scale disaster waste management under uncertain environment. *J. Clean. Prod.* **2019**, *212*, 200–222. [CrossRef]
33. Ullah, M.; Sarkar, B. Smart and sustainable supply chain management: A proposal to use rfid to improve electronic waste management. In Proceedings of the International Conference on Computers and Industrial Engineering, Auckland, New Zealand, 2–5 December 2018; Volume 2018.
34. Johansson, O.; Hellström, D. The effect of asset visibility on managing returnable transport items. *Int. J. Phys. Distrib. Logist. Manag.* **2007**, *37*, 799–815. [CrossRef]
35. Witt, C. Transport packaging: Neat and clean or down and dirty. *Mater. Hand. Eng.* **1999**, *54*, 71–73.
36. Livingstone, S.; Sparks, L. The new German packaging laws: Effects on firms exporting to Germany. *Int. J. Phys. Distrib. Logist. Manag.* **1994**, *24*, 15–25. [CrossRef]
37. Kroon, L.; Vrijens, G. Returnable containers: An example of reverse logistics. *Int. J. Phys. Distrib. Logist. Manag.* **1995**, *25*, 56–68. [CrossRef]
38. Glock, C.H. Decision support models for managing returnable transport items in supply chains: A systematic literature review. *Int. J. Prod. Econ.* **2017**, *183*, 561–569. [CrossRef]
39. Kim, T.; Glock, C.H.; Kwon, Y. A closed-loop supply chain for deteriorating products under stochastic container return times. *Omega* **2014**, *43*, 30–40. [CrossRef]
40. Glock, C.H.; Kim, T. Container management in a single-vendor-multiple-buyer supply chain. *Logist. Res.* **2014**, *7*, 112. [CrossRef]
41. Kelle, P.; Silver, E.A. Forecasting the returns of reusable containers. *J. Oper. Manag.* **1989**, *8*, 17–35. [CrossRef]
42. Goh, T.; Varaprasad, N. A statistical methodology for the analysis of the life-cycle of reusable containers. *IIE Trans.* **1986**, *18*, 42–47. [CrossRef]
43. Lee, H.L. Lot sizing to reduce capacity utilization in a production process with defective items, process corrections, and rework. *Manag. Sci.* **1992**, *38*, 1314–1328. [CrossRef]
44. Gupta, T.; Chakraborty, S. Looping in a multistage production system. *Int. J. Prod. Res.* **1984**, *22*, 299–311. [CrossRef]
45. Tayi, G.K.; Ballou, D.P. An integrated production-inventory model with reprocessing and inspection. *Int. J. Prod. Res.* **1988**, *26*, 1299–1315. [CrossRef]
46. Lee, H.H.; Chandra, M.J.; Deleveaux, V. Optimal batch size and investment in multistage production systems with scrap. *Prod. Plan. Cont.* **1997**, *8*, 586–596. [CrossRef]
47. Glock, C.H.; Jaber, M.Y. Learning effects and the phenomenon of moving bottlenecks in a two-stage production system. *Appl. Math. Model.* **2013**, *37*, 8617–8628. [CrossRef]
48. Salameh, M.; Jaber, M. Economic production quantity model for items with imperfect quality. *Int. J. Prod. Econ.* **2000**, *64*, 59–64. [CrossRef]
49. Jaber, M.Y.; Zanoni, S.; Zavanella, L.E. Economic order quantity models for imperfect items with buy and repair options. *Int. J. Prod. Econ.* **2014**, *155*, 126–131. [CrossRef]

50. Ouyang, L.Y.; Chen, C.K.; Chang, H.C. Quality improvement, setup cost and lead-time reductions in lot size reorder point models with an imperfect production process. *Comput. Oper. Res.* **2002**, *29*, 1701–1717. [CrossRef]
51. Cárdenas-Barrón, L.E. Economic production quantity with rework process at a single-stage manufacturing system with planned backorders. *Comput. Ind. Eng.* **2009**, *57*, 1105–1113. [CrossRef]
52. Eroglu, A.; Ozdemir, G. An economic order quantity model with defective items and shortages. *Int. J. Prod. Econ.* **2007**, *106*, 544–549. [CrossRef]
53. Ben-Daya, M. The economic production lot-sizing problem with imperfect production processes and imperfect maintenance. *Int. J. Prod. Econ.* **2002**, *76*, 257–264. [CrossRef]
54. Sarkar, B.; Majumder, A.; Sarkar, M.; Kim, N.; Ullah, M. Effects of variable production rate on quality of products in a single-vendor multi-buyer supply chain management. *Int. J. Adv. Manuf. Technol.* **2018**, *99*, 567–581. [CrossRef]
55. Sarkar, B. Mathematical and analytical approach for the management of defective items in a multi-stage production system. *J. Clean. Prod.* **2019**, *218*, 896–919. [CrossRef]
56. Kim, M.S.; Kim, J.S.; Sarkar, B.; Sarkar, M.; Iqbal, M.W. An improved way to calculate imperfect items during long-run production in an integrated inventory model with backorders. *J. Manuf. Syst.* **2018**, *47*, 153–167. [CrossRef]
57. Sett, B.K.; Sarkar, S.; Sarkar, B. Optimal buffer inventory and inspection errors in an imperfect production system with preventive maintenance. *Int. J. Adv. Manuf. Technol.* **2017**, *90*, 545–560. [CrossRef]
58. Sarkar, B.; Sett, B.K.; Sarkar, S. Optimal production run time and inspection errors in an imperfect production system with warranty. *J. Ind. Manag. Optim.* **2018**, *14*, 267–282. [CrossRef]
59. Kang, C.W.; Ullah, M.; Sarkar, B. Optimum ordering policy for an imperfect single-stage manufacturing system with safety stock and planned backorder. *Int. J. Adv. Manuf. Technol.* **2018**, *95*, 109–120. [CrossRef]
60. Khanna, A.; Kishore, A.; Sarkar, B.; Jaggi, C. Supply Chain with Customer-Based Two-Level Credit Policies under an Imperfect Quality Environment. *Mathematics* **2018**, *6*, 299. [CrossRef]
61. Tayyab, M.; Sarkar, B.; Ullah, M. Sustainable Lot Size in a Multistage Lean-Green Manufacturing Process under Uncertainty. *Mathematics* **2019**, *7*, 20. [CrossRef]
62. Tayyab, M.; Sarkar, B.; Yahya, B. Imperfect Multi-Stage Lean Manufacturing System with Rework under Fuzzy Demand. *Mathematics* **2019**, *7*, 13. [CrossRef]
63. Ullah, M.; Sarkar, B.; Asghar, I. Effects of Preservation Technology Investment on Waste Generation in a Two-Echelon Supply Chain Model. *Mathematics* **2019**, *7*, 189. [CrossRef]

9

Maintenance Optimization for Repairable Deteriorating Systems under Imperfect Preventive Maintenance

Juhyun Lee [1], Byunghoon Kim [2] and Suneung Ahn [2,*]

1 Department of Industrial and Management Engineering, Hanyang University, Seoul 04763, Korea
2 Department of Industrial and Management Engineering, Hanyang University ERICA, Ansan 15588, Korea
* Correspondence: sunahn@hanyang.ac.kr.

Abstract: This study deals with the preventive maintenance optimization problem based on a reliability threshold. The conditional reliability threshold is used instead of the system reliability threshold. Then, the difference between the two thresholds is discussed. The hybrid failure rate model is employed to represent the effect of imperfect preventive maintenance activities. Two maintenance strategies are proposed under two types of reliability constraints. These constraints are set to consider the cost-effective maintenance strategy and to evaluate the balancing point between the expected total maintenance cost rate and the system reliability. The objective of the proposed maintenance strategies is to determine the optimal conditional reliability threshold together with the optimal number of preventive maintenance activities that minimize the expected total maintenance cost per unit time. The optimality conditions of the proposed maintenance strategies are also investigated and shown via four propositions. A numerical example is provided to illustrate the proposed preventive maintenance strategies. Some sensitivity analyses are also conducted to investigate how the parameters of the proposed model affect the optimality of preventive maintenance strategies.

Keywords: preventive maintenance; imperfect repair; optimization; reliability threshold; repairable system

---

1. Introduction

Most engineering systems experience degradation influenced by the usage environment and operation time length such as aircrafts, high-speed railroads, nuclear power plants, and weapons systems, which leads to frequent failures [1]. Once a failure has occurred in the system, it decreases the availability and reliability of the system, which increase unexpected losses. For example, the availability degradation can increase downtime costs and decrease product quality. Additionally, the reliability degradation can affect the safety of systems and reduce the operating life. Therefore, in many industries, maintenance activities are performed to prevent decreasing availability and reliability of the systems due to severe degradation.

Maintenance activities play an important role in reliability theory because they decrease uncertainty of the failure, reduce the operation cost of the system, and increase the functioning life of the system [2]. In practice, due to increasing complexity of the systems, the cost related to the maintenance activities has increased constantly over the past decades, which attributes 15–70% of production costs [3–5]. Therefore, it is important to establish cost-effective maintenance planning for ensuring higher efficiency, quality production, availability of the system, and reliability of the system.

There are two general types of maintenance activities: corrective maintenance (CM) and preventive maintenance (PM). CM is unscheduled and performed when a system fails, and the system is restored from a failed state to a working state [6]. Most CM employs a minimal repair that returns the failed system to its previous working state after repairing the system. Therefore, the minimal repair has been widely used in numerous maintenance models to address maintenance problems such as the failure of a system. There are many studies covering the minimal repair model [7]. On the other hand, PM activities delay the degradation of a system and reduce operational stress. In traditional PM models, a repaired system can end up in the state of either "as good as new" (i.e., perfect PM) or "as bad as old" (i.e., minimal repair) after the PM activities [8]. Thereafter, a more generalized and realistic approach has been proposed, called the imperfect PM model. The imperfect PM model considers that the system state lies somewhere between "as good as new" and "as bad as old" after PM activities.

Many models describing the effect of imperfect PM activities have been discussed in many existing studies. Malik et al. [9] introduced an imperfect PM model based on the age reduction method. Lin et al. [10] proposed a hybrid hazard rate function that combines the age reduction factor and hazard increasing factor. This model can better reflect the general and realistic situations because the model restores the system effective age to a better state and degrades the hazard rate of the system, after PM activities. El-Ferik and Ben-Daya [11] developed an age-based imperfect PM model that uses the hybrid hazard rate model and considers two types of failure modes. Khatab and Abdelhakim [12] developed an imperfect PM model that is subject to random deterioration based on the hybrid hazard rate using a reliability threshold.

In addition, the PM activity is subcategorized into two types of activities: periodic and non-periodic. Periodic PM activities are repeatedly performed over a fixed period of time T. The periodic PM activity is still widely applied to many PM models because it is mathematically convenient [13]. Shue et al. [14] developed an extended periodic imperfect PM model by considering an age-dependent failure type. Some previous studies [15–17] proposed maintenance policies for a leased system using the periodic imperfect PM model. Moreover, other extended studies on periodic PM have been applied to various fields such as warranty policies, production systems, and used systems [18–21]. However, a periodic PM has a shortcoming in that it is hard to prevent failures that frequently occur in the late period of the life of a system based on the periodic PM. Non-periodic PM models have been proposed to handle this shortcoming, which can prevent the inefficient PM activities [22].

In non-periodic PM models, controlling system reliability is an important issue especially when dealing with repairable deteriorating systems. In practice, maintenance based on reliability threshold provides good guidance for decision-makers who want to attain high performance of maintenance for their operating systems [23]. In addition, reliability of an engineering system is a decision variable that can increase investment in production technology [24]. Zhao [25] proposed a PM policy based on the critical reliability level subject to degradation. Zhou et al. [26] proposed a reliability-centered predictive maintenance policy that is composed of scheduled PM and non-scheduled PM using the reliability threshold. They considered the situation in which the system was continuously monitored. Liao et al. [27] developed a sequential imperfect PM model with a reliability threshold for a continuously monitored system that considered both the operational cost and the breakdown cost. Schutz et al. [16] developed a maintenance strategy for leased equipment based on the reliability threshold. Doostparast et al. [28] developed an optimal PM scheduling considering a certain level of reliability for coherent systems. Lin et al. [29] developed an imperfect PM model with a reliability threshold that considered three reliability constraints to help evaluate the maintenance cost. This model used the system reliability threshold that maintains a survival probability until the system replacement. Khatab et al. [30] developed an imperfect PM model for failure-prone second-hand systems considering the conditional reliability threshold and upgrade action. Khatab [31] improved Liao's model [27] by adding mathematical properties and remedying inconsistencies in the maintenance optimization model.

Traditional PM models using the reliability threshold assume that PM activities are performed when the reliability reaches a threshold value, and that the system is constantly monitored and maintained [29]. However, the system reliability may not be maintained at a constant level when following the foregoing assumption. This is because there are two types of reliability threshold in the

PM model: the conditional reliability threshold and the system reliability threshold. The conditional reliability threshold refers to the survival probability of each PM cycle, whereas the system reliability threshold refers to the survival probability until the replacement. The performance of the PM model (e.g., the expected maintenance cost rate and the system operating time until replacement) is not robust to the selection of reliability threshold in constructing the PM model. Therefore, it is important to clearly differentiate between the conditional reliability threshold and the system reliability threshold. This is the motivation of this study.

In addition, it is important to consider the trade-off between reliability and cost when establishing PM models. Some existing models employ reliability thresholds to address the trade-off. However, the existing models assume that the threshold value is same for every PM cycle. This assumption is biased and unrealistic toward reliability. Therefore, the existing models find it difficult to reflect the cost-effectiveness accurately. This is another motivation of this study.

The main contributions of this study are (i) to identify and define the differences between the system reliability threshold and the conditional reliability threshold and (ii) to develop a novel PM model that minimizes the expected maintenance cost rate with consideration of two reliability constraints. Table 1 highlights the contributions of different authors in the field of PM model focusing on the reliability threshold. It can be observed in Table 1 that the literature uses the system reliability threshold more, rather than the conditional reliability threshold. In addition, the literature assumes a fixed level of the reliability threshold.

Table 1. Author contribution in the reliability-based PM model based on reliability constraints, maintenance costs, and imperfect maintenance model.

Research	Types of Reliability Threshold	Reliability Constraints		Maintenance Cost		Other Cost	Imperfect Maintenance Model		
		Fixed	Unfixed	CM	PM	Breakdown	Age Reduction Factor	Hazard Increasing Factor	Other
Zhao [25]	Conditional	√		√	√				√
Zhou et al. [26]	System	√		√	√		√	√	
Liao et al. [27]	Conditional	√		√	√	√	√	√	
Schutz et al. [16]	System	√		√	√		√		
Doostparast et al. [28]	System	√		√	√	√			√
Lin et al. [29]	System	√	√	√	√		√		
Khatab et al. [30]	Conditional	√		√	√	√	√	√	
Khatab [31]	Conditional	√		√	√	√	√	√	
Proposed model	Conditional	√	√	√	√	√	√	√	

In this study, we use the conditional reliability as a criterion of PM activities instead of using system reliability because the conditional reliability threshold can consider the situation in which the system reliability decreases as the frequency of PM activities increase. The hybrid failure rate model is used to represent the effect of imperfect PM activities. In addition, we analyzed the proposed model through two strategies that can help evaluate the trade-off between maintenance cost and system reliability. We discussed the optimal conditions for each strategy via four propositions and one lemma that show the existence and uniqueness of the optimal solution.

The remainder of this paper is organized as follows. Section 2 explains the assumptions to establish the proposed PM model, and derives the function of the expected maintenance cost rate. Section 3 defines two strategies and provides an algorithm to find the optimal solution. Section 4 provides a numerical example to illustrate the proposed PM model and conducts sensitivity analyses to investigate critical elements. Finally, Section 5 discusses the conclusions of this research.

2. Imperfect Preventive Maintenance Model

We consider the situation in which a new system is subject to stochastic deterioration over time and will be replaced at the end of the Nth PM cycle by a new one. The system experiences PM whenever the conditional reliability threshold reaches a certain level R_i. PM is imperfect and modeled via the hybrid hazard rate model of Lin et al. [10]. Minimal repair is taken for any failures prior to the PM actions. The goal of this study is to determine the optimal number of PM activities and the optimal conditional reliability thresholds for each PM cycle that minimizes the expected maintenance cost rate. To consider the above-mentioned goal and assumptions, we compose the long-term expected maintenance cost rate in this study, which is given as:

$$ECR(\underline{R} = (R_1, R_2, \ldots, R_N), N) = \frac{E[MC] + E[BC]}{E[T]}, \tag{1}$$

where $E[MC]$ denotes the expected total maintenance cost, $E[BC]$ denotes the expected total breakdown cost, and $E[T]$ denotes the expected system replacement cycle. It should be noted that $E[MC]$ is composed of the cost for minimal repair, PM, and replacement. These costs are assumed to be constant and can be obtained from a field investigation. As they may also be considered the expected values for the random variables, it is reasonable to assume $max\ (C_{MR}, C_{PM}, C_{BD}) < C_{RP}$.

2.1. Notations and Assumptions

The notations and detailed assumptions in this study are:

Notations

α	Scale parameter of hazard intensity functions where $\alpha > 0$
β	Deterioration parameter of hazard intensity functions where $\beta > 2$
a_i	Age reduction factor where $0 = a_0 < a_1 < \cdots < a_i < \cdots < 1$
b_i	Hazard increasing factor where $1 = b_0 \leq b_1 \leq \cdots \leq b_i \leq \cdots$
x_i	The ith PM interval, where $x_0 = 0$ and $x_1 = Y_1$.
Y_i	Effective age of the system before the ith PM activity
T_O	System lifetime until replacement
R_i	Conditional reliability threshold where $I = 1, 2, \ldots, N$
C_{MR}	Cost of minimal repair
C_{PM}	Cost of PM activity
C_{RP}	Cost of system replacement
C_{BD}	Cost of system breakdown
$R_{SYS,i}$	System reliability up to the ith PM cycle
$R_{CON,i}$	Conditional reliability at the ith PM cycle
$R_{RES,i}$	Restored reliability after the ith PM activity
$h_i\ (t)$	Hazard intensity functions during the ith PM cycle

Assumptions

1. The planning horizon is infinite.
2. The system is replaced at the Nth PM cycle, and $(N - 1)$ imperfect PMs are performed.
3. The durations of minimal repair, PM, and replacement are negligible.
4. The deterioration process of the system follows the non-homogeneous Poisson process (NHPP), and their hazard intensity function is given as

$$h(t) = \alpha\beta t^{\beta-1}, \text{ where } t \geq 0. \tag{2}$$

5. The hazard rate of the system varies with the hybrid model after imperfect PM activities that are performed whenever the conditional reliability threshold reaches a certain value.
6. Costs for minimal repair, PM, replacement, and breakdown are assumed constant.

2.2. *Hybrid Hazard Rate Model under Imperfect PM*

Now, we describe the effect of imperfect PM using the hybrid hazard rate model. This model adjusts the hazard rate of the system using the age reduction factor and the hazard increasing factor after PM activity. Let us consider the situation in which a system's age is reduced by the age reduction factor after PM. If the first PM activity is performed at the effective age Y_1, the age of the system is updated from Y_1 to a_1Y_1. The effective age Y_2 is obtained as the sum of a_1Y_1 and x_2, where x_2 denotes the length of the second PM cycle, and the effective age Y_i can then be generalized as

$$Y_i = x_i + a_{i-1}Y_{i-1}, \tag{3}$$

where $Y_0 = 0$, $x_1 = Y_1$, and a_i denotes the age reduction factor for $0 = a_0 < a_1 < \cdots < a_i < \cdots < 1$.

The hybrid hazard rate model adjusts both the effective age of the system and the hazard rate of the system after PM activity. Now, we describe the hazard rate of the system in each PM cycle. The hazard rate of the system in the first PM cycle is the same as Equation (2) within $[0, Y_1)$. After the first PM, the hazard rate of the system is adjusted from $h(Y_1)$ to $b_1h(a_1Y_1)$, and thus, the hazard rate in the second PM cycle becomes $b_1h(a_1Y_1{+}t)$ and $t \in [0, x_2)$. After the second PM, the hazard rate of the system is adjusted from $b_1h(Y_2)$ to $b_2b_1h(a_2Y_2)$. The hazard rate in the third PM cycle, then becomes $b_2b_1h(a_2Y_2{+}t)$ and $t \in [0, x_3)$. According to these procedures, the hazard rate in the ith PM cycle can be generalized as

$$h_i(t) = B_{i-1}h(a_{i-1}Y_{i-1} + t), \tag{4}$$

where $i \geq 1$, $t \in [0, x_i)$, and $B_{i-1} = \Pi_{j=1}^{i-1}b_j$ with $1 = b_0 \leq b_1 \leq \cdots \leq b_i \leq \cdots$.

2.3. *Conditional Reliability Threshold*

Here, we try to derive the system reliability under imperfect PM. According to the theory of reliability engineering, the reliability is defined as the probability that a system will survive over time period t [32]. The system reliability without PM activity can be defined as:

$$\begin{aligned} R_{SYS}(t) &= \Pr\{T \geq t\} \\ &= \exp\left(-\int_0^t h(u)du\right). \end{aligned} \tag{5}$$

Let $R_{SYS,i}(t)$ be the system reliability at the ith PM cycle, $R_{RES,i-1}(t)$ be the restored reliability after the $(i-1)$th PM, and $R_{CON,i}(t|a_{i-1}Y_{i-1})$ be the conditional reliability at the ith PM cycle. According to the hybrid hazard rate model, after the first PM activity, the effective age is adjusted from Y_1 to a_1Y_1 and the hazard rate of the system is adjusted from $h(Y_1)$ to $B_1h(a_1Y_1)$. Based on these procedures, the restored reliability after the first PM can be inferred as

$$R_{RES,1}(a_1Y_1) = \exp(-B_1H(a_1Y_1)). \tag{6}$$

The conditional reliability at the second PM cycle, $R_{CON,2}(t|a_1Y_1)$, is the survival probability of the system at time t given that the system undergoes the first PM at Y_1; it follows that

$$R_{CON,2}(t|a_1Y_1) = \exp(-B_1\int_{a_1Y_1}^{t} h(u)du), \tag{7}$$

where $t \in [a_1Y_1, Y_2)$. The system reliability after the first PM at time period t is the survival probability over the period $Y_1 + (t - a_1Y_1)$; it can then be computed as the product of the restored reliability and the conditional reliability, which is given as

$$R_{SYS,2}(t) = R_{RES,1}(a_1Y_1)R_{CON,2}(t|a_1Y_1). \tag{8}$$

Based on these procedures, the system reliability after the $(i-1)$th PM cycle can obviously be determined as:

$$R_{SYS,i}(t) = R_{RES,i-1}(a_{i-1}Y_{i-1})R_{CON,i}(t|a_{i-1}Y_{i-1}). \quad (9)$$

According to the aforementioned discussion, we can describe the difference between the conditional reliability threshold and the system reliability threshold through two definitions.

Definition 1. *If imperfect PM activities are performed under the following constraints*

$$R_i = R_{CON,i}(Y_i|a_{i-1}Y_{i-1}), \quad (10)$$

then R_i is called the conditional reliability threshold of the ith PM cycle.

Definition 2. *If imperfect PM activities are performed under the following constraints*

$$\begin{aligned} R_S &= R_{SYS,j}(Y_j) \\ &= R_{RES,j-1}(a_{j-1}Y_{j-1})R_{CON,j}(Y_j|a_{j-1}Y_{j-1}), \end{aligned} \quad (11)$$

then R_S is called a system reliability threshold.

Details of the PM model using the system reliability threshold have been reported in the literature [29]. As the conditional reliability threshold can reflect the situation in which the system reliability decreases as the frequency of PM activities increases while maintaining the reliability above a certain level, it is used to model the present work. Moreover, as the system undergoes PM whenever the conditional reliability threshold reaches the certain level, the corresponding reliability constraints can be given as:

$$R_i = \exp\left(-B_{i-1}\int_{a_{i-1}Y_{i-1}}^{Y_i} h(t)dt\right), \quad (12)$$

for $i = 1, 2, \ldots, N$. Equation (12) is inferred as:

$$\ln R_i = -B_{i-1}\int_{a_{i-1}Y_{i-1}}^{Y_i} h(t)dt. \quad (13)$$

Solving Equation (13) with respect to Y_i, the effective age at the ith PM activity can be determined as:

$$Y_i = \left(\frac{-B_{i-1}^{-1}\ln R_i - \sum_{j=1}^{i-1}\left(\prod_{k=j}^{i-1}\rho_k^{\beta}\right)B_{j-1}^{-1}\ln R_j}{\alpha}\right)^{1/\beta}, \quad (14)$$

where $i = 2, \ldots, N$ and $Y_1 = (-\ln R_1/\alpha)^{1/\beta}$.

2.4. Long-Term Expected Maintenance Cost Rate

The expected total maintenance cost until the Nth PM cycle is composed of the replacement cost, PM cost, and minimal repair cost, which is given as:

$$\begin{aligned} E[MC] &= C_{RP} + (N-1)C_{PM} + C_{MR}\sum_{i=1}^{N}\left(\int_0^{Y_i} h_i(t)dt\right) \\ &= C_{RP} + (N-1)C_{PM} + C_{MR}\sum_{i=1}^{N}\left(B_{i-1}\int_{a_{i-1}Y_{i-1}}^{Y_i} h(t)dt\right). \end{aligned} \quad (15)$$

Through Equation (13), Equation (15) becomes:

$$E[MC] = C_{RP} + (N-1)C_{PM} - C_{MR}\sum_{i=1}^{N} \ln R_i. \tag{16}$$

Real engineering systems suffer from frequent breakdowns after long-term use due to the severe degradation of the system. At this point, it may be more economical to replace the system rather than performing PM. Liao et al. [27] pointed out the importance of breakdown cost when constructing the PM models and used the breakdown cost in their PM model. They considered that the breakdown costs are charged once per PM cycle. Unlike the computation approach used in Liao et al. [27], Khatab [31] considered the breakdown cost that occurs whenever the system undergoes maintenance activities, such as a minimal repair, PM, and replacement. He discussed that the breakdown costs are computed as the product of the sum of all maintenance activities until the Nth PM cycle and the value of the breakdown cost. This study employs the approach proposed by Khatab [31] in calculating the breakdown cost. The total breakdown cost in this study is given as

$$\begin{aligned} E[BC] &= C_{BD}\left(N + \sum_{i=1}^{N} \int_0^{Y_i} h_i(t)dt\right) \\ &= C_{BD}\left(N - \sum_{i=1}^{N} \ln R_i\right). \end{aligned} \tag{17}$$

The replacement cycle of the system can be calculated by the sum of the durations up to the Nth PM cycle, which is given as:

$$T_O = Y_N + \sum_{j=1}^{N-1} (1-a_j)Y_j. \tag{18}$$

Through Equations (15)–(18), the long-term expected maintenance cost rate becomes:

$$ECR(\underline{R} = (R_1, R_2, \ldots, R_N), N) = \frac{C_{RP} + (N-1)C_{PM} - C_{MR}\sum_{i=1}^{N} \ln R_i + C_{BD}\left(N - \sum_{i=1}^{N} \ln R_i\right)}{Y_N + \sum_{j=1}^{N-1} (1-a_j)Y_j}. \tag{19}$$

3. Maintenance Optimization

In this section, we introduce the proposed PM model by discussing two strategies with different reliability constraints. The optimization problem of the proposed PM model is to find the decision variables that minimize the expected maintenance cost rate. In this optimization problem, the decision variables are defined as the optimal number of PM activities and the optimal conditional reliability thresholds.

Strategy 1: The PM activities are performed with the same level of conditional reliability threshold, which is given as:

$$R_1 = \exp\left(-B_{i-1} \int_{a_{i-1}Y_{i-1}}^{Y_i} h(t)dt\right), \text{ for } i = 1, 2, \ldots, N. \tag{20}$$

Strategy 2: The PM activities are performed at the different values of conditional reliability thresholds on all PM cycles. This PM strategy relaxes the reliability constraints of strategy 1 and is assumed to follow Equation (12). The system reliability of strategy 2 is lower than that of strategy 1, but it can lower the expected total maintenance cost rate.

3.1. Strategy 1: Same Level of Conditional Reliability Threshold

This strategy assumes that the conditional reliability thresholds on all PM cycles have the same value. The PM activity of this strategy is performed whenever the conditional reliability threshold reaches a certain level. The decision variables in this strategy are the optimal number of PM activities and the optimal conditional reliability threshold. These values minimize the expected total maintenance cost rate. In this strategy, the reliability constraints for performing PM activities follow Equation (20). Solving Equation (20) with respect to Y_i, the effective age at the ith PM activity can be determined as:

$$Y_i = \left(\frac{-B_{i-1}^{-1} \ln R_1 - \sum_{j=1}^{i-1} \left(\prod_{k=j}^{i-1} \rho_k^{\beta}\right) B_{j-1}^{-1} \ln R_1}{\alpha} \right)^{1/\beta}, \tag{21}$$

where $i = 2, \ldots, N$ and $Y_1 = (-\ln R_1/\alpha)^{1/\beta}$.

The expected total maintenance cost rate of strategy 1 becomes:

$$ECR_1(R_1, N) = \frac{C_{RP} + (N-1)C_{PM} - NC_{MR} \ln R_1 + C_{BD}(N - N \ln R_1)}{Y_N + \sum_{j=1}^{N-1} (1-a_j) Y_j}. \tag{22}$$

Now, we try to find the optimal number of PM activities that minimizes Equation (22). The inequalities are:

$$ECR_1(R_1, N-1) > ECR_1(R_1, N) \text{ and } ECR_1(R_1, N) \leq ECR_1(R_1, N+1), \tag{23}$$

if and only if

$$K_1(N-1) < C_{RP} - C_{PM} \text{ and } K_1(N) \geq C_{RP} - C_{PM}, \tag{24}$$

where

$$K_1(N) = \left(\frac{\sum_{i=1}^{N} x_i}{x_{N+1}} - N \right) (C_{PM} + C_{BD} - (C_{MR} + C_{BD}) \ln R_1). \tag{25}$$

Inequalities (23) and (24) are a necessary condition to find where Equation (22) is a convex function with respect to N when R_1 is fixed. Through the following Proposition 1, we can see that there exists an optimal finite and unique N^*.

Proposition 1. *When R_1 is fixed, if Inequalities (22) and (23) are satisfied, Equation (22) is a convex function and there exists an optimal finite and unique N^* that minimizes Equation (22).*

Proof. See Appendix A. □

Through following Proposition 2, we discuss the optimal value of the conditional reliability threshold when N is fixed.

Proposition 2. *When N is fixed, the optimal value for the conditional reliability threshold is given as:*

$$-\ln R_1^* = \frac{N(C_{PM} + C_{BD}) - C_{PM} + C_{RP}}{(\beta - 1)N(C_{MR} + C_{BD})}. \tag{26}$$

Proof. See Appendix A. □

The optimal solutions for strategy 1 can be obtained via the following algorithm that is given as the pseudo-code and is followed as Table 2.

Table 2. Algorithm to obtain the optimal solution for strategy 1.

Algorithm Pseudo-Code for the Minimization of Equation (22)
Input data: $C_M, C_R, C_P, C_{BD}, \alpha, \beta, a_i, b_i$;
Set $N = 1, M = 2$;
Compute $R_1(N), R_1(M)$ given in Equation (26);
Compute $ECR_1(N\vert R_1(N)), ECR_1(M\vert R_1(M))$ given in Equation (22)
while ($ECR_1(R_1(N), N) > ECR_1(R_1(M), M)$)
$N = N + 1, M = M + 1$;
Compute $R_1(N), R_1(M)$ given in Equation (26);
Compute $ECR_1(R_1(N), N), ECR_1(R_1(M), M)$ given in Equation (22);
end while
Set $N^* = N, R^*_1 = R_1(N^*)$;
Using N^* and R^*_1, compute $Y_i$given in Equation (21);

3.2. Strategy 2: Different Level of Conditional Reliability Threshold

Unlike strategy 1, this strategy assumes that the conditional reliability thresholds on all PM cycles have difference values. The structural properties of this strategy follow Section 2. The aim of this strategy is to find the optimal number of PM activities N^* and the optimal conditional reliability thresholds $\underline{R} = (R_1, R_2, \ldots, R_N)$. The expected total maintenance cost rate is the same as Equation (19). Solving Equation (14) with respect to Y_i, the effective age at the ith PM activity can be determined as:

$$Y_i = \left(\frac{-B_{i-1}^{-1}\ln R_i - \sum_{j=1}^{i-1}\left(\prod_{k=j}^{i-1}\rho_k^{\beta}\right)B_{j-1}^{-1}\ln R_j}{\alpha} \right)^{1/\beta}, \tag{27}$$

where $i = 2, \ldots, N$ and $Y_1 = (-\ln R_1/\alpha)^{1/\beta}$.

The result of following Lemma 1 will be used in Propositions 3 and 4. The lemma derives the N conditional reliability thresholds to a single conditional reliability threshold.

Lemma 1. *When N is fixed, the conditional reliability thresholds that minimize Equation (17) have the following relationship:*

$$-\ln R_i = \begin{cases} B_{i-1}\left(\frac{A_{i-1}}{A_i} - a_{i-1}^{\beta}\right)\left(\frac{A_1}{A_{i-1}}\right)(-\ln R_1) & ,\text{if } 2 \le i \le N-1, \\ B_{N-1}\left(A_{N-1} - a_{N-1}^{\beta}\right)\left(\frac{A_1}{A_{N-1}}\right)(-\ln R_1) & ,\text{if } i = N, \end{cases} \tag{28}$$

where

$$A_i = \left[\frac{1}{B_{N-1}} \left(\frac{B_{i-1} - a_i^{\beta} B_i}{1 - a_i} \right) \right]^{\beta/(\beta-1)} \quad \text{for all } i. \tag{29}$$

Proof. See Appendix A. □

Through the result of the following Lemma, Equation (19) becomes:

$$ECR_2(R_1, N) = \frac{(C_{RP} + C_{BD}) + (N-1)(C_{PM} + C_{BD}) - (C_{MR} + C_{BD})Mr(N)\ln R_1}{Y_N + \sum\limits_{j=1}^{N-1}(1 - a_j)Y_j}, \tag{30}$$

where

$$Mr(N) = 1 + \sum_{i=2}^{N-1} B_{i-1}\left(\frac{A_{i-1}}{A_i} - a_{i-1}^{\beta}\right)\left(\frac{A_1}{A_{i-1}}\right) + B_{N-1}(A_{N-1} - a_{N-1}^{\beta})\left(\frac{A_1}{A_{N-1}}\right), \tag{31}$$

where $Mr(1) = 1$ and $Mr(2) = 1 + B_1(A_1 - a_1^{\beta})$.

Through following Proposition 3, we can see that there exists an optimal finite and unique N^*.

Proposition 3. *When R_1 is fixed, N^* that satisfies Inequality (32) is the unique and finite optimal value that minimizes Equation (30).*

$$K_2(N-1) < C_{RP} - C_{PM} \text{ and } K_2(N) \geq C_{RP} - C_{PM}, \quad (32)$$

where

$$K_2(N) = \left(\frac{\sum_{i=1}^{N} x_i}{x_{N+1}} - N\right)(C_P + C_{BD}) - (C_M + C_{BD})\frac{\ln R_1}{x_{N+1}}\left(Mr(N+1)\sum_{i=1}^{N} x_i - Mr(N)\sum_{i=1}^{N+1} x_i\right). \quad (33)$$

Proof. See Appendix A. □

The following Proposition provides the optimal conditional reliability threshold of strategy 2.

Proposition 4. *When N is fixed, the solution to the optimal conditional reliability threshold that minimizes Equation (30) is given as:*

$$-\ln R_1^* = \frac{N(C_{PM} + C_{BD}) - C_{PM} + C_{RP}}{(\beta - 1)Mr(N)(C_{MR} + C_{BD})}. \quad (34)$$

Proof. See Appendix A. □

The optimal solutions for strategy 2 can be obtained via the following algorithm that is given as pseudo-code and is followed as Table 3.

Table 3. Algorithm to obtain the optimal solution for strategy 2.

Algorithm Pseudo-Code for the Minimization of Equation (30)
Input data: $C_M, C_R, C_P, C_{BD}, \alpha, \beta, a_i, b_i$;
Set $N = 1, M = 2$;
Compute $R_1(N)$, $R_1(M)$ given in Equation (34);
Compute $Mr(N)$, $Mr(M)$ given in Equation (31);
Compute $ECR_2(R_1(N), N)$, $ECR_2(R_1(M), M)$ given in Equation (30);
while $(ECR_2(R_1(N), N) > ECR_2(R_1(M), M))$
$N = N + 1, M = M + 1$;
Compute $R_1(N)$, $R_1(M)$ given in Equation (34);
Compute $Mr(N)$, $Mr(M)$ given in Equation (31);
Compute $ECR_2(R_1(N), N)$, $ECR_2(R_1(M), M)$ given in Equation (30);
end while
Set $N^* = N$, $R^*_1 = R_1(N^*)$;
Using N^* and R^*_1, compute $R_N^* = (R_1, \ldots, R_N^*)$ given in Equation (28);
Using N^* and R^*_1, compute $Y_i$given in Equation (27);

4. Numerical Illustration

In this section, we illustrate the proposed PM strategy based on a numerical example. To conduct the numerical illustration, we used Python 3.7. The procedures of the numerical example are followed as:

Step 1. Input data related to maintenance costs and parameters of hybrid hazard rate function.
Step 2. Compute the conditional reliability threshold and the expected maintenance cost rate.
Step 3. Repeat the step 2 until satisfying $ECR_k(R_1(N), N) > ECR_k(R_1(M), M)$ for $k = 1,2$.

Step 4. Compute PM schedule, the optimal number of PM activities, and the optimal conditional reliability threshold.

In the numerical illustration, two additional PM strategies are considered that can help evaluate the effect of the trade-off between reliability and cost. In the additional PM strategies, the conditional reliability threshold is provided to the decision maker, in which this is given as $R_1 = 0.9$ (PM strategy 3) and $R_1 = 0.3$ (PM strategy 4). Sensitivity analyses are also conducted to find how the parameters affect the proposed PM strategy. Table 4 sets out the parameters for conducting the numerical illustration. The functions of the age reduction factor and the hazard increasing factor are set as follows:

$$a_i = \frac{i}{2i+2},\ b_i = \frac{13i+4}{12i+4}. \tag{35}$$

Table 4. Setting of parameters.

C_{MR}/C_{PM}	C_{RP}/C_{PM}	C_{BD}/C_{PM}	α	β
3.0	5.0	0.3	2.6	3.2

4.1. Results of the Numerical Illustration

Figure 1 shows the curves for the expected total maintenance cost rate under the increasing PM activities. Table 5 summarizes the results of this study.

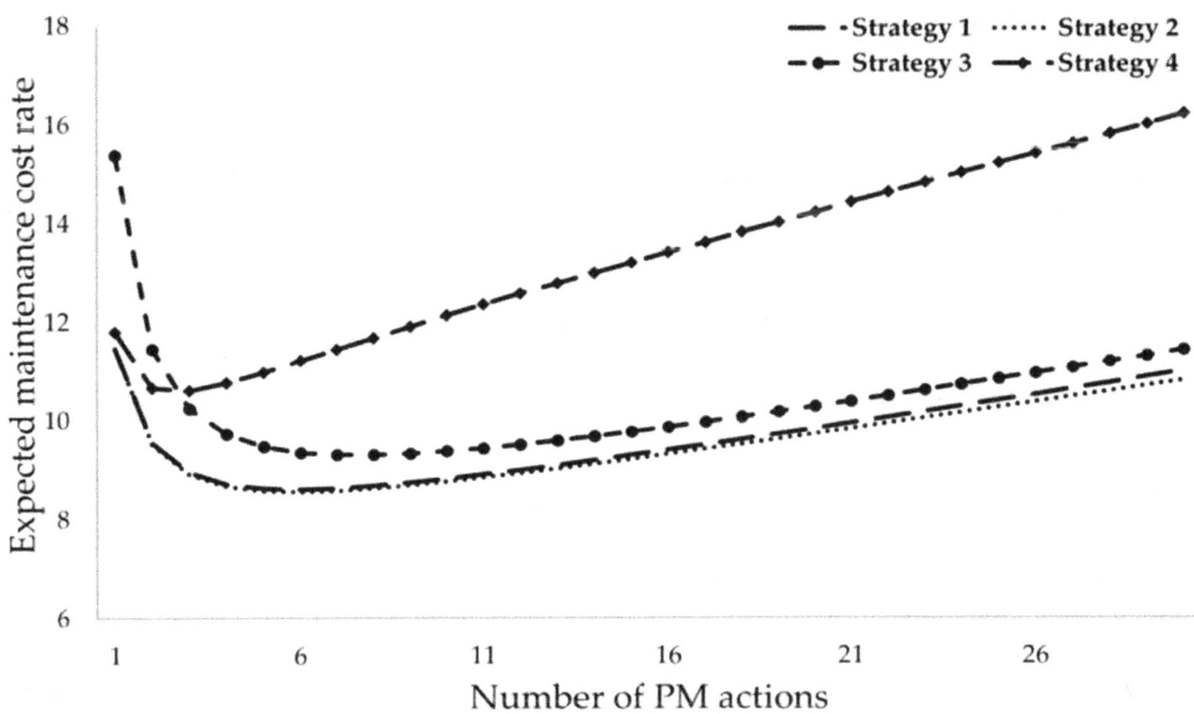

Figure 1. Cost curve under increasing the number of preventive maintenance (PM) activities.

Table 5. The optimal solutions and their system performances.

No. PM	PM Strategy 1		PM Strategy 2		PM Strategy 3		PM Strategy 4	
	x_i	R_S (R^*_1)	x_i	R_S (R^*_1)	x_i	R_S (R^*_1)	x_i	R_S (R^*_1)
1	0.4933	0.7627 (0.7627)	0.5194	0.7265 (0.7265)	0.3672	0.9000 (0.9000)	0.7862	0.3000 (0.3000)
2	0.3626	0.7601 (0.7627)	0.3534	0.7637 (0.7668)	0.2699	0.8988 (0.9000)	0.5779	0.2955 (0.3000)
3	0.3164	0.7561 (0.7627)	0.2968	0.7841 (0.7901)	0.2356	0.8969 (0.9000)	0.5043	0.2886 (0.3000)
4	0.2905	0.7528 (0.7627)	0.2655	0.7981 (0.8072)	0.2163	0.8955 (0.9000)		
5	0.2728	0.7504 (0.7627)	0.2441	0.8091 (0.8196)	0.2031	0.8943 (0.9000)		
6	0.2593	0.7485 (0.7627)	0.3271	0.6660 (0.6753)	0.1931	0.8934 (0.9000)		
7					0.1848	0.8928 (0.9000)		
8					0.1777	0.8922 (0.9000)		
N^*	6		6		8		3	
T_O	1.9950		2.0065		1.8477		1.8684	
R^*_1	0.7627		0.7265					
EMC (N^*, R^*_1)	8.6034		8.5542		9.2988		10.6077	

From the results of this study, we can see that the larger the conditional reliability threshold that we set, the shorter PM cycle and the more PM activities the optimal solution needs. As shown in Table 5, the optimal value N of PM strategy 3 is higher than the optimal ones obtained from strategies 1, 2, or 4. In addition, as R_1 for the PM strategy 3 is set to a high value, the reliability of the system remains high. However, this leads to high PM cost and breakdown cost, since there are frequent PM activities. As shown in Figure 1, the expected total maintenance cost rate of PM strategy 3 is higher than that of PM strategies 1 and 2.

In the cases where the conditional reliability threshold is set to a low value, the length of the PM cycle increases. The result of this is many system breakdowns and only minimal repairs are allowed, and thus, the expected total maintenance cost rate dramatically increases over time. Moreover, we can observe that a trade-off between the reliability threshold and the expected total maintenance cost rate. Therefore, appropriate conditional reliability thresholds are required to establish a cost-effective maintenance strategy.

Unlike the results of PM strategies 3 and 4, PM strategies 1 and 2 focus on finding a balance between the conditional reliability threshold and the expected total maintenance cost rate. Hence, we can see that the results of PM strategies 1 and 2 are more cost-effective than the results of PM strategies 3 and 4. Furthermore, we can see that PM strategy 2 is more cost-effective than PM strategy 1 because PM strategy 2 relaxed the reliability constraints of PM strategy 1. From Table 5, we can see that the optimal conditional reliability thresholds of PM strategy 2 tend to increase slightly except for the value that corresponds to N^*. This means that it is more reasonable to perform PM activities frequently than to replace the system up to ($N^* - 1$) PM cycles. Since the system is then replaced at the N^*th PM activity, the conditional reliability threshold of the last cycle decreases. This implies that operating the system for as long as possible in the last PM cycle is more efficient than performing imperfect PM activity.

4.2. Sensitivity Analyses

Several sensitivity analyses were conducted to investigate how the parameters affected the proposed models. First, to determine how the parameters of the hazard intensity function influence the proposed models, sensitivity analysis was conducted for the situation in which α was adjusted from 1.92 to 2.88 and β was adjusted from 2.88 to 4.32. The results are summarized in Table 6. It is shown that changing of α had no effect on the optimal solutions because the change in α could not reflect the trend in the deterioration process. However, since α was used to calculate the PM cycle, the system lifetime and the expected total maintenance cost rate were affected by the change in α.

The change in β affected the optimal solutions unlike the results of the sensitivity analyses for α because β reflects the shape of the hazard intensity. For example, when β received a low value, the hazard intensity increased from the initial period, resulting in shorter PM intervals. As shown in

Table 6, N^* decreased as the value of β was low. Moreover, $R_1{}^*$ increases as β increases, which implies that the system degradation accelerates when β is high.

Table 6. Results of sensitivity analysis related to the failure rate's parameters.

α	β	Strategy 1					Strategy 2				
		R^*_1	N^*	R_S	T_O	EMC_1 (N^*, R^*_1)	R^*_1	N^*	R_S	T_O	EMC_2 (N^*, R^*_1)
2.08	3.20	0.7627	6	0.7485	2.1391	8.0239	0.7265	6	0.6660	2.1514	7.7980
2.34	3.20	0.7627	6	0.7485	2.0618	8.3248	0.7265	6	0.6660	2.0736	8.2771
2.60	3.20	0.7627	6	0.7485	1.9550	8.6034	0.7265	6	0.6660	2.0065	8.5542
2.86	3.20	0.7627	6	0.7485	1.9364	8.8635	0.7265	6	0.6660	1.9476	8.8128
3.12	3.20	0.7627	6	0.7485	1.8845	9.1079	0.7265	6	0.6660	1.8953	9.0557
2.60	2.56	0.6650	5	0.6351	1.7113	10.0687	0.6251	5	0.5486	1.7213	10.0101
2.60	2.88	0.7128	5	0.6933	1.7342	9.2751	0.6831	5	0.6090	1.7441	9.2224
2.60	3.20	0.7627	6	0.7485	1.9550	8.6034	0.7265	6	0.6660	2.0065	8.5542
2.60	3.52	0.7894	6	0.7798	2.0492	8.0434	0.7595	6	0.7021	2.0604	7.9998
2.60	3.84	0.8190	7	0.8119	2.3400	7.5694	0.7857	7	0.7379	2.3523	7.5298

In addition, we conducted sensitivity analyses to investigate how the change in related costs for maintenance activity affected the proposed model. The results are summarized in Table 7. As the replacement cost increases or the PM cost decreases, the optimal number of PM activities increases. This implies that it is more economical to perform PM activities than to replace the system. As the minimal repair cost increases, the optimal conditional reliability threshold increases. However, the optimal number of PM activities remains unchanged. As the conditional reliability threshold increases, the length of the PM cycle shortens, thereby increasing the expected total maintenance cost rate. Indeed, as shown in Table 7, the optimal conditional reliability threshold and the expected total maintenance cost rate increases as the minimal repair cost increases. Moreover, for both PM strategies, the value of the optimal conditional reliability threshold is significantly sensitive to changes in the minimal repair cost. Therefore, decision-makers should pay close attention to the estimation of minimal repair cost.

Table 7. Results of sensitivity analysis related to maintenance costs.

C_{MR}	C_{PM}	C_{RP}	Strategy 1					Strategy 2				
			R^*_1	N^*	R_S	T_O	EMC_1 (N^*, R^*_1)	R^*_1	N^*	R_S	T_O	EMC_2 (N^*, R^*_1)
2.40	1.00	5.00	0.7181	6	0.7018	2.1241	8.0805	0.6766	6	0.6085	2.1363	8.0342
2.70	1.00	5.00	0.7423	6	0.7271	2.0553	8.3510	0.7036	6	0.6395	2.0671	8.3032
3.00	1.00	5.00	0.7627	6	0.7485	1.9550	8.6034	0.7265	6	0.6660	2.0065	8.5542
3.30	1.00	5.00	0.7801	6	0.7668	1.9415	8.8406	0.7461	6	0.6890	1.9526	8.7900
3.60	1.00	5.00	0.7952	6	0.7826	1.8935	9.0645	0.7631	6	0.7090	1.9044	9.0126
3.00	0.80	5.00	0.7912	7	0.7772	2.1434	8.0756	0.7472	7	0.6991	2.1562	8.0276
3.00	0.90	5.00	0.7715	6	0.7577	1.9682	8.3511	0.7364	6	0.6776	1.9795	8.3033
3.00	1.00	5.00	0.7627	6	0.7485	1.9550	8.6034	0.7265	6	0.6660	2.0065	8.5542
3.00	1.10	5.00	0.7406	5	0.7274	1.7924	8.8454	0.7169	5	0.6453	1.8023	8.7970
3.00	1.20	5.00	0.7325	5	0.7189	1.8127	9.0673	0.7082	5	0.6350	1.8227	9.0177
3.00	1.00	4.00	0.7697	5	0.7577	1.7170	8.0477	0.7482	5	0.6827	1.7265	8.0037
3.00	1.00	4.50	0.7592	5	0.7467	1.7448	8.3365	0.7368	5	0.6691	1.7544	8.2910
3.00	1.00	5.00	0.7627	6	0.7485	1.9550	8.6034	0.7265	6	0.6660	2.0065	8.5542
3.00	1.00	5.50	0.7652	7	0.7497	2.2347	8.8521	0.7168	7	0.6643	2.2481	8.7994
3.00	1.00	6.00	0.7577	7	0.7418	2.2601	9.0746	0.7081	7	0.6543	2.2736	9.0206

5. Conclusions

In this study, we established a cost-effective PM strategy for a repairable system subject to stochastic deteriorations. Two PM strategies were proposed and modeled by two reliability constraints that help evaluate the trade-off between the expected total maintenance cost rate and the system reliability. Strategy 1 constricted the conditional reliability threshold with the same level, whereas

Strategy 2 relaxed the reliability constraints of Strategy 1 by using different level of the conditional reliability threshold. Moreover, this study discussed the difference between the conditional reliability threshold and the system reliability threshold via two definitions. The conditional reliability threshold was used as a condition variable that represents the decreasing system reliability as the number of PM activities increases.

This study employed a hybrid hazard rate model to represent imperfect PM activities. The function of the expected total maintenance cost rate was determined, including the costs of PM activity, replacement, minimal repair, and breakdown. We provided the structural properties of the proposed PM strategies via four propositions and proved their existence and uniqueness of the propositions. Two algorithms were proposed to find optimal solutions. A numerical example was conducted to illustrate the proposed PM strategy. Sensitivity analyses were also conducted to investigate how the parameters of the proposed PM strategy affect the optimal solutions.

5.1. Contributions to Theory

The contributions of this study are as follows:

(1) This study identifies and defines the differences between the system reliability threshold and the conditional reliability threshold. This contributes to the reliability theory.
(2) This study develops a cost-effective PM model using the conditional reliability threshold that considers the decreasing system reliability as the number of PM activities increases, which can relax the trade-off between the reliability and the cost.
(3) This study provides the optimal conditions for each strategy through four propositions and one lemma that show the uniqueness and existence of the optimal solution.

5.2. Implications for the Decision-Makers

Decision makers need to decide whether they undertake all the maintenance activities to keep the system in good condition or reduce maintenance activities in planning PM policy to cut down the costs. For example, if they reduce the costs related to maintenance activities in a PM strategy, it is difficult to obtain high quality of the system due to the insufficient PM activities. On the other hand, frequent PM activities result in higher maintenance costs and poorer system availability. The proposed model can be a good alternative in solving these problems. Based on the proposed model, they can avoid both the extremely frequent maintenance activities and the insufficient maintenance ones effectively.

5.3. Limitations and Further Research

This study has some limitations. First, the parameters of hazards rate function are not estimated based on data but predetermined manually. Second, we need more constraints to apply the model in real word applications because the proposed model is applied to complex and expensive systems. For example, the optimal PM strategies in this work may be expanded to the system under lease, considering the various constraints for the reliability such as the maximum reliability, the minimum reliability, and the reliability at the end of lease. To tackle these limitations, we can estimate the parameters based on the data and add more constraints to handle the complex and difficult situations in real world applications.

For the next step of the research, we can also extend the proposed model to apply to many different areas that cover multi component systems. The expected result is a novel algorithm that not only solves the integrated optimization problem of the proposed PM strategies but also maintains production costs and requirements. In addition, another next step of this study includes solving the two dimensional warranty problem of the proposed model while considering the warranty cost and the numerous uncertainties.

Author Contributions: Conceptualization, writing—original draft preparation, formal analysis, visualization, J.L.; methodology, J.L and S.A.; supervision, writing—review and editing, B.K. and S.A.; data curation, B.K. and J.L.; funding acquisition, project administration, S.A.

Appendix A

Appendix A.1 Proof of Proposition 1

When R_1 is fixed, the expected total maintenance cost rate becomes a univariate function. Inequalities (23)–(24) are a necessary condition to obtain the optimal number of PM activities. Inequality (24) is obtained from substituting Equation (22) into Inequality (23). If N^* satisfies Inequality (24), it becomes a finite and unique optimal solution such that $Min(ECR_1(N), N = 1,2,\dots) = ECR_1(N^*)$. Now, to prove Proposition 1, we recall $K_1(N)$, which is as follows:

$$K_1(N) = \left(\frac{\sum_{i=1}^{N} x_i}{x_{N+1}} - N\right)(C_{PM} + C_{BD} - (C_{MR} + C_{BD}) \ln R_1). \tag{A1}$$

The left hand side of $K_1(N)$ increases as N increases because x_{N+1} decrease to zero. Moreover, the right hand side of $K_1(N)$ is a positive because $-\ln R_1$ is a positive value. Hence, $K_1(N)$ in Inequality (24) is the increasing function with respect to N, and N^* that satisfies Inequality (24) is then a finite and unique optimal solution that minimizes Equation (22).

Appendix A.2 Proof of Proposition 2

When the number of PM activities is fixed, the condition to obtain the optimal conditional reliability threshold that minimizes Equation (22) is as follows:

$$\frac{dECR_1(R_1|N)}{dR_1} = 0. \tag{A2}$$

Solving Equation (A2) with R_1, we obtained Equation (26). Moreover, Equation (26) is a finite and unique solution minimizes Equation (22) because $d^2ECR_1(R_1|N)/d^2R_1 > 0$.

Appendix A.3 Proof of Lemma 1

When the number of PM activities is fixed, the conditions to obtain the optimal conditional reliability threshold that minimizes Equation (17) are as follows:

$$\begin{cases} \frac{dECR_2(\underline{R}=(R_1,R_2,\cdots,R_N)|N)}{dR_1} = 0 \\ \vdots \\ \frac{dECR_2(\underline{R}=(R_1,R_2,\cdots,R_N)|N)}{dR_N} = 0 \end{cases}. \tag{A3}$$

Equation (A3) yields the following relationships:

$$\alpha\beta(C_m + C_{bd})B_{i-1}\left(\frac{dT_o}{dR_i}\left(Y_N + \sum_{i=1}^{N-1}(1 - a_i Y_i)\right)\right) = ECR_2(\underline{R} = (R_1, R_2, \cdots, R_N)|N), \tag{A4}$$

for $i = 1, 2, \ldots, N$. Solving the simultaneous equation in Equation (A4), we have the following relationships:

$$\begin{aligned} S(N) &= \left[\frac{1}{B_{N-1}} \frac{B_{N-2}-a_{N-1}^{\beta}B_{N-1}}{1-a_{N-1}}\right]^{\frac{\beta}{\beta-1}} S(N-1) \\ &\vdots \\ &= \left[\frac{1}{B_{N-1}} \frac{B_1-a_2^{\beta}B_2}{1-a_2}\right]^{\frac{\beta}{\beta-1}} S(2) \\ &= \left[\frac{1}{B_{N-1}} \frac{1-a_1^{\beta}B_1}{1-a_1}\right]^{\frac{\beta}{\beta-1}} S(1) \end{aligned} \tag{A5}$$

where $S(1) = -\ln R_1$ and

$$S(i) = -B_{i-1}^{-1}\ln R_i - \sum_{j=1}^{i-1}\left(\prod_{k=j}^{i-1}\rho_k^{\beta}\right)B_{j-1}^{-1}\ln R_j,\ \text{for } i = 2, 3, \ldots, N. \tag{A6}$$

Using the relationships in Equation (A5), we obtained Equation (28).

Appendix A.4 Proof of Proposition 3

Using the results of Lemma 1, Equation (17) becomes Equation (30). To prove Proposition 3, we redefine $K_2(N)$ in Inequality (33) as follows:

$$K_2(N) = A(N) + B(N)C(N), \tag{A7}$$

where

$$\begin{aligned} A(N) &= \left(\frac{\sum_{i=1}^{N} x_i}{x_{N+1}} - N\right)(C_{PM} + C_{BD}), \\ B(N) &= -(C_{MR} + C_{BD})\frac{\ln R_1}{x_{N+1}}, \\ C(N) &= \sum_{i=1}^{N} x_i(Mr(N+1) - Mr(N)) - Mr(N)x_{N+1}. \end{aligned} \tag{A8}$$

$A(N)$ and $B(N)$ increase as N increase, and $C(N)$ increases as N increases because x_{N+1} decreases to zero. Hence, there exists a finite and unique N^* that minimizes Equation (30).

Appendix A.5 Proof of Proposition 4

Using the results of Lemma 1, Equation (17) becomes Equation (30), and then, Equation (34) is a finite and unique solution that minimizes Equation (30) because $d^2ECR_2(R_1|N)/d^2R_1 > 0$.

References

1. Lee, J.; Park, J.; Ahn, S. On determining a non-periodic preventive maintenance schedule using the failure rate threshold for a repairable system. *Smart Struct. Syst.* **2018**, *22*, 151–159.
2. Kilic, E.; Ali, S.S.; Weber, G.W.; Dubey, R. A value-adding approach to reliability under preventive maintenance costs and its applications. *Optimization* **2014**, *63*, 1805–1816. [CrossRef]
3. Velmurugan, R.; Dhingra, T. Maintenance strategy selection and its impact in maintenance function: A conceptual framework. *Int. J. Oper. Prod. Manag.* **2015**, *35*, 1622–1661. [CrossRef]
4. Ilangkumaran, M.; Kumanan, S. Application of hybrid VIKOR model in selection of maintenance strategy. *Int. J. Inf. Syst. Supply Chain Manag.* **2012**, *5*, 59–81. [CrossRef]
5. Lee, J.; Ni, J.; Djurdjanovic, D.; Qiu, H.; Liao, H. Intelligent prognostics tools and e-maintenance. *Comput. Ind.* **2006**, *57*, 476–489. [CrossRef]
6. Ahn, S.; Kim, W. On determination of the preventive maintenance interval guaranteeing system availability under a periodic maintenance policy. *Struct. Infrastruct. Eng.* **2011**, *7*, 307–314. [CrossRef]
7. Murthy, D. A note on minimal repair. *IEEE Trans. Reliab.* **1991**, *40*, 245–246. [CrossRef]
8. Yanez, M.; Joglar, F.; Modarres, M. Generalized renewal process for analysis of repairable systems with limited failure experience. *Reliab. Eng. Syst. Saf.* **2002**, *77*, 167–180. [CrossRef]

9. Malik, M.A.K. Reliable preventive maintenance scheduling. *AIIE Trans.* **1979**, *11*, 221–228. [CrossRef]
10. Lin, D.; Zuo, M.J.; Yam, R.C. Sequential imperfect preventive maintenance models with two categories of failure modes. *Nav. Res. Logist.* **2001**, *48*, 172–183. [CrossRef]
11. El-Ferik, S.; Ben-Daya, M. Age-based hybrid model for imperfect preventive maintenance. *IIE Trans.* **2006**, *38*, 365–375. [CrossRef]
12. Khatab, A. Hybrid hazard rate model for imperfect preventive maintenance of systems subject to random deterioration. *J. Intell. Manuf.* **2015**, *26*, 601–608. [CrossRef]
13. Huang, Y.-S.; Chen, E.; Ho, J.-W. Two-dimensional warranty with reliability-based preventive maintenance. *IEEE Trans. Reliab.* **2013**, *62*, 898–907. [CrossRef]
14. Sheu, S.-H.; Chang, C.-C. An extended periodic imperfect preventive maintenance model with age-dependent failure type. *IEEE Trans. Reliab.* **2009**, *58*, 397–405. [CrossRef]
15. Yeh, R.H.; Kao, K.-C.; Chang, W.L. Preventive-maintenance policy for leased products under various maintenance costs. *Expert Syst. Appl.* **2011**, *38*, 3558–3562. [CrossRef]
16. Schutz, J.; Rezg, N. Maintenance strategy for leased equipment. *Comput. Ind. Eng.* **2013**, *66*, 593–600. [CrossRef]
17. Mabrouk, A.B.; Chelbi, A.; Radhoui, M. Optimal imperfect maintenance strategy for leased equipment. *Int. J. Prod. Econ.* **2016**, *178*, 57–64. [CrossRef]
18. Jung, K.M.; Park, M.; Park, D.H. Cost optimization model following extended renewing two-phase warranty. *Comput. Ind. Eng.* **2015**, *79*, 188–194. [CrossRef]
19. Liao, G.-L. Optimal economic production quantity policy for a parallel system with repair, rework, free-repair warranty and maintenance. *Int. J. Prod. Econ.* **2016**, *54*, 6265–6280. [CrossRef]
20. Park, M.; Jung, K.M.; Park, D.H. Optimization of periodic preventive maintenance policy following the expiration of two-dimensional warranty. *Reliab. Eng. Syst. Saf.* **2018**, *170*, 1–9. [CrossRef]
21. Sett, B.K.; Sarkar, S.; Sarkar, B. Optimal buffer inventory and inspection errors in an imperfect production system with preventive maintenance. *Int. J. Adv. Manuf. Technol.* **2017**, *90*, 545–560. [CrossRef]
22. Nakagawa, T. Sequential imperfect preventive maintenance policies. *IEEE Trans. Reliab.* **1988**, *37*, 295–298. [CrossRef]
23. Sarkar, B.; Sana, S.S.; Chaudhuri, K. Optimal reliability, production lotsize and safety stock: An economic manufacturing quantity model. *Int. J. Ind. Eng. Manag. Sci.* **2010**, *5*, 192–202. [CrossRef]
24. Sarkar, B.; Sana, S.S.; Chaudhuri, K. Optimal reliability, production lot size and safety stock in an imperfect production system. *Int. J. Math. Oper. Res.* **2010**, *2*, 467–490. [CrossRef]
25. Zhao, Y. On preventive maintenance policy of a critical reliability level for system subject to degradation. *Reliab. Eng. Syst. Saf.* **2003**, *79*, 301–308. [CrossRef]
26. Zhou, X.; Xi, L.; Lee, J. Reliability-centered predictive maintenance scheduling for a continuously monitored system subject to degradation. *Reliab. Eng. Syst. Saf.* **2007**, *92*, 530–534. [CrossRef]
27. Liao, W.; Pan, E.; Xi, L. Preventive maintenance scheduling for repairable system with deterioration. *J. Intell. Manuf.* **2010**, *21*, 875–884. [CrossRef]
28. Doostparast, M.; Kolahan, F.; Doostparast, M. A reliability-based approach to optimize preventive maintenance scheduling for coherent systems. *Reliab. Eng. Syst. Saf.* **2014**, *126*, 98–106. [CrossRef]
29. Lin, Z.-L.; Huang, Y.-S.; Fang, C.-C. Non-periodic preventive maintenance with reliability thresholds for complex repairable systems. *Reliab. Eng. Syst. Saf.* **2015**, *136*, 145–156. [CrossRef]
30. Khatab, A.; Diallo, C.; Sidibe, I. Optimizing upgrade and imperfect preventive maintenance in failure-prone second-hand systems. *J. Manuf. Syst.* **2017**, *43*, 58–78. [CrossRef]
31. Khatab, A. Maintenance optimization in failure-prone systems under imperfect preventive maintenance. *J. Intell. Manuf.* **2018**, *29*, 707–717. [CrossRef]
32. Ebeling, C.E. *An Introduction to Reliability and Maintainability Engineering*; Tata McGraw-Hill Education: New York, NY, USA, 2004.

10

Application of Optimization to Select Contractors to Develop Strategies and Policies for the Development of Transport Infrastructure

Chia-Nan Wang [1,2], Tien-Muoi Le [1,*] and Han-Khanh Nguyen [3]

1 Department of Industrial Engineering and Management, National Kaohsiung University of Sciences and Technology, Chien Kung Campus, 415 Chien Kung Road, Kaohsiung 807, Taiwan; cn.wang@fotech.edu.tw
2 Department of Industrial Engineering and Management, Fortune Institute of Technology, Kaohsiung 83160, Taiwan
3 Faculty of Economics, Thu Dau Mot University, Number 6, Tran Van On Street, Phu Hoa Ward, Thu Dau Mot City 820000, Binh Duong Province, VietNam; khanhnh@tdmu.edu.vn
* Correspondence: letienmuoinkust2018@gmail.com.

Abstract: Many factors influence the efficiency and quality of transport works. In particular, consultants and construction contractors of these works play important roles, and critical factors directly affect the quality of traffic works. If the quality of consultancy and construction is good, the project will reduce the total investment; if the contractor is good, the completion time of the new project is guaranteed, thus reducing construction costs. The longer the construction time is, the higher the cost of the project. In this study, the authors used optimal algorithms to evaluate past, present, and future contractors' technical, technological, and performance effectiveness. Research results show that bidders are divided into three groups: highly effective bidders, stable contractors, and inefficient groups. Research results for this subject will help the government, regulatory agencies, and investors select good contractors as the basis for developing strategies and policies for the development of transport infrastructure.

Keywords: transport infrastructure; develop strategies; optimal algorithms; business performance

1. Introduction

Contractor Selection Procedure: There are currently many models for selecting contractors for transport infrastructure projects, in which the best bidder selection model by Jyh-Bin Yang and Wei-Chih Wang is utilized (Figure 1).

Choosing a contractor for construction of traffic works is an important issue. Therefore, the contractor must have sufficient professional capacity, human resources, and financial capacity. Depending on the scale, the nature and sources of capital for the construction of traffic works are considered when selecting contractors. In particular, the selection of contractors for transport works is divided into three main forms as follows [2]:

Open bidding: Used to select contractors for construction works in the form of unlimited number of participating contractors.

Restricted bidding: To be used for the selection of contractors for construction of transport works; there are only a number of contractors that meet all the conditions in regard to the capability of construction activities and the capability of practicing the construction of works. Traffic is invited to participate in bidding.

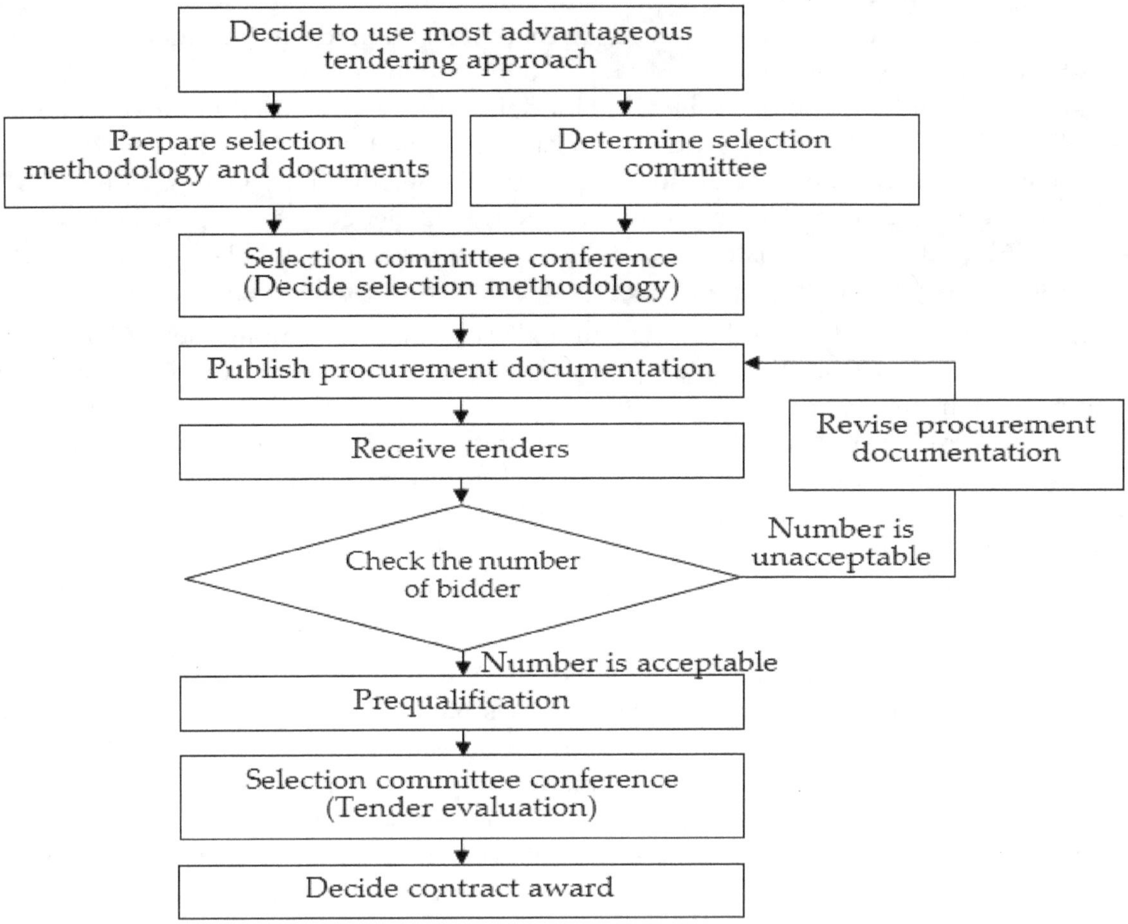

Figure 1. Contractor selection procedure [1].

Appointment of contractor: Investment deciders or investors of transport works are entitled to designate contractors directly qualified for construction activities and are capable of practicing construction of traffic works at a reasonable cost.

Each traffic works in different areas require different qualifications, construction techniques, and experience. In order to ensure that construction works are designed to bring economic efficiency for the investor in particular and the community in general, the selection of contractors is made in accordance with the complexity of each project. In this study, the authors use optimal algorithms to forecast and evaluate the economic, technical, and technological efficiency of contractors in the field of construction of traffic works during the period from 2014–2021. This research provides government, managers, and investors with a solid foundation in the development of strategies and policies for the development of transport infrastructure. The results of this research are important to consider when selecting contractors for construction of traffic works in the future.

2. Literature Review

2.1. Related Research

Nowadays, optimal mathematical models have become popular tools for researchers. In particular, there have been many high-impact research projects serving the macroeconomic management decisions of governments and decisions of investors in strategies and policies for developing products for businesses. In particular, research uses forecasting methods and modern data analysis techniques to find the best solutions for businesses that are of special interest to researchers worldwide. In particular, in 2006, Zhou, Ang, and Poh [3] used the Grey model to forecast electricity demand in Singapore. In 2012, Ze Zhao, Jianzhou Wang, Jing Zhao, and Zhongyue Su [4] used the Grey model to forecast annual net income per capita of rural households in China. In 2010, Guang-Ming Shi, Jun Bi, and Jin-Nan Wang [5] evaluated China's industrial energy efficiency based on DEA models. In 2007, Matthias Staat [6] used DEA models to evaluate the performance of hospitals in Germany. There are also many researchers who combine predictive models and performance assessments.

These studies have yielded good results and high applicability. Among them, Chia-Nan Wang, Han-Khanh Nguyen, and Ruei-Yuan Liao [7] combined the Grey model and DEA model to find a strategic partner in the supply chain of the textile and garment industry in Vietnam. In 2011, Jia-Jane Shuai and Wei-Wen Wu [8] used the DEA model and Grey model to evaluate marketing effectiveness through the website of hotels in Taiwan. In those studies, DEA models were used to analyze and evaluate the business situation of enterprises and economic sectors of countries, together with the forecasting models to forecast the development trend of these enterprises, from which to make appropriate decisions. The results show that these studies have yielded good results and achievements, making an important contribution to the economic growth of businesses and countries.

In the construction industry, the application of optimal mathematical models to find solutions to minimize costs and maximize profits for businesses has also been made by many researchers. In this study, the authors used the GM (1, 1) model to forecast the business situation of contractors of traffic works in Vietnam. Optimal mathematical models are used to simultaneously evaluate the three indicators of technical efficiency, technology efficiency, and business performance of past, present, and future contractors, which give the government, regulators, investors, and business leaders the basis for developing strategies and policymaking for the development of transport infrastructure. This is a new and important application in the selection of qualified contractors for the construction of highly efficient and economically efficient transport works.

2.2. Overview of Transport Infrastructure in Vietnam

Roads: Vietnam's transport infrastructure has changed dramatically in recent years. Many large and modern works have been put into operation. Especially, the road infrastructure has been built and improved. The overview of road transport infrastructure in Vietnam is shown in Table 1.

Table 1. Road infrastructure.

Order	Type of Roads	Total Route	Total Length (km)
1	Highways	13	745
2	Route	146	23,816
3	Provincial road	998	27,176
4	District roads	8680	57,294
5	Commune roads	61,402	173,294
6	Urban road	23,495	27,910
7	Other rural roads	168,888	256,377
8	Specialized road	2476	8528

Source: [9].

Railway: The Vietnam national railway infrastructure is as follows [9]:

- Total length of railway: 3161 km.
- Terminal area: 2,029,837 m^2.
- Leased area: 1,316,175 m^2.

Seaway: Maritime transport infrastructure is as follows: There are 44 seaports in the country (14 seaports of type I and IA, 17 seaports of type II, 13 offshore ports of type III). There are 254 harbors with 59.4 km long wharfs and total design capacity of about 500 million tons per year [9].

Inland waterways: At present, Vietnam has 45 inland waterways with a total length of about 7075 km [9].

- Signal systems on the route: 12,539 signaling towers, 18,458 signaling beams, 3070 signaling beams, and 9153 signaling lights.
- Bridges over the route: 251/532 bridges and river crossings located on national inland waterway lanes with less than technical specifications as approved.
- Inland waterway system: By the end of August 2017, Vietnam had 277 ports, of which 220 ports are on national inland waterways and 57 on local inland waterways.

By air: Currently, Vietnam has 21 airports in operation, including eight international airports and 13 domestic airports [9].

3. Data and Methodology

3.1. Data

3.1.1. Contractors Collection

The data collection of contractors to carry out this study plays an important and practical role in the development of strategies and policies for the development of transport infrastructure. Therefore, the authors pay special attention to the conditions of the use of optimal mathematical models. The selected bidders must have complete, continuous data, along with extensive experience in the design and construction of traffic works. These contractors must meet the human, financial, and property requirements. Through the Website of the General Statistics Office, the authors compiled 17 contractors to satisfy the conditions of this study (Table 2).

Table 2. List of contractors.

I_n	Companies
I_1	Construction joint stock company No. 1
I_2	HUD3 investment and construction joint stock company
I_3	Construction joint stock company No. 9
I_4	Coteccons construction joint stock company
I_5	Hoa Binh construction group joint stock company
I_6	Binhduong trade and development joint stock company
I_7	Development investment construction Hoi An joint stock company
I_8	Licogi 14 joint stock company
I_9	Tien Giang investment and construction joint stock company
I_{10}	Construction joint stock company No. 5
I_{11}	Chuong Duong joint stock company
I_{12}	Constrexim No 8 investment and construction joint stock company
I_{13}	Vietnam Construction joint stock company No. 2
I_{14}	Binh Duong construction and civil engineering joint stock company
I_{15}	HUD1 investment & construction joint stock company
I_{16}	Vinaconex 6 joint stock company
I_{17}	Sci E&C joint stock company

Source: [10].

3.1.2. Elements Selection

Considering the characteristics in the business of construction contractors of transport infrastructure with long production cycles, the place of production is frequently changed according to the works, and the production value of the design activities and construction is large. However, contractors still follow basic economic models. Therefore, in this study, the authors use four inputs and two outputs to analyze the economic, technical, and technological performance of the contractor. These inputs include

Total assets (F_1) are the value of all assets of an enterprise, including tangible assets such as buildings, machinery, equipment, materials, products, and goods, along with other intangible assets such as computer software, patent, commercial advantage, copyright, etc.

Equity (F_2) is the source of capital that makes up the assets of an enterprise that is contributed by the owner and investor or formed from the results of its business.

Cost of sales (F_3) is the total cost of producing a finished product. For a contractor to build consultancy and construction of transportation works, the cost of goods sold is the total cost needed to complete the works or services that the contractor receives for consultancy, design, and construction (purchase prices from suppliers, shipping, insurance, etc.).

Enterprise cost management (F_4) reflects the total administrative expense allocated to the finished product and goods sold in the enterprise's reporting year.

The outputs include

After-tax profit (F_5): The total value of a business's annual turnover, determined by the net of operating profit and other profits subtracting the cost corporate income tax.

Net sales (F_6): Represents the total sales of products, goods, and services in the reporting year of the enterprise.

The inputs and outputs used by these authors fully reflect the assets, costs, and profits of the contractors. After compiling data in local currency (VND), the authors convert to international currency (USD). All data are summarized in Tables 3–6.

Table 3. Data in 2014.

I_n	F_1	F_2	F_3	F_4	F_5	F_6
I_1	27,619.71	10,642.98	18,457.90	937.68	515.42	20,056.04
I_2	28,709.94	7639.48	13,874.53	829.40	555.53	15,705.79
I_3	69,303.22	8152.75	25,489.89	1355.45	492.23	29,433.26
I_4	213,612.52	111,010.37	310,891.73	9433.09	15,701.84	335,310.80
I_5	254,920.91	43,753.02	136,519.46	12,303.89	3021.07	154,538.01
I_6	275,092.39	51,467.31	61,864.00	2470.02	4951.20	76,694.83
I_7	8965.29	2328.10	11,151.77	415.67	132.48	11,794.35
I_8	23,157.62	2239.67	7996.86	179.96	830.19	9456.81
I_9	17,820.80	6940.10	19,197.12	1141.27	747.30	23,063.05
I_{10}	96,207.60	14,653.46	57,300.18	956.57	1465.09	61,143.84
I_{11}	35,509.26	11,122.29	9447.72	526.14	348.99	11,005.28
I_{12}	4032.71	1130.60	5697.05	100.94	57.37	5866.17
I_{13}	70,769.76	11,952.00	23,135.13	2340.04	729.60	26,622.03
I_{14}	41,732.93	15,413.33	25,252.74	273.30	941.02	26,854.88
I_{15}	32,607.10	7802.01	30,375.07	986.96	334.76	31,927.61
I_{16}	23,968.93	4944.83	24,042.54	874.82	342.88	25,566.41
I_{17}	12,052.33	2929.92	8139.35	582.72	357.69	9618.24

Source: [10].

Table 4. Data in 2015.

I_n	F_1	F_2	F_3	F_4	F_5	F_6
I_1	25,427.87	10,544.98	15,046.55	546.92	524.69	16,143.51
I_2	27,240.89	7804.60	16,308.62	935.70	690.64	18,736.61
I_3	58,661.12	8387.84	26,583.99	1288.07	486.56	33,167.85
I_4	343,282.14	142,430.07	551,576.25	15,936.92	32,188.75	600,414.22
I_5	320,271.10	47,193.66	210,066.59	5260.52	3638.61	223,054.97
I_6	323,038.50	52,527.06	50,531.61	2336.35	4789.37	68,593.61
I_7	9782.04	2316.94	9064.47	454.06	131.64	9783.93

Table 4. *Cont.*

I_n	F_1	F_2	F_3	F_4	F_5	F_6
I_8	21,538.00	2982.72	26,301.02	372.01	896.48	28,106.62
I_9	26,452.39	8868.83	24,122.09	1319.39	2460.31	30,724.63
I_{10}	99,017.47	13,671.11	59,662.10	788.29	1571.26	62,866.42
I_{11}	35,416.75	11,524.74	12,789.79	594.80	753.06	15,497.41
I_{12}	3941.09	1127.26	4227.08	115.57	43.88	4402.08
I_{13}	68,716.47	12,154.76	26,534.48	1686.70	649.75	29,570.57
I_{14}	78,950.45	15,484.49	20,196.87	307.35	1107.01	22,297.69
I_{15}	27,798.57	7888.80	26,135.77	892.70	368.32	27,642.07
I_{16}	23,237.08	4712.94	22,131.74	746.43	102.21	23,132.23
I_{17}	15,926.25	5473.47	10,984.45	731.45	617.42	12,838.24

Source: [10].

Table 5. Data in 2016.

I_n	F_1	F_2	F_3	F_4	F_5	F_6
I_1	35,109.27	10,485.68	22,603.28	1134.68	564.14	24,390.61
I_2	26,796.59	7970.64	22,050.17	1076.09	703.64	24,479.21
I_3	60,403.74	8407.83	34,716.18	1268.74	609.55	37,280.20
I_4	515,723.84	273,815.34	833,852.12	13,056.99	62,468.41	912,891.79
I_5	502,938.31	80,390.64	418,775.39	16,586.71	24,950.27	472,911.18
I_6	319,702.35	53,657.30	45,777.73	2306.88	5634.81	63,677.86
I_7	10,255.38	2320.89	9588.46	493.99	147.59	10,267.46
I_8	19,557.49	3960.06	7722.72	374.11	1183.62	9963.19
I_9	30,463.49	12,107.59	28,276.70	1286.45	3799.47	36,441.09
I_{10}	87,299.68	14,039.25	61,030.95	945.63	1841.62	64,615.23
I_{11}	32,274.41	12,500.76	11,130.81	568.00	710.45	12,491.93
I_{12}	4615.74	1132.18	5321.93	125.01	42.96	5497.76
I_{13}	111,536.69	12,839.03	39,513.77	2294.98	1322.29	45,818.27
I_{14}	55,524.62	15,639.86	36,510.14	341.96	678.34	37,868.28
I_{15}	28,725.27	7715.34	15,890.36	485.60	142.01	16,929.51
I_{16}	30,738.16	4943.11	30,534.96	847.72	332.38	31,949.23
I_{17}	24,888.24	5590.88	13,841.11	1338.85	177.02	15,710.31

Source: [10].

Table 6. Data in 2017.

I_n	F_1	F_2	F_3	F_4	F_5	F_6
I_1	35,716.50	10,548.01	24,608.44	1811.97	666.61	27,375.57
I_2	32,613.60	8584.32	14,505.26	1242.48	842.67	17,305.95
I_3	74,012.57	8369.17	43,573.89	1199.83	553.81	46,708.38
I_4	697,419.42	320,950.06	1,104,166.34	17,331.82	72,594.78	1,192,729.49
I_5	614,888.03	108,362.54	630,370.59	20,780.77	37,798.35	704,445.13
I_6	345,556.56	54,166.49	40,539.08	2031.60	5903.72	59,339.64
I_7	10,254.63	2289.53	7539.38	476.64	105.86	8201.51
I_8	13,150.60	6368.54	9517.91	658.88	2767.79	14,748.52
I_9	37,817.72	14,129.26	30,026.08	2179.23	4033.59	39,965.81
I_{10}	88,450.18	15,173.54	81,247.45	2393.50	2634.74	86,402.59
I_{11}	41,387.15	12,037.26	8806.98	599.72	1272.61	10,050.29
I_{12}	4187.99	1135.78	5589.87	135.07	43.22	5772.25
I_{13}	99,261.08	13,428.69	81,743.76	3821.78	1348.60	92,106.15
I_{14}	50,846.69	15,659.84	24,190.22	410.00	1286.71	25,434.63
I_{15}	42,474.18	7645.81	22,175.67	749.19	230.35	23,825.16
I_{16}	35,662.03	4940.52	25,106.20	795.23	329.79	26,310.29
I_{17}	32,817.24	6018.81	35,321.30	1315.83	673.69	38,384.97

Source: [10].

3.2. Methodology

3.2.1. Grey Forecasting Model

Deng [11] first introduced the grey system theory in 1982. Since its introduction, grey theories have been used extensively in statistics. Models in this theory require only a limited amount of data in the past to predict and estimate future values [11]. So far, grey system theory has been applied in almost all fields of economics, finance, science and industry, and transportation.

In models, the grey system theory is referred to as GM (1, 1); GM (1, *n*); GM (2, 1); GM (2, *n*). The general form is GM (*n*, *m*), where *n* is the order of the differential equations, and *m* is the number of variables. There are many models, but the most commonly used model is GM (1, 1), as this model is an accurate prediction model.

In this study, the steps in the GM (1, 1) model were carried out in the following steps [11]:
From the original data range:

$$\mathrm{X}^{(0)} = \left(x_{(1)}^{(0)}, x_{(2)}^{(0)}, x_{(3)}^{(0)}, \ldots, x_{(n)}^{(0)}\right); n \geq 4. \tag{1}$$

Use the accumulating generation operation (AGO) method to compute $\mathrm{X}^{(1)}$ values [11]:

$$\mathrm{X}^{(1)} = \left(x_{(1)}^{(1)}, x_{(2)}^{(1)}, x_{(3)}^{(1)}, \ldots, x_{(n)}^{(1)}\right); n \geq 4. \tag{2}$$

In which,

$$\begin{cases} x_{(1)}^{(0)} = x_{(1)}^{(1)} \\ x_{(k)}^{(1)} = \sum\limits_{i=1}^{k} x_{(i)}^{(0)} \end{cases}, (k = 1,\ 2,\ 3, \ldots,\ n) \tag{3}$$

Calculate the mean values $\mathrm{Z}^{(1)}$:

$$\mathrm{Z}^{(1)} = \left(z_{(2)}^{(1)}, z_{(3)}^{(1)}, z_{(4)}^{(1)}, \ldots, z_{(n)}^{(1)}\right); n \geq 4. \tag{4}$$

In which,

$$\mathrm{Z}_{(n)}^{(1)} = \frac{x_{(k)}^{(1)} + x_{(k-1)}^{(1)}}{2}, k = 1,\ 2,\ 3,\ 4 \ldots; n = \ 2,\ 3,\ 4 \tag{5}$$

From the values of $\mathrm{X}^{(0)}$, $\mathrm{Z}^{(1)}$ the authors obtain the following system of equations:

$$\begin{cases} x_{(2)}^{(0)} + a \times z_{(2)}^{(1)} = b \\ x_{(3)}^{(0)} + a \times z_{(3)}^{(1)} = b \\ x_{(4)}^{(0)} + a \times z_{(4)}^{(1)} = b \end{cases} \tag{6}$$

From the above equations, transformed into the matrix form as follows:

$$B = \begin{bmatrix} -z_{(2)}^{(1)} \\ \ldots\ \ldots \\ -z_{(n)}^{(1)} \end{bmatrix}; Y_N = \begin{bmatrix} x_{(2)}^{(0)} \\ \ldots.. \\ x_{(n)}^{(0)} \end{bmatrix} \tag{7}$$

$$\hat{a} = \begin{bmatrix} a \\ b \end{bmatrix} = \frac{1}{(B^T B)} B^T Y_N \tag{8}$$

Use the values a and b to find the differential equation:

$$\frac{dx_{(k)}^{(1)}}{dt} + ax^{(0)} = b \tag{9}$$

Equation prediction:

$$\hat{x}_{(k+1)}^{(1)} = (\hat{x}_{(1)}^{(0)} - \frac{b}{a}) \times e^{-(ak)} + \frac{b}{a} \tag{10}$$

Use the accumulating generation operation method to calculate predictive values:

$$\hat{x}_{(k+1)}^{(0)} = \hat{x}_{(k+1)}^{(1)} - \hat{x}_{k}^{(1)} \tag{11}$$

The GM (1, 1) model is used in this study to predict the business performance of contractors consulting, designing, and executing traffic infrastructure works for the period 2018–2021.

3.2.2. Mean Absolute Percentage Error

To test the predictive accuracy, the authors used the mean absolute error (ε) calculated according to the formula and convention below [12].

$$\varepsilon = \frac{1}{n}\sum_{i=1}^{n} \frac{|\lambda_i - \hat{\lambda}_i|}{\lambda_i} \tag{12}$$

$$\begin{cases} (\varepsilon) \leq 10\% : \text{Excellent}, 10\% < (\varepsilon) \leq 20\% : \text{Good} \\ 20\% < (\varepsilon) \leq 50\% : \text{Qualified}, (\varepsilon) > 50\% : \text{Unqualified} \end{cases}$$

3.2.3. Malmquist Model

Many researchers have used the Malmquist productivity index (EMPI) to evaluate the effectiveness of many areas of the world [13–18]. In this study, the authors evaluated the technical, technology, and business efficiency of the consultants, design, and construction contractors of traffic infrastructure between the two periods $P(x_k^t, y_k^t)$ and $Q(x_k^{t+1}, y_k^{t+1})$ (as shown in Figure 2). In it, the catch-up index (ECA) is used to evaluate the technical efficiency of the contractor. The frontier-shift index (EFR) is used to evaluate the technology efficiency of the contractors. The Malmquist productivity index (EMPI) is used to evaluate the performance of contractors.

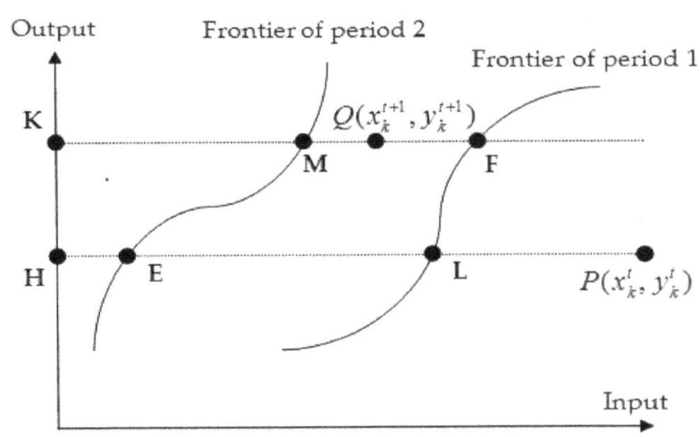

Figure 2. Catch-up terms [18].

The ECA, EFR, and EMPI indexes are used to evaluate the technical efficiency, technology efficiency, and business performance of investors at time (t + 1) is the point $Q(x_k^{t+1}, y_k^{t+1})$, score for where t is the point $P(x_k^t, y_k^t)$ is calculated by the following formula.

$$\text{ECA} = \frac{\text{KM}}{\text{KQ}} \times \frac{\text{HP}}{\text{HL}} = \left[\frac{D_k^{t+1}(x_k^{t+1}, y_k^{t+1})}{D_k^t(x_k^t, y_k^t)}\right] \tag{13}$$

$$\text{EFR} = \sqrt{\frac{\text{HL}}{\text{HE}} \times \frac{\text{KF}}{\text{KM}}} = \left[\frac{D_k^t(x_k^{t+1}, y_k^{t+1})}{D_k^{t+1}(x_k^{t+1}, y_k^{t+1})} \times \frac{D_k^t(x_k^t, y_k^t)}{D_k^{t+1}(x_k^t, y_k^t)}\right]^{\frac{1}{2}} \tag{14}$$

$$\begin{aligned} \text{EMPI} &= \text{ECA} \times \text{EFR} \\ &= \left[\frac{D_k^{t+1}(x_k^{t+1}, y_k^{t+1})}{D_k^t(x_k^t, y_k^t)}\right] \times \left[\frac{D_k^t(x_k^{t+1}, y_k^{t+1})}{D_k^{t+1}(x_k^{t+1}, y_k^{t+1})} \times \frac{D_k^t(x_k^t, y_k^t)}{D_k^{t+1}(x_k^t, y_k^t)}\right]^{\frac{1}{2}} \end{aligned} \tag{15}$$

In particular:
The efficiency of investors at point (x_k^t, y_k^t) period t is determined by the following formula [19]:

$$D_k^t(x_k^t, y_k^t) = \text{Max}\varphi_k^t \tag{16}$$

$$\sum_{j=1}^{n} \lambda_j x_{ij}^t \le x_{ik}^t,\ j = 1, 2, 3, \ldots, m \tag{17}$$

$$\sum_{j=1}^{n} \lambda_j y_{rj}^t \ge \varphi_k^t y_{rk}^t,\ r = 1, 2, 3, \ldots, s \tag{18}$$

$$\lambda_j \ge 0 \tag{19}$$

The efficiency of investors at point (x_k^t, y_k^t) period $t + 1$ is determined by the following formula [19]:

$$D_k^{t+1}(x_k^t, y_k^t) = \text{Max}\varphi_k^{t+1} \tag{20}$$

$$\sum_{j=1}^{n} \lambda_j x_{ij}^{t+1} \le x_{ik}^t,\ j = 1, 2, 3, \ldots, m \tag{21}$$

$$\sum_{j=1}^{n} \lambda_j y_{rj}^{t+1} \ge \varphi_k^{t+1} y_{rk}^t,\ r = 1, 2, 3, \ldots, s \tag{22}$$

$$\lambda_j \ge 0 \tag{23}$$

The efficiency of investors at point (x_k^{t+1}, y_k^{t+1}) period t is determined by the following formula [19]:

$$D_k^t(x_k^{t+1}, y_k^{t+1}) = \text{Max}\varphi_k^t \tag{24}$$

$$\sum_{j=1}^{n} \lambda_j x_{ij}^{t+1} \le x_{ik}^t,\ j = 1, 2, 3, \ldots, m \tag{25}$$

$$\sum_{j=1}^{n} \lambda_j y_{rj}^t \ge \varphi_k^t y_{rk}^{t+1},\ r = 1, 2, 3, \ldots, s \tag{26}$$

$$\lambda_j \ge 0 \tag{27}$$

The efficiency of investors at point (x_k^{t+1}, y_k^{t+1}) period $t + 1$ is determined by the following formula [19]:

$$D_k^{t+1}(x_k^{t+1}, y_k^{t+1}) = \text{Max}\varphi_k^{t+1} \tag{28}$$

$$\sum_{j=1}^{n} \lambda_j x_{ij}^{t+1} \le x_{ik}^{t+1},\ j = 1, 2, 3, \ldots, m \tag{29}$$

$$\sum_{j=1}^{n} \lambda_j y_{rj}^{t+1} \geq \varphi_k^{t+1} y_{rk}^{t+1},\ r = 1,\ 2,\ 3, \ldots,\ s \quad (30)$$

$$\lambda_j \geq 0 \quad (31)$$

If the ECA, EFR, EMPI indexes are less than 1 (<1), this reflects that the technical, technological, and financial investment of the consultants, and design and construction contractors in the traffic layer during the assessment period were not effective. If these indicators are greater than 1 (>1), it is clear that the bids achieved the above three indicators during this period. These indicators are equal to 1 (=1) reflecting the performance in the comparison period corresponding to the previous period [20].

3.2.4. Correlation Coefficients

The correlation coefficient (k), first introduced in 1895, is a statistical tool used to test correlations between variables. The formula for calculating the correlation coefficient is as follows [21]:

$$k = \frac{\sum_{i=1}^{n} (\alpha_i - \overline{\alpha})(\delta_i - \overline{\delta})}{\left[\sum_{i=1}^{n} (\alpha_i - \overline{\alpha})^2 \sum_{i=1}^{n} (\delta_i - \overline{\delta})^2\right]^{\frac{1}{2}}} \quad (32)$$

The value of k depends on the range (−1; 1). The value of negative k denotes the inverse relationship between the two factors (if the value of this factor increases, then the other factor decreases and vice versa). The value of positive k denotes the positive correlation between the two factors (if the value of this factor increases, the value of the other factor increases; if this value decreases, the value of the other factor will decrease accordingly). The value of $k = 0$ denotes independent factors [22].

4. Results

4.1. Forecasting

Designing and executing traffic infrastructure works for the period 2018–2021, the authors use data from factor F_1 of I_{12} (Table 7) to explain the computational steps in the GM (1, 1) model used in this study (as follows).

Table 7. Data of I_{12}.

Year	F_1	F_2	F_3	F_4	F_5	F_6
2014	4032.71	1130.60	5697.05	100.94	57.37	5866.17
2015	3941.09	1127.26	4227.08	115.57	43.88	4402.08
2016	4615.74	1132.18	5321.93	125.01	42.96	5497.76
2017	4187.99	1135.78	5589.87	135.07	43.22	5772.25

From the original data range: $X^{(0)} = (4032.71; 3941.09; 4615.74; 4187.99)$
Use the accumulating generation operation method to compute $X^{(1)}$ values:

$$X^{(1)} = (4032.71; 7973.80; 12589.54; 16777.53)$$

Calculate the mean values $Z^{(1)}$: $Z^{(1)} = (6003.26; 10281.67; 14683.53)$
From the values of $X^{(0)}$, $Z^{(1)}$ the authors obtain the following system of equations:

$$\begin{cases} 3941.0853 + a \times 6003.26 = b \\ 4615.7374 + a \times 10281.67 = b \\ 4187.9907 + a \times 14683.53 = b \end{cases}$$

From the above equations, transformed into the matrix form as follows:

$$B = \begin{bmatrix} -6003.26 \\ -10281.67 \\ -14683.53 \end{bmatrix}; \; Y_N = \begin{bmatrix} 3941.0853 \\ 4615.7374 \\ 4187.9907 \end{bmatrix}$$

$$\begin{bmatrix} a \\ b \end{bmatrix} = \begin{bmatrix} -0.0278 \\ 3960.8791 \end{bmatrix}$$

Use the values a and b to find the differential equation:

$$\frac{dx^{(1)}_{(k)}}{dt} - 0.0278x^{(0)} = 3960.8791$$

Equation prediction:

$$\begin{aligned} \hat{x}^{(1)}_{(k+1)} &= (4032.71 + \tfrac{3960.8791}{0.0278}) \times e^{0.0278k} - \tfrac{3960.8791}{0.0278} \\ &= 146510.3755 \times e^{0.0278k} - 142477.6655 \end{aligned}$$

In turn, the values for $k = 0, 1, 2, \ldots, 7$ are given, $\hat{x}^{(1)}_{(k+1)}$ by the following values:

$$\hat{x}^{(1)}_{(k+1)} = \{4032.71; 8163.09; 12410.08; 16776.97; 21267.14; 25884.08; 30631.36; 35512.66\} \quad (33)$$

Use the accumulating generation operation method to calculate the predicted values by the formula: $\hat{x}^{(0)}_{(k+1)} = \hat{x}^{(1)}_{(k+1)} - \hat{x}^{(1)}_{k}$ obtains the predicted values for the years in Table 8.

Table 8. Forecast data F_1 of I_{12}.

Year	2014	2015	2016	2017	2018	2019	2020	2021
Forecast	4032.71	4130.38	4246.99	4366.89	4490.17	4616.94	4747.28	4881.30

As calculated above, the authors obtained predictive values for all elements of bidders for the period 2018–2021 which are summarized in Tables 9–12 below:

Table 9. Forecast data in 2018.

I_n	F_1	F_2	F_3	F_4	F_5	F_6
I_1	43,148.97	10,529.26	31,583.34	3002.34	743.78	35,590.60
I_2	34,876.68	8937.01	16,084.48	1427.70	914.95	18,918.35
I_3	81,722.91	8369.65	55,330.65	1166.97	616.75	54,842.26
I_4	970,992.39	464,410.49	1,528,230.15	17,007.63	105,075.32	1,645,842.91
I_5	837,908.16	158,089.65	1,010,373.13	35,425.99	80,927.61	1,146,236.67
I_6	352,912.66	55,106.43	36,489.40	1941.59	6626.52	55,128.51
I_7	10,574.94	2281.91	7373.88	497.44	106.19	8000.74
I_8	11,396.45	9067.29	2843.32	861.83	4424.96	6251.14
I_9	44,823.14	17,802.49	33,790.76	2746.82	5195.65	45,802.04
I_{10}	81,267.23	15,874.31	92,890.06	3753.24	3361.32	99,281.31
I_{11}	43,105.18	12,531.78	7512.81	592.52	1623.51	8090.65
I_{12}	4490.17	1140.29	6514.66	145.95	42.70	6697.85
I_{13}	124,451.91	14,130.21	132,385.34	5576.55	1904.47	149,909.49
I_{14}	37,470.96	15,770.61	30,520.06	469.61	1254.61	31,322.63
I_{15}	51,484.78	7509.46	17,234.57	555.00	116.78	18,751.01
I_{16}	44,165.08	5095.48	28,770.26	844.86	526.54	30,174.41
I_{17}	45,949.20	6265.79	55,965.60	1772.24	573.39	60,413.53

Table 10. Forecast data in 2019.

I_n	F_1	F_2	F_3	F_4	F_5	F_6
I_1	50,268.76	10,530.78	39,335.96	5135.68	841.40	45,135.29
I_2	38,405.69	9380.65	15,372.27	1645.09	1016.03	18,323.39
I_3	92,406.66	8360.35	70,525.62	1126.83	653.54	65,455.69
I_4	1,362,850.22	652,348.51	2,123,825.36	17,859.22	147,332.03	2,274,415.58
I_5	1,128,340.52	231,141.55	1,650,867.09	57,935.96	161,765.74	1,892,158.68
I_6	365,367.83	55,956.38	32,713.77	1815.37	7327.12	51,272.07
I_7	10,823.35	2268.44	6785.52	509.16	96.74	7383.38
I_8	9129.50	13,497.17	1389.64	1206.79	8428.18	3906.00
I_9	53,806.69	22,183.54	37,561.50	3691.75	6451.81	52,051.24
I_{10}	76,607.59	16,738.56	109,851.23	7293.26	4420.41	117,992.54
I_{11}	47,016.95	12,796.42	6273.92	595.05	2228.47	6521.79
I_{12}	4616.94	1144.59	7428.14	157.77	42.37	7609.37
I_{13}	144,387.22	14,849.66	241,883.95	8566.41	2532.30	272,267.60
I_{14}	29,524.41	15,859.39	32,487.51	544.20	1392.92	32,828.68
I_{15}	65,218.21	7392.40	15,507.31	492.67	82.55	17,042.17
I_{16}	54,172.45	5215.07	30,323.31	870.30	770.08	31,838.47
I_{17}	64,451.77	6575.94	111,382.72	2242.08	621.99	115,378.84

Table 11. Forecast data in 2020.

I_n	F_1	F_2	F_3	F_4	F_5	F_6
I_1	58,563.35	10,532.30	48,991.59	8784.90	951.83	57,239.69
I_2	42,291.79	9846.31	14,691.59	1895.59	1128.26	17,747.14
I_3	104,487.13	8351.07	89,893.44	1088.07	692.53	78,123.11
I_4	1,912,847.87	916,341.45	2,951,541.13	18,753.45	206,582.56	3,143,049.80
I_5	1,519,441.39	337,950.12	2,697,381.85	94,748.97	323,352.64	3,123,494.98
I_6	378,262.58	56,819.44	29,328.81	1697.36	8101.78	47,685.40
I_7	11,077.60	2255.05	6244.11	521.17	88.13	6813.66
I_8	7313.48	20,091.29	679.17	1689.81	16,053.07	2440.64
I_9	64,590.75	27,642.71	41,753.01	4961.74	8011.67	59,153.09
I_{10}	72,215.11	17,649.85	129,909.41	14,172.20	5813.20	140,230.21
I_{11}	51,283.70	13,066.65	5239.32	597.59	3058.84	5257.15
I_{12}	4747.28	1148.90	8469.72	170.56	42.05	8644.94
I_{13}	167,515.87	15,605.75	441,951.10	13,159.30	3367.09	494,496.04
I_{14}	23,263.11	15,948.68	34,581.79	630.64	1546.49	34,407.14
I_{15}	82,615.00	7277.16	13,953.15	437.35	58.35	15,489.06
I_{16}	66,447.39	5337.48	31,960.20	896.51	1126.29	33,594.31
I_{17}	90,404.86	6901.43	221673.89	2836.48	674.71	220,352.59

Table 12. Forecast data in 2021.

I_n	F_1	F_2	F_3	F_4	F_5	F_6
I_1	68,226.59	10,533.82	61,017.34	15,027.11	1076.76	72,590.24
I_2	46,571.11	10,335.09	14,041.06	2184.23	1252.90	17,189.01
I_3	118,146.88	8341.79	114,580.07	1050.65	733.84	93,242.00
I_4	2,684,804.91	1,287,167.27	4,101,841.54	19,692.46	289,661.07	4,343,428.75
I_5	2,046,104.07	494,114.03	4,407,301.41	154,953.26	646,347.78	5,156,132.51
I_6	391,612.42	57,695.82	26,294.10	1587.02	8958.35	44349.63
I_7	11,337.82	2241.74	5745.90	533.45	80.28	6287.90
I_8	5858.70	29,907.01	331.94	2366.17	30,576.11	1525.03
I_9	77,536.16	34,445.35	46,412.26	6668.62	9948.66	67,223.91
I_{10}	68,074.49	18,610.76	153,630.09	27,539.30	7644.83	166,658.94
I_{11}	55,937.66	13,342.59	4375.34	600.14	4198.64	4237.74
I_{12}	4881.30	1153.24	9657.34	184.37	41.73	9821.43
I_{13}	194,349.37	16,400.33	807,497.87	20,214.66	4477.08	898,110.26
I_{14}	18,329.65	16,038.46	36,811.08	730.80	1716.98	36,061.50
I_{15}	104,652.34	7163.71	12,554.75	388.24	41.24	14,077.50
I_{16}	81,503.71	5462.76	33,685.45	923.51	1647.25	35,446.97
I_{17}	126,808.59	7243.04	441,175.36	3588.46	731.90	420,833.36

To test the accuracy of the predicted values, the authors used MAPE to ensure that the results of this study were highly reliable. The results in Table 13 are as follows:

Table 13. Errors results.

I_n	MAPE (%)	I_n	MAPE (%)
I_1	3.82	I_{10}	5.12
I_2	5.01	I_{11}	3.73
I_3	1.92	I_{12}	1.78
I_4	5.38	I_{13}	6.85
I_5	15.13	I_{14}	9.90
I_6	0.99	I_{15}	13.16
I_7	3.35	I_{16}	8.08
I_8	13.55	I_{17}	16.47
I_9	3.80	Average $_{\text{all } In}$ (%)	6.94

As shown in Table 13, average $_{\text{all } In}$ = 6.94% indicates that the forecasts for contractors' business performance for the period 2018–2021 are high. In particular, the predicted data for 4/17 of contractors ranged from 10 to 20%, while the predicted data for 13 of 17 contractors had errors of less than 10%. This confirms that the GM (1, 1) model used to predict the contractor's business performance in this study is consistent with high reliability.

4.2. Correlation Coefficient

The correlation coefficients are shown in Table 14. Based on the convention mentioned in Section 3.2.4, it was found that the factors used in this study were positive (the increase in inputs would lead to the increase in output). This is in line with economic law. All coefficients are greater than 0.6, indicating that these factors have a strong correlation.

Table 14. Correlation coefficient.

	Time Period 2014						Time Period 2015					
	F_1	F_2	F_3	F_4	F_5	F_6	F_1	F_2	F_3	F_4	F_5	F_6
F_1	1.00	0.81	0.71	0.79	0.67	0.73	1.00	0.85	0.75	0.75	0.68	0.76
F_2	0.81	1.00	0.96	0.77	0.97	0.96	0.85	1.00	0.96	0.97	0.96	0.96
F_3	0.71	0.96	1.00	0.83	0.96	1.00	0.75	0.96	1.00	0.99	0.96	1.00
F_4	0.79	0.77	0.83	1.00	0.67	0.84	0.75	0.97	0.99	1.00	0.97	0.99
F_5	0.67	0.97	0.96	0.67	1.00	0.96	0.68	0.96	0.96	0.97	1.00	0.96
F_6	0.73	0.96	1.00	0.84	0.96	1.00	0.76	0.96	1.00	0.99	0.96	1.00
	Time Period 2016						Time Period 2017					
F_1	1.00	0.83	0.87	0.92	0.85	0.87	1.00	0.89	0.92	0.93	0.92	0.92
F_2	0.83	1.00	0.97	0.77	0.99	0.97	0.89	1.00	0.97	0.81	0.98	0.96
F_3	0.87	0.97	1.00	0.88	0.99	1.00	0.92	0.97	1.00	0.92	0.99	1.00
F_4	0.92	0.77	0.88	1.00	0.84	0.89	0.93	0.81	0.92	1.00	0.90	0.93
F_5	0.85	0.99	0.99	0.84	1.00	0.99	0.92	0.98	0.99	0.90	1.00	0.99
F_6	0.87	0.97	1.00	0.89	0.99	1.00	0.92	0.96	1.00	0.93	0.99	1.00

This result shows that the data used in this study are consistent with the conditions of use of the optimal mathematical models used in this study. The conclusions of the study have sufficient grounds for assessing the technical, technological, and economic efficiency of contractors in the field of construction of transport infrastructure works.

4.3. Catch-Up Index

The catch-up index reflects the technical efficiency of the consultants, design, and construction contractors for transport infrastructure in Vietnam for the period 2014–2021, as shown in Table 15.

Table 15. The catch-up index.

(ECA)	'14–'15	'15–'16	'16–'17	'17–'18	'18–'19	'19–'20	'20–'21	Average
I_1	0.9956	0.9211	1.0104	1.0768	1.0419	1.0380	1.0180	1.0146
I_2	0.9718	0.9855	0.9172	1.0726	1.0179	1.0063	0.9963	0.9954
I_3	1.0000	0.9404	1.0634	1.0000	0.9599	0.8961	0.9620	0.9745
I_4	1.0000	1.0000	1.0000	1.0000	1.0000	1.0000	1.0000	1.0000
I_5	1.0000	1.0000	1.0000	1.0000	1.0000	1.0000	1.0000	1.0000
I_6	1.0000	1.0000	1.0000	1.0000	1.0000	1.0000	1.0000	1.0000
I_7	0.9546	1.0145	0.9616	1.0450	1.0276	1.0000	1.0000	1.0005
I_8	1.0000	1.0000	1.0000	1.0000	1.0000	1.0000	1.0000	1.0000
I_9	1.0000	1.0000	1.0000	1.0000	1.0000	1.0000	1.0000	1.0000
I_{10}	1.0000	1.0000	0.9769	0.9571	0.9936	1.0014	1.0088	0.9911
I_{11}	0.9935	0.8856	0.8911	0.9988	1.0654	1.2272	1.0000	1.0088
I_{12}	1.0000	1.0000	1.0000	1.0000	1.0000	1.0000	1.0000	1.0000
I_{13}	0.9167	1.0966	1.0490	1.0000	1.0000	1.0000	1.0000	1.0089
I_{14}	1.0000	1.0000	1.0000	0.9437	0.9841	0.9882	0.9828	0.9855
I_{15}	0.9256	0.9967	0.9960	1.0410	1.0017	1.0041	1.0132	0.9969
I_{16}	0.9121	1.0963	0.9424	1.0088	0.9947	0.9936	0.9950	0.9919
I_{17}	0.9588	0.9491	1.0989	1.0000	1.0000	1.0000	1.0000	1.0010
Average	0.9782	0.9933	0.9945	1.0085	1.0051	1.0091	0.9986	0.9982
Max	1.0000	1.0966	1.0989	1.0768	1.0654	1.2272	1.0180	1.0146
Min	0.9121	0.8856	0.8911	0.9437	0.9599	0.8961	0.9620	0.9745
SD	0.0323	0.0523	0.0499	0.0342	0.0232	0.0626	0.0120	0.0092

The results in Table 15 and Figure 3 show that the consultants, design, and construction of transport infrastructure have not achieved technical efficiency in the period 2013–2017. Specifically, the catch-up index in this phase is less than 1 (average ECA 2014–2015 = 0.9782; average ECA 2015–2016 = 0.9933; average ECA 2016–2017 = 0.9945). However, according to the results above, in the period 2017–2020, the contractor has made changes and achieved technical efficiency. In the period 2017–2020, the catch-up index is greater than 1 (average ECA 2017–2018 = 1.0085; average ECA 2018–2019 = 1.0051; average ECA 2019–2020 = 1.0091). In particular, there are strong changes and high efficiency of three contractors (I_1, I_{11}, I_{15}). These bidders have not really achieved technical efficiency in the period 2014–2017, but, from the period 2017–2021, there are solutions to create efficiency (catch-up index is greater than 1). In addition, nine out of 17 contractors have maintained stable technical performance in the period 2017–2021, including I_4, I_5, I_6, I_7, I_8, I_9, I_{12}, I_{13}, I_{17}. However, many contractors have not yet achieved technical efficiency (I_1, I_2, I_{10}, I_{14}, I_{16}). Managers rely on this result to set up timely measures, in line with the actual situation, to improve the technical efficiency of their businesses in the future.

In construction, machinery is always a decisive factor in productivity, quality, and efficiency. Therefore, the frequent research and development of new high-tech techniques play a decisive role. On the other hand, the machinery and equipment in construction are typically of great economic value; further, the maintenance of machinery and equipment, ensuring the machinery is always operating properly and takes advantage of the capacity of machinery equipment, plays a huge role in improving the technical efficiency of contractors.

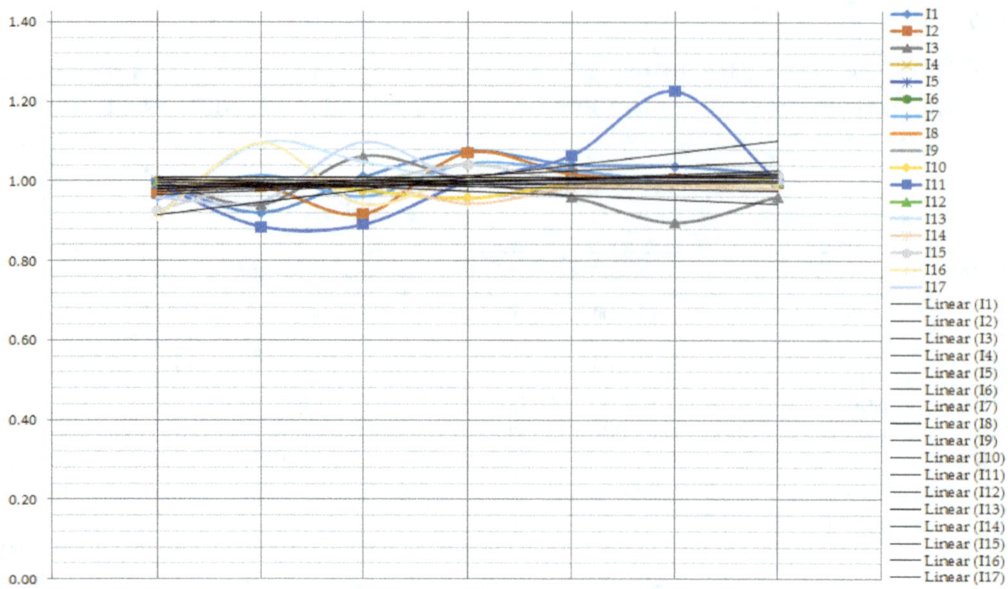

Figure 3. Catch-up index.

4.4. Frontier Index

The CFR index reflects the technological efficiency of construction contractors for transport infrastructure development. The results in Table 16 show that the technology efficiency scores of past contractors (2014–2017) and future predictions (2018–2021) are good. Specifically, the average technology efficiency scores for bidders for the period 2014–2021 are average 2014–2015: 1.0744; average 2015–2016: 1.0764; average 2016–2017: 1.0656; average 2017–2018:1.0872; average 2018–2019: 1.0928; average 2019–2020: 1.0936; average 2020–2021: 1.0955.

Table 16. Frontier index.

(EFR)	'14–'15	'15–'16	'16–'17	'17–'18	'18–'19	'19–'20	'20–'21	Average
I_1	1.0368	1.0291	1.0220	0.9833	1.0017	0.9990	1.0124	1.0120
I_2	1.0407	1.0087	1.0753	0.9297	0.9813	1.0006	1.0128	1.0070
I_3	1.0844	0.9756	1.0096	1.1078	1.1737	1.2170	1.1320	1.1000
I_4	1.4438	1.4294	1.0633	1.2018	1.1780	1.1831	1.1799	1.2399
I_5	1.1949	1.1485	1.1439	1.3345	1.3750	1.4227	1.4220	1.2917
I_6	1.0948	1.1329	1.0830	1.0965	1.0616	1.0301	1.0293	1.0755
I_7	1.0183	0.9836	1.0256	0.9536	0.9802	1.0020	1.0005	0.9948
I_8	0.7654	1.1597	1.7164	0.9066	1.0000	1.0000	1.0000	1.0783
I_9	1.3344	1.2253	1.0100	1.0820	1.0204	1.0138	1.0119	1.0997
I_{10}	1.1063	0.9491	0.9558	1.0619	1.0943	1.0855	1.0147	1.0382
I_{11}	1.0329	1.0480	1.1554	1.1902	0.9499	0.8692	1.0352	1.0401
I_{12}	1.0000	1.1096	1.0000	1.0743	1.0621	1.0599	1.0562	1.0517
I_{13}	1.0637	1.0108	1.0723	1.2639	1.3339	1.3365	1.3396	1.2030
I_{14}	0.8733	1.1119	0.7263	1.0872	1.0017	0.9991	1.0193	0.9741
I_{15}	1.0672	1.0071	1.0069	0.9640	1.0043	1.0025	0.9947	1.0067
I_{16}	1.0757	0.9580	0.9814	1.0202	1.0112	1.0106	1.0087	1.0094
I_{17}	1.0321	1.0111	1.0685	1.2259	1.3474	1.3590	1.3545	1.1998
Average	1.0744	1.0764	1.0656	1.0872	1.0928	1.0936	1.0955	1.0836
Max	1.4438	1.4294	1.7164	1.3345	1.3750	1.4227	1.4220	1.2917
Min	0.7654	0.9491	0.7263	0.9066	0.9499	0.8692	0.9947	0.9741
SD	0.1523	0.1210	0.1920	0.1231	0.1389	0.1545	0.1415	0.0946

Based on the results of the technology efficiency score of the contractors, as shown in Table 16 and Figure 4, the authors created three groups:

1. Highly efficient bidders in the future, including I_3, I_4, I_5, I_6, I_9, I_{10}, I_{12}, I_{13}, I_{16}, I_{17}. These bidders are expected to have a technology efficiency score greater than 1 (>1) over the period 2018–2021.
2. Contractors that maintain stable technology in the future, including I_8. This contractor is expected to be technologically stable (=1) over the period 2018–2021.
3. Contractors not effective in terms of technology in the future, including I_1, I_2, I_7, I_{11}, I_{14}, I_{15}. These bidders are predicted to be technologically efficient with high volatility during the period 2018–2021.

In general, construction contractors of transport works in Vietnam have approached and mastered modern technologies, improving the capacity of construction of transport works to meet the requirements of construction works. Along with that is the use of high-quality materials and application of new technology to bring high economic efficiency and longevity for the works.

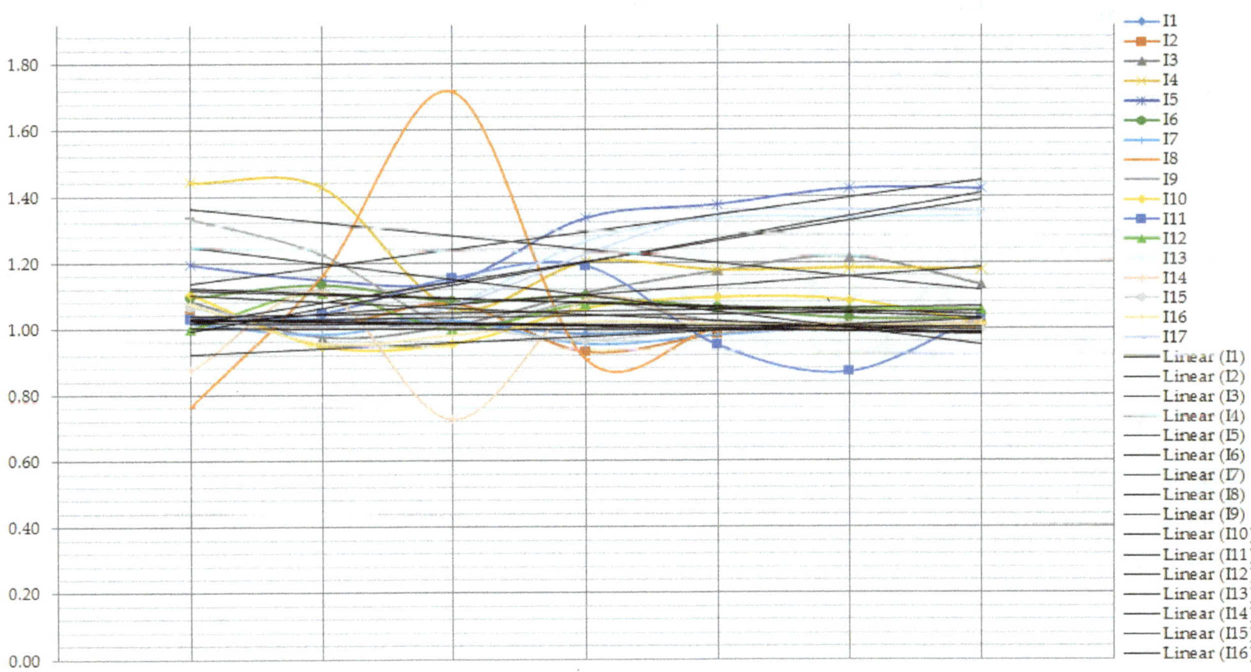

Figure 4. Frontier index.

4.5. Malmquist Index

As shown in Table 17 and Figure 5, EMPI average 2014–2018 – 1.0807, which reflects the business performance of the contractor of traffic works in general and is quite satisfactory. Specifically, according to the forecast results, in the period 2018–2021, 13 out of 17 contractors have achieved high business results, including I_1, I_3, I_4, I_5, I_6, I_9, I_{10}, I_{11}, I_{12}, I_{13}, I_{15}, I_{16}, I_{17}. In addition, 2/17 contractors are expected to maintain stable business performance during this period, including I_7, I_8. However, 2/17 of the contractors whose business results are forecasted to be ineffective include I_2, I_{14}.

Table 17. Malmquist index.

(EMPI)	'14–'15	'15–'16	'16–'17	'17–'18	'18–'19	'19–'20	'20–'21	Average
I_1	1.0322	0.9480	1.0326	1.0588	1.0436	1.0370	1.0307	1.0261
I_2	1.0114	0.9941	0.9863	0.9972	0.9989	1.0070	1.0090	1.0005
I_3	1.0844	0.9174	1.0736	1.1078	1.1266	1.0905	1.0890	1.0699
I_4	1.4438	1.4294	1.0633	1.2018	1.1780	1.1831	1.1799	1.2399
I_5	1.1949	1.1485	1.1439	1.3345	1.3750	1.4227	1.4220	1.2917
I_6	1.0948	1.1329	1.0830	1.0965	1.0616	1.0301	1.0293	1.0755
I_7	0.9721	0.9979	0.9862	0.9965	1.0072	1.0020	1.0005	0.9946
I_8	0.7654	1.1597	1.7164	0.9066	1.0000	1.0000	1.0000	1.0783
I_9	1.3344	1.2253	1.0100	1.0820	1.0204	1.0138	1.0119	1.0997
I_{10}	1.1063	0.9491	0.9338	1.0163	1.0872	1.0870	1.0236	1.0290
I_{11}	1.0262	0.9281	1.0296	1.1888	1.0120	1.0667	1.0352	1.0409
I_{12}	1.0000	1.1096	1.0000	1.0743	1.0621	1.0599	1.0562	1.0517
I_{13}	0.9751	1.1085	1.1248	1.2639	1.3339	1.3365	1.3396	1.2118
I_{14}	0.8733	1.1119	0.7263	1.0260	0.9858	0.9872	1.0018	0.9589
I_{15}	0.9878	1.0038	1.0029	1.0035	1.0060	1.0066	1.0078	1.0026
I_{16}	0.9812	1.0503	0.9250	1.0292	1.0058	1.0041	1.0037	0.9999
I_{17}	0.9895	0.9597	1.1741	1.2259	1.3474	1.3590	1.3545	1.2014
Average	1.0513	1.0691	1.0595	1.0947	1.0972	1.0996	1.0938	1.0807
Max	1.4438	1.4294	1.7164	1.3345	1.3750	1.4227	1.4220	1.2917
Min	0.7654	0.9174	0.7263	0.9066	0.9858	0.9872	1.0000	0.9589
SD	0.1590	0.1316	0.1974	0.1128	0.1318	0.1397	0.1407	0.0973

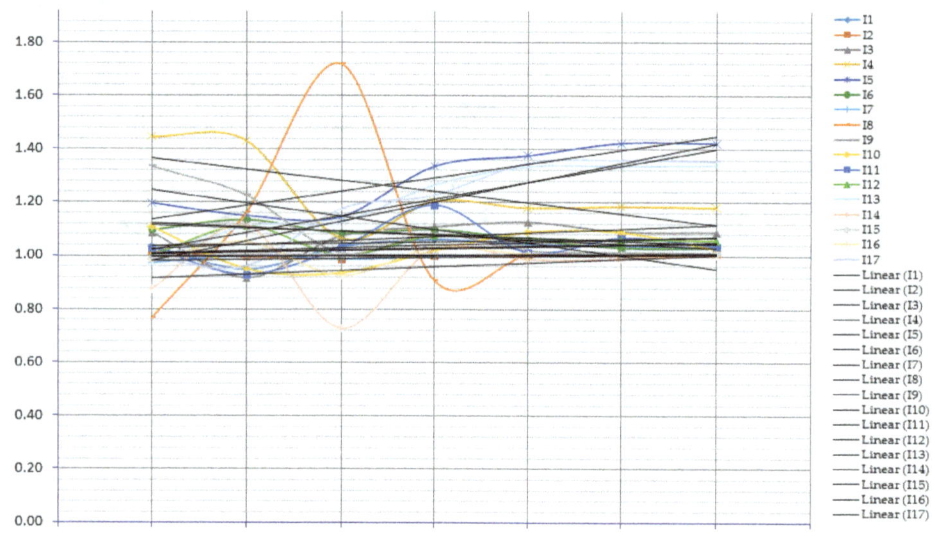

Figure 5. Malmquist index.

In order to maintain and improve business efficiency, contractors themselves must actively create, overcome difficulties, promote advantages, and exploit and make use of favorable conditions and factors of the environment and geographic location, thus combining multiple measures to maximize the use of resources and businesses to achieve optimal efficiency. The economic efficiency of production and business activities in consultancy, design, and construction of traffic works is an integrated category in many fields. In order to improve the economic efficiency of production and business activities, contractors must use the combination of measures from raising the management capacity and managing production and business activities of enterprises in the office to work. In addition to enhancing and improving all the activities within the enterprise, enterprises are always adapting to the changes of the market, i.e., adaptation to each project in different localities. Bidders must regularly maintain and ensure the balance of the relationship between parties from the construction site, so the new office can enhance the sense of responsibility of each person, i.e., enhance the initiative in production to bring high economic efficiency.

4.6. Contractor Selection

The results of forecasting and evaluating the technical, technological, and business efficiency of the consultants, design, and construction contractors of the transport infrastructure are shown in Table 18.

Table 18. Classify contractor.

	Good Efficiency	Efficiency	Inefficiency
ECA	I_1, I_{11}, I_{15}	$I_4, I_5, I_6, I_7, I_8, I_9, I_{12}, I_{13}, I_{17}$	$I_2, I_3, I_{10}, I_{14}, I_{16}$
EFR	$I_3, I_4, I_5, I_6, I_9, I_{10}, I_{12}, I_{13}, I_{16}, I_{17}$	I_8	$I_1, I_2, I_7, I_{11}, I_{14}, I_{15}$
EMPI	$I_1, I_3, I_4, I_5, I_6, I_9, I_{10}, I_{11}, I_{12}, I_{13}, I_{15}, I_{16}, I_{17}$	I_7, I_8	I_2, I_{14}

Based on these results, the government, regulatory authorities in strategy formulation, policy development, selection of consultants, and design and construction contractors have good capacity to implement projects. It is of decisive importance in ensuring the progress and quality of traffic works. The result of this business performance assessment is also a good basis for all self-revising subjects, to see where their businesses are in the overall picture of the construction investment sector to provide timely solutions to improve capacity and more effective implementation of assigned tasks.

5. Conclusions

At present, with the rapid development of the economy, demand for transportation and transportation of goods and people remains large. As a result, road, rail, waterway, and air infrastructure works are being increasingly built. Therefore, the role of contractor consulting, design, and construction of transport infrastructure has become more important and maintains a special position. The main contractor is the decisive factor affecting the quality and progress of construction of the transportation infrastructure. In this study, the authors used a modern, highly accurate forecasting method to forecast contractors' business, design, and construction contractor performance. Also, in this study, the authors used optimized mathematical models to evaluate past, present, and future contractors' technical, technological, and performance effectiveness. The result of this study is a solid basis for the government, regulatory agencies, and investors to use for strategic planning and policy development of transport infrastructure with high efficiency. The best way to do this is through the selection of contractors who have the human, financial, technical, and technological capabilities to meet the requirements of managers and investors.

In addition to the results achieved, this research still has certain limitations: It does not combine with the qualitative factors, weather factors, and government policies. In addition, many of the optimal mathematical models have not been considered in this study. The authors will continue to address these issues in subsequent studies.

Author Contributions: C.-N.W. developed and checked theoretical calculations. T.-M.L. wrote and formatted the manuscript. H.-K.N. contributed to analyzing the data. All authors have read and accepted this research.

References

1. Yang, J.B.; Wang, W.C. Contractor selection by the most advantageous tendering approach in Taiwan. *J. Chin. Inst. Eng.* **2003**, *26*, 381–387. [CrossRef]
2. Holt, G.D. Which contractor selection methodology? *Int. J. Proj. Manag.* **1998**, *16*, 153–164. [CrossRef]

3. Zhou, P.A.B.W.; Ang, B.W.; Poh, K.L. A trigonometric grey prediction approach to forecasting electricity demand. *Energy* **2006**, *31*, 2839–2847. [CrossRef]
4. Zhao, Z.; Wang, J.; Zhao, J.; Su, Z. Using a Grey model optimized by Differential Evolution algorithm to forecast the per capita annual net income of rural households in China. *Omega* **2012**, *40*, 525–532. [CrossRef]
5. Shi, G.M.; Bi, J.; Wang, J.N. Chinese regional industrial energy efficiency evaluation based on a DEA model of fixing non-energy inputs. *Energy Policy* **2010**, *38*, 6172–6179. [CrossRef]
6. Pilyavsky, A.; Staat, M. Efficiency and productivity change in Ukrainian health care. *J. Prod. Anal.* **2008**, *29*, 143–154. [CrossRef]
7. Wang, C.N.; Nguyen, H.K.; Liao, R.Y. Partner Selection in Supply Chain of Vietnam's Textile and Apparel Industry: The Application of a Hybrid DEA and GM (1,1) Approach. *Math. Probl. Eng.* **2017**. [CrossRef]
8. Shuai, J.J.; Wu, W.W. Evaluating the influence of E-marketing on hotel performance by DEA and grey entropy. *Expert Syst. Appl.* **2011**, *38*, 8763–8769. [CrossRef]
9. Minh-Hue, N. Logistics Vietnam report 2017. Available online: http://nhaxuatbancongthuong.com (accessed on 21 May 2018).
10. Statistics. General Statistics Office of Vietnam. Available online: https://www.gso.gov.vn (accessed on 25 May 2018).
11. Deng, J. Control problems of grey systems. *Syst. Control Lett.* **1982**, *1*, 288–294.
12. Wang, C.N.; Nguyen, H.K. Enhancing Urban Development Quality Based on the Results of Appraising Efficient Performance of Investors—A Case Study in Vietnam. *Sustainability* **2017**, *9*, 1397. [CrossRef]
13. Boles, J.-S.; Donthu, N.; Ritu, L. Salesperson evaluation using relative performance efficiency: The application of data envelopment analysis. *J. Pers. Sell. Sales Manag.* **1995**, *15*, 31–49.
14. Charnes, A.; Cooper, W.-W.; Rhodes, E. Evaluating Program and Managerial Efficiency: An Application of Data Envelopment Analysis to Program Follow Through. *Manag. Sci.* **1981**, *27*, 668–697. [CrossRef]
15. Banker, R.-D.; Charnes, A.; Cooper, W.-W. Some models for estimating technical and scale inefficiencies in data envelopment analysis. *Manag. Sci.* **1984**, *30*, 1078–1092. [CrossRef]
16. Mao, W.; Koo, W.W. Productivity growth, technological progress, and efficiency change in chinese agriculture after rural economic reforms: A DEA approach. *China Econ. Rev.* **1997**, *8*, 157–174. [CrossRef]
17. Macpherson, A.-J.; Principe, P.-P.; Mehaffey, M. Using Malmquist Indices to evaluate environmental impacts of alternative land development scenarios. *Ecological Indicators* **2013**, *34*, 296–303. [CrossRef]
18. Paradi, J.C.; Asmild, M.; Aggarwall, V.; Schaffnit, C. *Performance Evaluation in an Oligopoly Environment: Combining DEA Window Analysis with the Malmquist Index Approach—A Study of the Canadian Banking Industry*; Centre for Management of Technology and Entrepreneurship University of Toronto, University of Toronto: Toronto, ON, Canada, 2011; pp. 31–47.
19. Odeck, J. Identifying traffic safety best practice: An application of DEA and Malmquist indices. *Omega* **2006**, *34*, 28–40. [CrossRef]
20. Wang, C.N.; Nguyen, N.T.; Tran, T.T. Integrated DEA Models and Grey System Theory to Evaluate Past-to-Future Performance: A Case of Indian Electricity Industry. *Sci. World J.* **2015**. [CrossRef] [PubMed]
21. Wang, J. Pearson Correlation Coefficient. In *Encyclopedia of Systems Biology*; Springer: New York, NY, USA, 2013; pp. 97–141.
22. Nahler, G. Pearson correlation coefficient. In *Dictionary of Pharmaceutical Medicine*; Springer: Vienna, Austria, 2017.

11

A Single-Stage Manufacturing Model with Imperfect Items, Inspections, Rework and Planned Backorders

Chang Wook Kang [1], Misbah Ullah [2], Mitali Sarkar [1], Muhammad Omair [3] and Biswajit Sarkar [1,*]

[1] Department of Industrial and Management Engineering, ERICA Campus, Hanyang University, Ansan, Gyeonggi-do 15588, Korea; cwkang57@hanyang.ac.kr (C.W.K.); mitalisarkar.ms@gmail.com (M.S.)

[2] Department of Industrial Engineering, University of Engineering and Technology, Peshawar 25000, Pakistan; misbah@uetpeshawar.edu.pk

[3] Department of Industrial Engineering, Jalozai Campus, University of Engineering and Technology, Peshawar 25000, Pakistan; muhamad.omair87@gmail.com

* Correspondence: bsbiswajitsarkar@gmail.com.

Abstract: Each industry prefers to sell perfect products in order to maintain its brand image. However, due to a long-run single-stage production system, the industry generally obtains obstacles. To solve this issue, a single-stage manufacturing model is formulated to make a perfect production system without defective items. For this, the industry decides to stop selling any products until whole products are ready to fulfill the order quantity. Furthermore, manufacturing managers prefer product qualification from the inspection station especially when processes are imperfect. The purpose of the proposed manufacturing model considers that the customer demands are not fulfilled during the production phase due to imperfection in the process, however customers are satisfied either at the end of the inspection process or after reworking the imperfect products. Rework operation, inspection process, and planned backordering are incorporated in the proposed model. An analytical approach is utilized to optimize the lot size and planned backorder quantities based on the minimum average cost. Numerical examples are used to illustrate and compare the proposed model with previously developed models. The proposed model is considered more beneficial in comparison with the existing models as it incorporates imperfection, rework, inspection rate, and planned backorders.

Keywords: imperfect manufacturing system; backordering; defective products; rework; inspection

1. Introduction

Economic and production order quantity models have been extensively used in real industrial life for calculation of the optimum lot size. However, these models are based on the assumption that production processes result in perfect quality products always. There are various factors, which affect the manufacturing processes to make that assumption highly unrealistic. Engineers and technical experts reprocess these imperfect products and deliver them as per customer requirements after its qualification from the quality control section. For example, manufacturing processes involved in ultra-precision manufacturing industries including the automobile sector, ship manufacturing, defense and aerospace related products manufacturing industries, and tool manufacturing industries are a few of the common areas where the production of reworked and non-reworked products cannot be eliminated at all. Such products need qualification from the quality control department of the said organization in order to ensure that the product has been manufacturing as per customer requirements. In such setups, it is very rare that the product is delivered without its confirmation from inspection stations, i.e., demands fulfillment during the production stage. This proposed model re-considers the concept of Cárdenas-Barrón [1] to develop rework and backordering for the single-stage manufacturing setup.

Jamal et al. [2] introduced the rework operation to develop an optimal batch size for a single-stage production system. Two models were developed with the assumption that the rework operation can be performed either immediately or at the end of N-cycles. The optimal batch size was calculated based on the average minimum cost function. In the model, it is assumed that customer demands are met during the production process where demands are fulfilling during the production stage although the process is imperfect. Later, Cárdenas-Barrón [1] extended this model by incorporating the concept of planned backorders for a single-stage manufacturing setup. The rework operation was performed immediately at the end of the production process. Cárdenas-Barrón [1] obtained a closed-form solution for this model based on the average system cost. The optimum lot size and backorder quantity was calculated. Recently, Sarkar et al. [3] assumed that defective items may follow three kinds of distribution functions, i.e., uniform, triangular, and beta distribution. The analytical method has been used to calculate the optimal lot size and backorder quantity. These models assume imperfect processes during the production phase on one side and customer demands fulfillment on the other side. Therefore, inventory modeling in most of the models is based on $(p(1-\gamma)-d)$. The production model based on $(p(1-\gamma)-d)$ is based on the assumption that processes deviate from ideal conditions and are unable to produce quality products each and every time. This deviation from the ideal situation results in imperfection in processes that ultimately affect the products produced (p) by these processes. This imperfection results in the production of γ that need rectification / rework. Therefore, this number of items cannot be added to the total inventory buildup rather they need rectification. Hence, the net inventory is $(p(1-\gamma)-d)$ instead of $(p-d)$, which is applicable in the case of the perfect manufacturing environment. Whereas simultaneous demands fulfillment during the production phase is relatively unrealistic. Products need qualification from an inspection station before they are delivered to the customer or warehouse especially when processes are imperfect. Optimum lot sizes were calculated for real life scenarios, resulting in extension to the basic economic order quantity (EOQ) and production quantity models. Lee and Rosenblatt [4] introduced an idea of larger investment in quality control approaches to obtain the benefit of lower defective rate, reduced setup, and holding costs. Ben-Daya and Hariga [5] developed an economic lot size model with an important realistic assumption that perfect production processes deviate at random rate and results in some non-conforming products. Salameh and Jaber [6] extended the EOQ model for electronic industry items in particular with the assumption that 100% screening is performed of all incoming units. They assumed that low quality products could be sold out at the reduced price in a single batch. Later, Goyal and Cárdenas-Barrón [7] used a simple algebraic approach for optimum lot size calculation. The results were almost similar when other approaches were utilized in an imperfect production environment.

Biswas and Sarker [8] developed an optimal lot size for a lean manufacturing system where scrap items were identified 'before', 'during' and 'after' the rework process. Shortages in the production process were met through a buffer station of the qualified finish products. Chiu et al. [9] initiated a random defective rate within an imperfect production system. Pal et al. [10] worked on the production model with a stochastic demand not a stochastic defective rate. Ojha et al. [11] considered Chiu et al.'s [9] model with a quality assurance and rework. Chiu et al. [12] made their own model [9] with service level constraint. Rahim and Al-Hajailan [13] worked on Ojha et al.'s [11] model with a time-varying defective rate. Sarkar [14] considers delay-in-payments and stock-dependent demand. Sarkar et al. [15] converted a single-stage production system into a multi-stage production system when the defective item's reworking would be after n cycles or in each cycle that would be decided, whereas they considered a constant defective rate. Sarkar et al. [16] incorporated lead-time demand in the integrated production model. Ullah and Kang [17] mathematically managed the rejections of defective products and inspection on work-in-process lot-size. Cárdenas-Barrón et al. [18–20] covered three production models with discrete deliveries, multiple shipments, and partial reworks. Wee et al. [21] proposed an alternative solution approach of the production model. Sarkar [22] worked on the improved production model with the concept of reliability without any defective products. Tayyab and Sarkar [23] extended Sarkar et al.'s [3] model with multi-stage but with constant defective products. Sarkar and Moon [24] introduced the model the under inflation and time value of money. Sarkar et al. [25] extended Sarkar's [22] model with time-dependent demand under the reliable production system. Sarkar et al. [26] developed an optimal reliable model with defective products. However, none of the researchers consider the demand after production and full inspection to fulfill the customer's demand. Wee et al. [27] used the renewal reward theorem (RRT) to calculate the average total profit for optimization of the economic production quantity lot size by considering the imperfect production environment and screening constraint. Researchers (Sana [28,29] and Chaudhuri [30]) concluded that imperfect quality items can be reworked for their practical utilization and significance.

The basic objective of the study is to convert and model the imperfect production system into a mathematical form. The purpose of the proposed production model considers the scenario where the customer demands are not fulfilled during the production phase due to imperfection in the process, and customers are satisfied either at the end of the inspection process or after reworking the imperfect products. The rework operation, inspection process, and planned backordering are incorporated in the proposed model. An analytical approach is utilized to optimize the lot size and planned backorder quantities based on the minimum average cost. The introduction has been given in this section. The rest of the paper is modeled as follows: The next section is related to the past and current research studies related to the imperfection in the production model. Section 3 describes the model development with notation, assumptions, and equations required for optimal solutions. In Section 4, numerical examples are used to evaluate and discuss the proposed model's results. Section 5 includes the results and discussion of the numerical experiments performed using the proposed methodology. The sensitivity analysis has been given in detailed form in Section 6. Conclusions and future directions of the research are presented in the last section.

2. Literature Review

Most of the research carried out in the past concentrated on the production process itself. Defective products are produced in an out-of-control status of the process. These products are either scrapped or reworked in most of the research literature. One major aspect of the production setup is the product qualification via the inspection process. Industries including aerospace, automobile, and pharmaceutical cannot deliver the product until and unless it is qualified by the inspection stage. Inspection is considered as a process that consumes resources (time, material, techniques) in a similar way as are consumed by other processes. Relatively less attention has been given to the inspection rate in conjunction with the manufacturing process. However, in most of the inventory related literature, it is assumed that inspection takes either minimal time or is performed in parallel to the manufacturing processes. Whereas inspection rate importance is much required in environments where production processes are imperfect as the probability of non-conforming products increases. Ben-Salem et al. [31,32] considered an environmental issue and sub-control issues within two alternative production maintenance systems where the production degradation is considered without stopping the demanded products for the market. Purohit et al. [33] extended a basic production model with maintenance planning within an integrated approach, even though they considered production and demand in parallel. Ullah et al. [34] explained the inspection errors within the inspection system within a tradition production system. Shin et al. [35] extended Ullah et al.'s [34] model with trade-credit financing. Kim and Sarkar [36] converted the basic single-stage production system to the multi-stage production system where they did not consider initial production and then demand. Diabat et al. [37] extended the basic production model with partial downstream down payment. Kang et al. [38] explained the effect of smart manufacturing technology based on the human quality control system. Sarkar et al. [39] extended the field of the imperfect production with a distribution free approach to calculate shortages during lead-time demand. Omair et al. [40] and Sarkar et al. [41] extended the production model by considering the imperfect production with the inspection process, rework, and rejected products. Kang et al. [42] extended Cárdenas-Barrón's [1] model with safety stock and planned backorders whereas they did not think about imperfect products.

Researchers highlighted the impact of the inspection process on the optimum lot size during the last few decades. Zhang and Gerchak [43] introduced a model based on the joint lot size in addition with the inspection policy. The single period manufacturing problem was considered with uncertainty in demand and inconstant yield. Inspection errors effect on total cost and lot size was highlighted by Ben-Daya and Rahim [44] in a multistage production setup. They assumed that inspection is performed at the end of every stage. Ben-Daya et al. [45] developed models for an integrated system highlighting the impact of different inspection policies including 'no inspection', 'sample based inspection', and 100 percent inspection. Khan et al. [46] presented an extensive review of EOQ models focusing on the model developed by the Salameh and Jaber model [6]. The review classified various extended models based on imperfect items, shortages backordering, quality, and supply chain. Razaei [47] incorporated backorder into the inventory model for an imperfect manufacturing setup. Analytical approaches were used to get the optimal lot size and shortages level. Wee et al. [48] extended Salameh and Jaber's model [6] for the imperfect production setup incorporating the backorders. Taleizadeh et al. [49] developed an economic production order quantity model with disturbance in processes that result in scrap and rework. They considered the backorder with cycle length and optimal backorder quantity as the decision variable. Hu et al. [50] introduced an inventory model that highlighted that the backorder cost per unit is increasing linearly with shortage time. Ganguly et al. [51] considered partial backorders in supply chain management to meet the demand in the market.

Table 1 highlights contributions made during the last several years in the field of inventory management focusing on imperfect production processes. It can be observed in Table 1 that the literature assumes imperfect production processes for a production quantity model along with rework operation, inspection process, and backorder process. Most of the considered research works focuses on the objective to minimize the cost of the production system under the optimization process except the work of Wee et al. [27] (maximizing total profit of the production system). All these extended models assume that customer demands are fulfilled in the production phase. However, products need qualification from an inspection station before they are delivered to the customer especially when the process is imperfect. Therefore, it is possible that demands fulfillment during the production phase is not appropriate as the probability of delivering imperfect quality products to the customer rises. Research gap exists in the available literature, to the best of our knowledge, to ensure qualified products delivery to the end user while working in random imperfect production setup. This paper is an approach towards fulfillment of this gap in the existing literature. From Table 1, it is found that several authors considered production and supply in parallel when imperfect products are in the system, which is impractical to maintain the brand image of the industry. The most benefitted way is that during production, the products supply should not be allowed to meet the demand as imperfect products are there. This is a major research gap within all studies in this direction. Thus, this proposed model is considered to solve this issue.

This model assumes that products can only be consumed at a demand rate when they are either qualified or processes are perfect. The management delivers the manufactured products to the inspection station where products are classified as either qualified, rejected or to be reworked. It is assumed that imperfect products are rewardable. However, the probability of rejected products is low and can be ignored. It may further be added that the backordering process has also been taken into consideration to calculate the optimum backorder quantity in an imperfect manufacturing environment. It is hoped that this model integrates all major aspects (rework, inspection rate and backordering) of the production setup to help managers in making decisions regarding the inventory level and backorder quantity for an imperfect production setup focusing on qualified products delivery to the customer.

Table 1. Author contribution in the production model on the basis of process, rework, inspection, backordering and demand.

Author (s)	Process			Rework Process			Inspection Process			Backordering Process			Demands Fulfillment (d)	
	Imperfect	Re-Workable	Scrap	Perfect	Not Allowed	Imperfect	No	Sampling	100%	No Back-Ordering	Lost Sale	Back-Ordering	During Production Phase	After Inspection
Cárdenas-Barrón [1]	√	√		√					√		√	√	√	
Salameh and Jaber [6]	√		√		√		√			√				
Ojha et al. [11]	√	√		√					√	√				
Chiu et al. [12]	√	√	√			√	√					√	√	
Sarkar et al. [15]	√	√		√			√				√		√	
Wee et al. [27]	√		√		√				√			√	√	
Sana [29]	√	√		√			√			√				
Ben-Daya and Rahim [44]	√		√	√				√				√	√	
Ben-Salem et al. [31]	√		√		√									
Ullah et al. [34]	√	√					√			√	√			
Kim and Sarkar [36]	√			√								√	√	
Kang et al. [38]	√			√			√			√			√	
Sarkar et al. [39]	√	√	√	√	√	√	√	√	√	√	√	√	√	
This Paper	√	√	√	√					√			√		√

3. Mathematical Model

The problem starts with the assumption that the optimal lot size and backorder quantity are derived for an imperfect manufacturing environment. Lot size Q arrives at the manufacturing station during each cycle. Due to process imperfection, the γ percent of total products needs rectification as the process goes out-of-control status during the production phase. All these imperfect products are reworkable. Due to imperfection in the process, manufactured products are not delivered to the customer directly, rather these products are processed through an inspection stage with a known inspection rate (M) in order to know about product qualification. On the other side, all reworkable products are re-processed after the inspection process. It is assumed that reworked products are qualified after reprocessing and no re-inspection is required.

Other assumptions are as follows:

1. Demand and production rates are known and constant. During production, no product is to be sold. After production completion, the demand is started as the production system contains the defective product.
2. Production rate is higher than the demand rate.
3. Reworkable products produced during the manufacturing phase are known in percentage.
4. Imperfect products produced during the production processes are re-processed on the same machine with 100% qualification.
5. Customer receives qualified perfect products only to maintain the good brand image of the industry.
6. Qualified products are delivered to the customer after the inspection process.
7. Planned backordering is allowed and the inventory storage space is unlimited.

The following notation has been used to develop the model.

Parameters

d	demand rate (units/unit time)
p	production rate (units/unit time)
h	inventory holding cost per unit time ($/unit/unit time)
t_1	time required for planned backorder units to be fulfilled when process starts again.
t_2	time required by machining station to manufacture lot size Q
t_3	time required to inspect manufactured units
t_4	time required to rework imperfect units
t_5	time required to consume on hand inventory
t_6	time required to build the backorder inventory
I_1	average inventory developed during machining of lot size Q (units/unit time)
I_2	average inventory produced during rework operation (units/unit time)
I_{max}	maximum average inventory produced during one cycle time (units/unit time)
γ	percentage of imperfect products produced during the production phase (%)
Z	backorder cost per unit product per unit time ($/unit backorder/ unit time)
c	cost of manufacturing per unit product ($/unit)
k	setup cost per lot size ($/setup)
C_i	inventory holding cost per unit of time ($/unit time)
C_b	backorder cost per unit time ($/unit time)
C_s	setup cost per unit time ($/unit time)
C_m	manufacturing cost per unit time ($/unit time)
$TC(Q,B)$	total cost per unit time ($/unit time)

Variables

Q	lot size (units)
B	backorder quantity (units)

The inventory over time for an economic order production quantity model for a complete cycle is shown below in Figure 1. The process considers planned backorders, manufacturing of lot size, inspection process, and rework process of imperfect products. As stated earlier, customer demands are not fulfilled during the production phase due to the process imperfection. The objective is to ensure qualified products for delivery to the customer. The inventory builds up during the production phase. Demands are not fulfilled in the production phase. Customer demands are fulfilled at the rate d during the rework process only. Some percentages of products are qualified at the rework stage and can be delivered to the customer in order to minimize inventory cost.

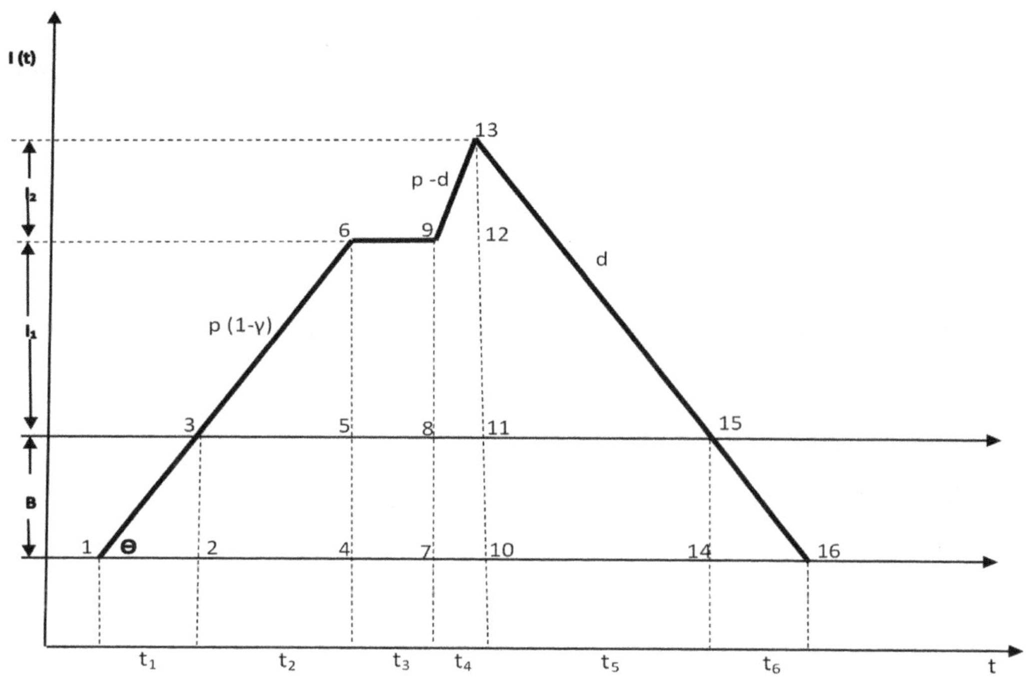

Figure 1. Inventory over time in an imperfect manufacturing environment with inspection and planned backordering.

Considering Figure 1, it is found that the time required to produce lot size Q at the production rate p is the sum of the time required by different processes to produce Q units. These processes include the time required for planned backorder units to be fulfilled when the process starts again (t_1), time required to manufacture lot size Q (t_2), time required to inspect the manufactured units (t_3), and time required to rework imperfect units (t_4), i.e.,

$$t_p = t_1 + t_2 + t_3 + t_4 \tag{1}$$

Obtaining these times in terms of the demand rate (d) and inventory lot size Q,

From Figure 1, the area of Δ 146 is the accumulated by production and backorder inventory, one can obtain

$$\tan\,\theta = \frac{I_1 + B}{t_1 + t_2}$$

$$p(1-\gamma) = \frac{I_1 + B}{t_1 + t_2}$$

$$t_1 + t_2 = \frac{I_1 + B}{p(1-\gamma)} \tag{2}$$

The time required for the inspection of manufactured products (t_3) is given by the following relation at the rate M,

$$t_3 = \frac{I_1 + B}{M} \quad (3)$$

Similarly, t_4 is the time required to rework imperfect products,

$$t_4 = \frac{\gamma Q}{p}$$

In addition, the area of Δ 91213 is the accumulated inventory during reworking, then the time can be obtained as follows:

$$t_4 = \frac{I_2}{(p - d)}$$

i.e.,

$$I_2 = \gamma Q\left(1 - \frac{d}{p}\right) \quad (4)$$

Therefore, (1) becomes

$$\frac{Q}{p} = \frac{I_1 + B}{p(1-\gamma)} + \frac{I_1 + B}{M} + \frac{\gamma Q}{p}$$

After some algebraic rearrangements and simplification, one can obtain

$$I_1 = \frac{MQ(1-\gamma)^2}{M + p(1-\gamma)} - B \quad (5)$$

Therefore, from (4) and (5), the maximum inventory is given by

$$I_{max} = I_1 + I_2 = \frac{QM\,(1-\gamma)^2}{M + p(1-\gamma)} - B + \gamma Q\left(1 - \frac{d}{p}\right) \quad (6)$$

Assuming

$$\frac{(1-\gamma)^2}{M + p(1-\gamma)} = \theta_1 \text{ and } (1 - \frac{d}{p}) = \theta_2, \text{ then}$$

$I_1 = MQ\theta_1 - B$
$I_2 = \gamma Q\theta_2$
$I_{max} = MQ\theta_1 + \gamma Q\theta_2 - B$

The average inventory can be found by taking the sum of the area of triangles and rectangles as shown in Figure 1 divided by the cycle time. The cycle time is defined as the time in which Q units inventory is consumed at the demand rate (d).

From the area Δ 356, the inventory for accruing Q quantity during inspection, area A_1 is given by

$$A_1 = \frac{I_1 \times t_2}{2}$$

where $t_2 = \frac{I_1}{p(1-\gamma)}$, putting in the above equation

$$A_1 = \frac{(MQ\theta_1 - B)^2}{2p(1-\gamma)} \quad (7)$$

The area of Δ 5698 is the area for holding inventory Q quantity during inspection, A_2 is as follows:

$$A_2 = I_1 \times t_3$$

where $t_3 = \frac{(I_1+B)}{M}$, thus

$$A_2 = \frac{(MQ\theta_1 - B)^2 + (B(MQ\theta_1 - B))}{M} \tag{8}$$

Similarly,

$A_3 = I_1 \times t_4$, where $t_4 = \frac{\gamma Q}{p}$

$$A_3 = \left(\frac{MQ^2\theta_1\gamma}{p} - \frac{BQ\gamma}{p}\right) \tag{9}$$

The area of Δ 91213 is the area for reworking, which gives the building of inventory. One can have

$$A_4 = \frac{I_2 \times t_4}{2}$$

$$A_4 = \frac{Q^2\gamma^2\theta_2}{2p} \tag{10}$$

The area A_5 of Δ111315 is the area for demand of products without any production. Thus, the area can be found as follows:

$$A_5 = \frac{I_{max} \times t_5}{2}, \text{ where } t_5 = \frac{I_{max}}{d}$$

Therefore,

$$A_5 = \frac{(MQ\theta_1 + \gamma Q\theta_2 - B)^2}{2d} \tag{11}$$

Therefore, the average inventory $\left(I_{avg}\right)$ per unit cycle time is the summation of (7–10), and (11) divided by the cycle time (t), i.e.,

$$I_{avg} = \left\{ \frac{\left(\frac{(MQ\theta_1 - B)^2}{2p(1-\gamma)}\right) + \left(\frac{(MQ\theta_1 - B)^2 + (B(MQ\theta_1 - B))}{M}\right) + \left(\frac{MQ^2\theta_1\gamma}{p} - \frac{BQ\gamma}{p}\right) + \left(\frac{Q^2\gamma^2\theta_2}{2p}\right) + \left(\frac{(MQ\theta_1 + \gamma Q\theta_2 - B)^2}{2d}\right)}{\frac{Q}{d}} \right\} \tag{12}$$

Algebraic simplification results in the following expression:

$$\begin{aligned} I_{avg} = \{(&\frac{(M^2Q^2\theta_1^2 + B^2 - 2\,BMQ\theta_1)}{2p(1-\gamma)} + MQ^2\theta_1^2 - BQ\theta_1 + \frac{MQ^2\theta_1\gamma}{p} + \frac{Q^2\gamma^2\theta_2}{2p} - \frac{BQ\gamma}{p} + \\ &\frac{1}{2d}(M^2Q^2\theta_1^2 + B^2 + Q^2{\theta_2}^2\gamma^2 - 2BMQ\theta_1 - 2\,BQ\gamma\theta_2 + 2MQ^2\theta_1\theta_2\gamma))/(\tfrac{Q}{d})\} \end{aligned} \tag{13}$$

Hence, the total inventory holding cost per unit cycle time is given by

$$\begin{aligned} C_i = h\{(&\frac{(M^2Q^2\theta_1^2 + B^2 - 2\,BMQ\theta_1)}{2p(1-\gamma)} + MQ^2\theta_1^2 - BQ\theta_1 + \frac{MQ^2\theta_1\gamma}{p} + \frac{Q^2\theta_2\gamma^2}{2p} - \frac{BQ\gamma}{p} \\ &+ \frac{1}{2d}(M^2Q^2\theta_1^2 + B^2 + Q^2{\theta_2}^2\gamma^2 - 2BMQ\theta_1 - 2\,BQ\gamma\theta_2 + 2MQ^2\theta_1\theta_2\gamma))/(\tfrac{Q}{d})\} \end{aligned} \tag{14}$$

The average backordering cost can be computed by calculating the area of Δ 123 and Δ 141516 which represent the average inventory in the form of backordered products per unit time (l).

Therefore,

$$\begin{aligned} \text{Backorder cost } (C_b) &= (\text{Average backorder inventory}) \times (\text{Unit cos t per unit backorder product per unit of time}) \\ &= \frac{(\text{Area of } (\Delta\ 123) + \text{Area of} (\Delta\ 141516))}{t} \times (z) \\ &= \frac{\left(\frac{B \times t_1}{2} + \frac{B \times t_6}{2}\right)}{\frac{Q}{d}} (z) \end{aligned}$$

where $t_1 = \frac{B}{p(1-\gamma)}$, $t_6 = \frac{B}{d}$, which gives

$$C_b = \left(\frac{B^2(p(1-\gamma)+d)}{2p(1-\gamma)Q}\right)(z) \tag{15}$$

Other important costs taken into consideration are the setup cost (C_s) and the manufacturing cost (C_m).

$$C_s = \frac{kd}{Q} \tag{16}$$

$$C_m = cd(1+\gamma) \tag{17}$$

Therefore, following the relation for all associated costs including the inventory-holding cost, setup cost, backorder cost and manufacturing cost per unit of time.

$$TC\ (Q,B) = h\{(\tfrac{M^2Q^2\theta_1^2+B^2-2\ BMQ\theta_1}{2p(1-\gamma)} + MQ^2\theta_1^2 - BQ\theta_1 + \tfrac{MQ^2\theta_1\gamma}{p} + \tfrac{Q^2\theta_2\gamma^2}{2p} - \tfrac{BQ\gamma}{p} + \\ \tfrac{1}{2d}(M^2Q^2\theta_1^2 + B^2 + Q^2{\theta_2}^2\gamma^2 - 2BMQ\theta_1\ - 2\ BQ\gamma\theta_2 + \ 2MQ^2\theta_1\theta_2\gamma))/(\tfrac{Q}{d})\} \\ + (\tfrac{B^2(p(1-\gamma)+d)}{2p(1-\gamma)Q})(z) + cd(1+\gamma) + \tfrac{kd}{Q} \tag{18}$$

After algebraic simplification, one obtains:

$$TC\ (Q,B) = \left(\frac{dhM^2\theta_1^2}{2p(1-\gamma)} + dhM\theta_1^2 + \frac{dh\theta_2\gamma^2}{2p} + \frac{dhM\theta_1\gamma}{p} + \frac{M^2h{\theta_1}^2}{2} + \frac{h{\theta_2}^2\gamma^2}{2} + hM\theta_1\theta_2\gamma\right)Q$$

$$\left(\frac{dh}{p(1-\gamma)} + h + \frac{(p(1-\gamma)+d)z}{p(1-\gamma)}\right)\frac{B^2}{2Q} - \left(\frac{dhM\theta_1}{p(1-\gamma)} + dh\theta_1 + \frac{dh\gamma}{p} + hM\theta_1 + h\theta_2\gamma\right)B + cd(1+\gamma) + \frac{kd}{Q}$$

The following symbols are used to simplify the above expression,

$$R_1 = \left(\frac{dhM^2\theta_1^2}{2p(1-\gamma)} + dhM\theta_1^2 + \frac{dh\theta_2\gamma^2}{2p} + \frac{dhM\theta_1\gamma}{p} + \frac{M^2h{\theta_1}^2}{2} + \frac{h{\theta_2}^2\gamma^2}{2} + hM\theta_1\theta_2\gamma\right)$$

$$R_2 = \left(\frac{dh}{p(1-\gamma)} + h + \frac{(p(1-\gamma)+d)z}{p(1-\gamma)}\right)$$

$$R_3 = \left(\frac{dhM\theta_1}{p(1-\gamma)} + dh\theta_1 + \frac{dh\gamma}{p} + hM\theta_1 + h\theta_2\gamma\right)$$

thus,

$$TC\ (Q,B) = Q(R_1) + \left(\frac{B^2}{2Q}\right)(R_2) - B(R_3) + \frac{kd}{Q} + cd(1+\gamma) \tag{19}$$

The optimum lot size (Q^*) and backorder quantity (B^*) can be found by minimization of the total cost function assuming Q and B as continues function.

Theorem 1. *$TC^*(Q,B)$ will have global maximum values at Q^* and B^* if* $Q^* = \sqrt{\frac{2kd(R_2)}{2R_1R_2-(R_3)^2}}$ *and* $B^* = \frac{R_3}{R_2}\left(\sqrt{\frac{2kd(R_2)}{2R_1R_2-(R_3)^2}}\right)$.

Proof. The following conditions need to be fulfilled in order to verify that (19) is a convex function.

$$\text{I)}\quad \frac{\partial^2 TC\ (Q,B)}{\partial Q^2} > 0,\ \frac{\partial^2 TC\ (Q,B)}{\partial B^2} > 0 \\ \text{II)}\quad \left(\frac{\partial^2 TC\ (Q,B)}{\partial Q^2}\right)\left(\frac{\partial^2 TC\ (Q,B)}{\partial B^2}\right) - \left(\frac{\partial^2 TC\ (Q,B)}{\partial Q \partial B}\right)^2 > 0 \tag{20}$$

□

Assuming Q and B as continues, first partial derivatives with respect to Q and B of (19) are shown in (21) and (22), respectively.

$$\frac{\partial TC\,(Q,\,B)}{\partial Q} = (R_1) - \left(\frac{B^2}{2Q^2}\right)(R_2) - \frac{kd}{Q^2} \tag{21}$$

$$\frac{\partial TC\,(Q,\,B)}{\partial B} = \left(\frac{B}{Q}\right)(R_2) - (R_3) \tag{22}$$

Similarly,

$$\frac{\partial^2 TC\,(Q,\,B)}{\partial Q^2} = \left(\frac{B^2}{Q^3}\right)(R_2) + \frac{2kd}{Q^3} > 0 \tag{23}$$

$$\frac{\partial^2 TC\,(Q,\,B)}{\partial B^2} = \left(\frac{R_2}{Q}\right) > 0 \tag{24}$$

$$\left(\frac{\partial^2 TC\,(Q,\,B)}{\partial Q^2}\right)\left(\frac{\partial^2 TC\,(Q,\,B)}{\partial B^2}\right) - \left(\frac{\partial^2 TC\,(Q,\,B)}{\partial Q \partial B}\right)^2 = \frac{2kd}{Q^4}(R_2) \tag{25}$$

The necessary and sufficient conditions for (19) to be convex have been fulfilled as shown in (23), (24), and (25). Therefore, an optimal solution exists at which the total cost function will be minimum. Thus, (21) and (22) are simultaneously solved to obtain the optimal lot size (Q) and backorder quantity (B) which are given by:

$$Q^* = \sqrt{\frac{2kd(R_2)}{2R_1R_2 - (R_3)^2}} \tag{26}$$

$$B^* = \frac{R_3}{R_2}\left(\sqrt{\frac{2kd(R_2)}{2R_1R_2 - (R_3)^2}}\right) \tag{27}$$

The optimal total cost can be obtained by substituting (26) and (27) in (19) as follows:

$$TC^*(Q^*, B^*) = \sqrt{\frac{2kd\left(2R_1R_2 - R_3^2\right)}{R_2}} + cd(1+\gamma)$$

Therefore, $TC^*(Q^*, B^*)$ is the global minimum total cost at Q^* and B^*.

This proposed model could be reduced to the basic economic production quantity model (EPQ) if processes are assumed perfect, demand is fulfilled during the production phase and shortages are not expected. In addition, the optimum lot size and optimum backorder quantity obtained by (26) and (27) can be reduced to the mathematical model developed by Cárdenas-Barrón [1] if the assumption of meeting customer demands during the production phase is made.

4. Numerical Examples

Two numerical examples have been used to understand the proposed model utilization for practical engineering problems and its comparison with previously developed models. Data has been taken from the Cárdenas-Barrón's model [1]. Numerical examples illustrate the effect of imperfect products and the inspection rate on the optimum lot size and backorder quantity. The total cost, optimal lot size and optimal backorder quantity at different values of the defect rate and inspection rate are calculated.

4.1. Example 1

The parametric data used for the mathematical experiment of the proposed model include the demand rate (d) = 300 units per year, inspection rate (M) = 550 units per year, production rate (p) = 550 units per year, holding cost (h) = \$50 per unit per year, shortage cost per unit (z) = \$10 per unit short per year, manufacturing cost (c) = \$7 per unit, setup cost (k) = \$50 per unit per year.

4.2. Example 2

The similar type of analysis is performed for another example whose data set is relatively for higher demand and production rate. In this example, the demand rate (d) = 4800 units per year, production rate (p) = 24,000 units per year, setup cost (k) = \$120 per lot size, holding cost (h) = \$0.6 per unit per year, shortages cost (z) = \$14 per unit short per year and manufacturing cost (c) = \$3 per unit. The inspection rate is assumed as 36,000 units per year and defective products produced during the process are varied gradually over an interval (0, 40%).

5. Results and Discussion

Table 2 shows the impact of the defective rate on the optimal lot size (Q) and optimal backorder quantity (B). It can be observed that lot size increases with an increase in the percentage of the defective rate. Similarly, the total cost variation is also significant. Total cost is increased with an increase in imperfection during the manufacturing process. However, an increase in the optimal backorder quantity level is relatively low with an increase in the defective rate. Comparing these results with the results obtained by Cárdenas-Barrón [1] (Example 1), it can be observed that the total cost of our proposed model is comparable for the range of imperfection (0%, 40%). However, the lot size has been increased significantly. The backorder quantity increases up to 35% of the defective rate and then decreases with increases in defective products.

Table 2. Change in optimal lot size and backorder quantity with variation in the defective rate (Example 1).

γ(%)	θ_1	Total Cost (\$/Year)	Q^*(Units)	B^*(Units)
0	0.000909091	2423.44	93	52
1	0.000895477	2437.49	95	53
5	0.000841492	2493.71	104	57
10	0.00077512	2564.18	118	62
15	0.000710074	2635.20	136	69
20	0.000646465	2707.40	160	79
25	0.000584416	2782.06	191	90
30	0.000524064	2861.69	228	104
35	0.000465565	2950.74	259	113
40	0.000409091	3054.67	262	109

It can be observed in Table 3 that the total cost and optimal lot size increase with an increase in the defective rate significantly at higher demand and production rate. The higher the defective products produced, the higher will be the lot size to be ordered. However, the change in backorder units is less significant for higher production rate and demand rate when the holding cost per unit product is comparatively low. Comparing these results with the results obtained by Cárdenas-Barrón [1] (Example 2), the total cost of our proposed model has not been increased significantly although the inspection process and carrying inventory over a longer period have been incorporated. Relatively larger lot sizes have been proposed by our model.

The change in optimal lot size with the change in the inspection rate has been shown in Figure 2 below. Example 2 data has been used for further analysis. It can be observed that the optimal lot size goes on decreasing with an increase in the inspection rate. It may be noted that the inspection rate is increased from 24,000 units (equal to the production rate) to 42,000 units per year (one and half time higher than the production rate). Defective products rate remain fixed at 20%. All other data remain the same. The total cost and backorder quantity level has been shown in Table 4. It may be noted that the backorder quantity level doesn't change with significant change in the inspection rate at the fixed level of defective products produced.

Table 3. Change in optimal lot size and backorder quantity with variation in the defective rate (Example 2).

γ(%)	θ_1	Total Cost ($/Year)	Q^* (Units)	B^*(Units)
0	0.000016	14,991.78	1947	52
1	0.000016	15,133.44	1954	52
5	0.000015	15,700.49	1985	52
10	0.000014	16,410.33	2020	52
15	0.000012	17,121.43	2052	52
20	0.000011	17,833.88	2080	52
25	0.000010	18,547.80	2103	51
30	0.000009	19,263.32	2120	51
35	0.000008	19,980.57	2131	51
40	0.000007	20,699.69	2135	50

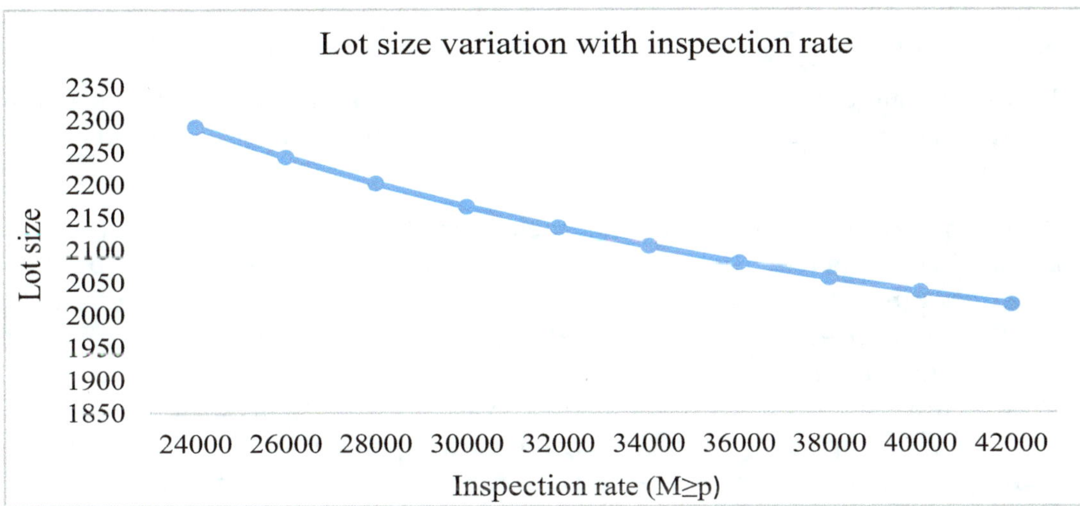

Figure 2. Change in optimal lot size with change in the inspection rate.

Table 4. Change in optimal lot size and backorder quantity with variation in the inspection rate (Example 2).

M (Units Per Year)	θ_1	Total Cost ($/Year)	Q^*(Units)	B^* (Units)
24,000	0.0000148	17,783.27	2289	52
26,000	0.0000141	17,793.67	2243	52
28,000	0.0000135	17,803.15	2202	52
30,000	0.0000130	17,811.81	2166	52
32,000	0.0000125	17,819.77	2134	52
34,000	0.0000120	17,827.10	2106	52
36,000	0.0000115	17,833.88	2080	52
38,000	0.0000112	17,840.17	2057	52
40,000	0.0000108	17,846.02	2035	52
42,000	0.0000104	17,851.47	2016	52

At the end, both the inspection rate and defective products production rate are changed simultaneously with an ascending order and their impact on the lot size, backorder level, and total cost is evaluated. The data has been shown in Table 5. It can be observed that total cost increases with an increase in the inspection and defective rate. Moreover, it can be observed that for higher inspection rate relative to the production process (M>>>p), the optimal lot size calculated by our proposed model approaches the lot size calculated by using the economic order quantity model under ideal conditions.

Table 5. Change in optimal lot size and backorder quantity with variation in the defective rate and inspection rate (Example 2).

M (Units Per Year)	γ(%)	θ_1	Total Cost ($/Year)	Q^* (Units)	B^* (Units)
24,000	0	0.000020	14,914.97	2237	52
26,000	5	0.000018	15,644.61	2196	52
28,000	10	0.000016	16,371.63	2167	52
30,000	15	0.000014	17,096.52	2147	52
32,000	20	0.000125	17,819.77	2134	52
34,000	25	0.000011	18,541.88	2126	52
36,000	30	0.0000092	19,263.32	2120	51
38,000	35	0.0000078	19,984.59	2115	50
40,000	40	0.0000066	20,706.15	2109	50
42,000	45	0.0000054	21,428.45	2100	49

6. Sensitivity Analysis

In order to evaluate the impact of important parameters on the optimal lot size and optimal backorder quantity as proposed by the developed model, sensitive analysis is carried out. It is based on the data shown in Example 2.

Table 6 summarizes the effect of key parameters on the optimal lot size, optimal backorder quantity, and total cost function. Parameters' values have been changed from −50% to +50%. It can be observed that the lot size increases with the increase in values for parameters k, d, p, and γ. Furthermore, the lot size decreases for increased values for parameters h M, *and* z. Similarly, the optimal backorder quantity increases for increased values of parameters k, d, h, and p. The impact of parameters' M, γ, and z is reverse, i.e., optimal backorder quantity decreases for higher values of M and z. Furthermore, the impact of positive change in parameters k, d, h, M, γ, and z values increases the total cost function value. However, the total cost function value decreases with increased values for parameter p. The setup cost and holding cost are key costs for any production system. If all costs are fixed and these costs are increased, the total cost must be increased. Within these costs, it is found that the setup cost is more sensitive than the holding cost. It means that the industry can reduce further the total cost by reducing the setup cost using some initial investment. It is found that the production rate is increased whereas the total cost is reduced. In the proposed model, the production rate is constant and it's increasing value gives decreasing total cost. This implies that the industry can think about the controllable production rate within a certain limit of minimum production rate and maximum production rate. Then, the total cost of the production system can be reduced more. As it is a traditional production system, thus the inspection cost is controlled through human inspection. With the increasing value of this cost, the total cost increases; this implies that the industry manager should think about the smart production system where all inspections controlled through online by the machine and production rate is flexible. The rejection cost is the most sensitive cost as all types of efforts are there with all procedures but there are no products that can be sold to obtain revenue. Thus, the machinery system should be perfect always to produce perfect product always such that there should not be any rejection cost. The shortage cost should not be in the production system as the production rate is greater than the demand but as during production no demanded products are given to the market. Thus, this cost has to incorporate. Even though, the shortage cost is there, the total cost is increasing very less comparing to the other costs.

Table 6. Sensitivity analysis (Example 2).

Parameter	Changes (%)	Parameter Value	Q (Units)	B (Units)	Total Cost ($/Year)
Setup cost (k)	−50%	60	1471	37	17,671.65
	−25%	90	1801	45	17,759.68
	+25%	150	2325	58	17,899.26
	+50%	180	2547	63	17,958.36
Holding cost (h)	−50%	0.3	2910	37	17,675.81
	−25%	0.45	2389	45	17,762.20
	+25%	0.75	1870	58	17,896.04
	+50%	0.9	1716	63	17,951.37
Production rate (p)	−50%	12,000	1720	47	17,949.81
	−25%	18,000	1915	50	17,881.72
	+25%	30,000	2229	53	17,796.91
	+50%	36,000	2366	54	17,766.95
Inspection rate (M)	−50%	18,000	2479	53	17,744.79
	−25%	27,000	2222	52	17,798.52
	+25%	45,000	1990	51	17,858.99
	+50%	54,000	1927	51	17,877.76
Rejection (%)	−50%	0.1	2020	52	16,410.33
	−25%	0.15	2052	52	17,121.43
	+25%	0.25	2103	51	18,547.80
	+50%	0.3	2120	51	19,263.32
Shortage cost (z)	−50%	7.2	2123	102	17,822.63
	−25%	10.8	2094	69	17,830.06
	+25%	18	2071	42	17,836.21
	+50%	21.6	2065	35	17,837.78

7. Conclusions

In this paper, a single stage manufacturing model for the imperfect production setup with rework, inspection, and backordering has been developed. It is assumed that inventory is carried over a longer period of time in the production phase, as the process is imperfect. Due to process imperfection, products are not delivered directly to the customer as observed in most of the conventional production order quantity models. This model optimized the lot size and backorder quantity in an imperfect manufacturing setup when demands are fulfilled after the production process. This approach is against the conventional production order quantity models where processes are imperfect and demands are fulfilled simultaneously. Moreover, products were passed through an inspection station for their qualification. The results were compared via numerical examples to the previously developed model from literature. The numerical examples highlighted the importance of the proposed model as compared to the previous model by considering the quality of the product through the inspection process, however, the cost is slightly been increased. Furthermore, as the demand rate and production rate were increased, the total cost of this model has been reduced in a significant way. The model provides insights to manufacturing engineers working in an imperfect manufacturing setup. It can be best utilized when manufacturing industries deliver products after the inspection process. Total cost is increased when demand is not fulfilled during the production process due to process imperfection under given conditions. Furthermore, the proposed model approaches the production quantity model to calculate the optimal lot size, when the inspection rate is much higher in comparison to the production rate under ideal situations. The model can be extended to incorporate the partial backordering and multistage manufacturing environment to obtain an optimal lot size and backorder level. The consideration of the suppliers and buyers with the imperfection process and sample based inspection can be the extension of the study with the assumption that demands are fulfilled after the inspection process. This production can be incorporated within an integrated inventory model or within a supply chain model to extend it further. The production rate can be considered as variable to make it a smart production system to produce smart products.

Author Contributions: C.W.K. supervised the research work and presented an idea; B.S. conceived and designed the research plan by the support of mathematical modeling; M.U. performed the numerical experiments and wrote the original draft; M.S. analyzed the data; M.O. contributed by computer software tool to get numerical results. All authors worked equally in reporting revisions and updating the manuscript.

References

1. Cárdenas-Barrón, L.-E. Economic production quantity with rework process at a single-stage manufacturing system with planned backorders. *Comput. Ind. Eng.* **2009**, *57*, 1105–1113. [CrossRef]
2. Jamal, A.; Sarker, B.R.; Mondal, S. Optimal manufacturing batch size with rework process at a single-stage production system. *Comput. Ind. Eng.* **2004**, *47*, 77–89. [CrossRef]
3. Sarkar, B.; Cárdenas-Barrón, L.-E.; Sarkar, M.; Singgih, M.L. An economic production quantity model with random defective rate, rework process and backorders for a single stage production system. *J. Manuf. Syst.* **2014**, *33*, 423–435. [CrossRef]
4. Rosenblatt, M.-J.; Lee, H.-L. Economic production cycles with imperfect production processes. *IIE Trans.* **1986**, *18*, 48–55. [CrossRef]
5. Ben-Daya, M.; Hariga, M. Economic lot scheduling problem with imperfect production processes. *J. Oper. Res. Soc.* **2000**, *51*, 875–881. [CrossRef]
6. Salameh, M.; Jaber, M. Economic production quantity model for items with imperfect quality. *Int. J. Prod. Econ.* **2000**, *64*, 59–64. [CrossRef]
7. Goyal, S.-K.; Cárdenas-Barrón, L.-E. Note on: Economic production quantity model for items with imperfect quality—A practical approach. *Int. J. Prod. Econ.* **2002**, *77*, 85–87. [CrossRef]
8. Biswas, P.; Sarker, B.-R. Optimal batch quantity models for a lean production system with in-cycle rework and scrap. *Int. J. Prod. Res.* **2008**, *46*, 6585–6610. [CrossRef]
9. Chiu, S.-W.; Gong, D.-C.; Wee, H.-M. Effects of random defective rate and imperfect rework process on economic production quantity model. *Jpn. J. Ind. Appl. Math.* **2004**, *2*, 375–389. [CrossRef]
10. Pal, B.; Sana, S.-S.; Chaudhuri, K. A mathematical model on EPQ for stochastic demand in an imperfect production system. *J. Manuf. Syst.* **2013**, *32*, 260–270. [CrossRef]
11. Ojha, D.; Sarker, B.; Biswas, P. An optimal batch size for an imperfect production system with quality assurance and rework. *Int. J. Prod. Res.* **2007**, *45*, 3191–3214. [CrossRef]
12. Chiu, S.-W.; Ting, C.-K.; Chiu, Y.-S.-P. Optimal production lot sizing with rework, scrap rate, and service level constraint. *Math. Comput. Model.* **2007**, *46*, 535–549. [CrossRef]
13. Rahim, M.; Al-Hajailan, W. An optimal production run for an imperfect production process with allowable shortages and time-varying fraction defective rate. *Int. J. Adv. Manuf. Technol.* **2006**, *27*, 1170–1177. [CrossRef]
14. Sarkar, B. An EOQ model with delay in payments and stock dependent demand in the presence of imperfect production. *Appl. Math. Comput.* **2012**, *21*, 8295–8308. [CrossRef]
15. Sarker, B.-R.; Jamal, A.-M.-M.; Mondal, S. Optimal batch sizing in a multi-stage production system with rework consideration. *Eur. J. Oper. Res.* **2008**, *184*, 915–929. [CrossRef]
16. Sarkar, B.; Gupta, H.; Chaudhuri, K.; Goyal, S.-K. An integrated inventory model with variable lead time, defective units and delay in payments. *Appl. Math. Comput.* **2014**, *237*, 650–658. [CrossRef]
17. Ullah, M.; Kang, C.-W. Effect of rework, rejects and inspection on lot size with work-in-process inventory. *Int. J. Prod. Res.* **2014**, *5*, 2448–2460. [CrossRef]
18. Cárdenas-Barrón, L.E.; Taleizadeh, A.A.; Treviño-Garza, G. An improved solution to replenishment lot size problem with discontinuous issuing policy and rework, and the multi-delivery policy into economic production lot size problem with partial rework. *Expert Syst. Appl.* **2012**, *39*, 13540–13546. [CrossRef]
19. Cárdenas-Barrón, L.-E.; Sarkar, B.; Treviño-Garza, G. An improved solution to the replenishment policy for the EMQ model with rework and multiple shipments. *Appl. Math. Model.* **2013**, *37*, 5549–5554. [CrossRef]
20. Cárdenas-Barrón, L.-E.; Treviño-Garza, G.; Widyadana, G.A.; Wee, H.-M. A constrained multi-products EPQ inventory model with discrete delivery order and lot size. *Appl. Math. Comput.* **2014**, *230*, 359–370. [CrossRef]
21. Wee, H.-M.; Wang, W.-T.; Cárdenas-Barrón, L.-E. An alternative analysis and solution procedure for the EPQ model with rework process at a single-stage manufacturing system with planned backorders. *Comput. Ind. Eng.* **2013**, *64*, 748–755. [CrossRef]

22. Sarkar, B. An inventory model with reliability in an imperfect production process. *Appl. Math. Comput.* **2012**, *218*, 4881–4891. [CrossRef]
23. Tayyab, M.; Sarkar, B. Optimal batch quantity in a cleaner multi-stage lean production system with random defective rate. *J. Clean. Prod.* **2016**, *139*, 922–934. [CrossRef]
24. Sarkar, B.; Moon, I. An EPQ model with inflation in an imperfect production system. *Appl. Math. Comput.* **2011**, *217*, 6159–6167. [CrossRef]
25. Sarkar, B.; Mandal, P.; Sarkar, S. An EMQ model with price and time dependent demand under the effect of reliability and inflation. *Appl. Math. Comput.* **2014**, *231*, 414–421. [CrossRef]
26. Sarkar, B.; Sana, S.-S.; Chaudhuri, K. Optimal reliability, production lot size and safety stock in an imperfect production system. *Int. J. Math. Oper. Res.* **2010**, *2*, 467–490. [CrossRef]
27. Wee, H.-M.; Wang, W.-T.; Yang, P.-C. A production quantity model for imperfect quality items with shortage and screening constraint. *Int. J. Prod. Res.* **2012**, *51*, 1869–1884. [CrossRef]
28. Sana, S.-S. A production–inventory model in an imperfect production process. *Eur. J. Oper. Res.* **2010**, *200*, 451–464. [CrossRef]
29. Sana, S.-S. An economic production lot size model in an imperfect production system. *Eur. J. Oper. Res.* **2010**, *201*, 158–170. [CrossRef]
30. Sana, S.-S.; Chaudhuri, K. An EMQ model in an imperfect production process. *Int. J. Syst. Sci.* **2010**, *41*, 635–646. [CrossRef]
31. Ben-Salem, A.; Gharbi, A.; Hajji, A. Environmental issue in an alternative production-maintenance control for unreliable manufacturing system subject to degradation. *Int. J. Adv. Manuf. Technol.* **2015**, *77*, 383–398. [CrossRef]
32. Ben-Salem, A.; Gharbi, A.; Hajji, A. Production and uncertain green subcontracting control for an unreliable manufacturing system facing emissions. *Int. J. Adv. Manuf. Technol.* **2016**, *83*, 1787–1799. [CrossRef]
33. Purohit, B.-S.; Lad, B.K. Production and maintenance planning: An integrated approach under uncertainties. *Int. J. Adv. Manuf. Technol.* **2016**, *86*, 3179–3191. [CrossRef]
34. Ullah, M.; Kang, C.-W.; Sarkar, B. Human errors incorporation in work-in-process group manufacturing system. *Sci. Iran.* **2017**, *24*, 2050–2061.
35. Shin, D.; Mittal, M.; Sarkar, B. Effects of human errors and trade-credit financing in two-echelon supply chain model. *Eur. J. Ind. Eng.* **2018**, *12*, 465–503. [CrossRef]
36. Kim, M.; Sarkar, B. Multi-stage cleaner production process with quality improvement and lead time dependent ordering cost. *J. Clean. Prod.* **2017**, *144*, 572–590. [CrossRef]
37. Diabat, A.; Taleizadeh, A.-A.; Lashgari, M. A lot sizing model with partial downstream delayed payment, partial upstream advanced payment, and partial backordering for deteriorating items. *J. Manuf. Syst.* **2017**, *45*, 322–342. [CrossRef]
38. Kang, C.-W.; Babar, B.-M.; Sarkar, B.; Imran, M. Effect of inspection performance in smart manufacturing system based on human quality control system. *Int. J. Adv. Manuf. Technol.* **2018**, *94*, 4351–4364. [CrossRef]
39. Sarkar, B.; Kim, J.-S.; Kim, M.; Sarkar, M.; Iqbal, W. An application of distribution free approach through an integrated inventory model with imperfect production and backorders. *J. Manuf. Syst.* **2018**, *4*, 153–167.
40. Omair, M.; Sarkar, B.; Cárdenas-Barrón, L.E. Minimum quantity lubrication and carbon footprint: A step towards sustainability. *Sustainability* **2017**, *9*, 714. [CrossRef]
41. Sarkar, B.; Omair, M.; Choi, S.-B. A multi-objective optimization of energy, economic, and carbon emission in a production model under sustainable supply chain management. *Appl. Sci.* **2018**, *8*, 1744. [CrossRef]
42. Kang, C.-W.; Ullah, M.; Sarkar, B. Optimum ordering policy for an imperfect single-stage manufacturing system with safety stock and planned backorder. *Int. J. Adv. Manuf. Technol.* **2018**, *95*, 109–120. [CrossRef]
43. Zhang, X.; Gerchak, Y. Joint lot sizing and inspection policy in an EOQ model with random yield. *IIE Trans.* **1990**, *22*, 41–47. [CrossRef]
44. Ben-Daya, M.; Rahim, A. Optimal lot-sizing, quality improvement and inspection errors for multistage production systems. *Int. J. Prod. Res.* **2003**, *41*, 65–79. [CrossRef]
45. Ben-Daya, M.; Noman, S.M.; Hariga, M. Integrated inventory control and inspection policies with deterministic demand. *Comput. Oper. Res.* **2006**, *33*, 1625–1638. [CrossRef]
46. Khan, M.; Jaber, M.-Y.; Guiffrida, A.-L.; Zolfaghari, S. A review of the extensions of a modified EOQ model for imperfect quality items. *Int. J. Prod. Econ.* **2011**, *132*, 1–12. [CrossRef]

47. Rezaei, J. Economic order quantity model with backorder for imperfect quality items. In Proceedings of the 2005 IEEE International Engineering Management Conference, St. John's, NL, Canada, 11–14 September 2005; pp. 466–470.
48. Wee, H.-M.; Yu, J.; Chen, M.-C. Optimal inventory model for items with imperfect quality and shortage backordering. *Omega* **2007**, *35*, 7–11. [CrossRef]
49. Taleizadeh, A.-A.; Cárdenas-Barrón, L.-E.; Mohammadi, B. A deterministic multi product single machine EPQ model with backordering, scraped products, rework and interruption in manufacturing process. *Int. J. Prod. Econ.* **2014**, *150*, 9–27. [CrossRef]
50. Hu, W.-T.; Kim, S.-L.; Banerjee, A. An inventory model with partial backordering and unit backorder cost linearly increasing with the waiting time. *Eur. J. Oper. Res.* **2009**, *197*, 581–587. [CrossRef]
51. Ganguly, B.; Pareek, S.; Sarkar, B.; Sarkar, M.; Omair, M. Influence of controllable lead time, premium price, and unequal shipments under environmental effects in a supply chain management. *RAIRO-Oper. Res.* **2018**. [CrossRef]

12

Stochastic-Petri Net Modeling and Optimization for Outdoor Patients in Building Sustainable Healthcare System Considering Staff Absenteeism

Chang Wook Kang [1], Muhammad Imran [2], Muhammad Omair [3], Waqas Ahmed [2], Misbah Ullah [4] and Biswajit Sarkar [1,*]

1 Department of Industrial & Management Engineering, Hanyang University, Ansan, Gyeonggido 155 88, Korea; cwkang57@hanyang.ac.kr
2 Department of Management & HR', NUST Business School, National University of Sciences & Technology, Islamabad 44000, Pakistan; imran.ime13@gmail.com (M.I.); engrwaqas284@gmail.com (W.A.)
3 Department of Industrial Engineering, Jalozai Campus, University of Engineering & Technology, Peshawar 25000, Pakistan; muhamad.omair87@gmail.com
4 Department of Industrial Engineering, University of Engineering & Technology, Peshawar 25000, Pakistan; misbah@uetpeshawar.edu.pk
* Correspondence: bsbiswajitsarkar@gmail.com.

Abstract: Sustainable healthcare systems are gaining more importance in the era of globalization. The efficient planning with sustainable resources in healthcare systems is necessary for the patient's satisfaction. The proposed research considers performance improvement along with future sustainability. The main objective of this study is to minimize the queue of patients and required resources in a healthcare unit with the consideration of staff absenteeism. It is a resource-planning model with staff absenteeism and operational utilization. Petri nets have been integrated with a mixed integer nonlinear programming model (MINLP) to form a new approach that is used as a solution method to the problem. The Petri net is the combination of graphical, mathematical technique, and simulation for visualizing and optimization of a system having both continuous and discrete characteristics. In this research study, two cases of resource planning have been presented. The first case considers the planning without absenteeism and the second incorporates planning with the absenteeism factor. The comparison of both cases showed that planning with the absenteeism factor improved the performance of healthcare systems in terms of the reduced queue of patients and improved operational sustainability.

Keywords: stochastic-petri net modeling; sustainable healthcare system; staff absenteeism; patient's queue; mixed integer nonlinear programming

1. Introduction

Sustainability was first introduced by Brundtland in 1987 [1] and it deals with economic, environmental, and social dimensions [2–6]. Efficient healthcare systems ensure social and economic sustainability. The management of healthcare systems involves the management of operations within the facility. Healthcare organizations should encourage social sustainability to set values for corporate social responsibility. To attain this, they should hunt for a balance between patient comfort, care, and economic need to guarantee sustainable development [7]. The patient's satisfaction and comfort are considered as a primary social indicator in the healthcare industry. The rapid increase in diseases has diverged the direction of researchers to sustainable healthcare systems. Promising innovations and sustainable operations are necessary for the healthcare system to achieve the patient's satisfaction level [8]. A healthcare system may be in a hospital or clinic or some temporary medical camp.

Patients visit hospitals for their treatment, but the limited resources and absence of doctors, dispensers, and nurses at healthcare systems creates issues. The severity of disease also varies from patient to patient. Therefore, sometimes, longer wait times in queue results in death. In order to avoid such problems, there should be some planning of resources by building sustainable healthcare systems for a patient's satisfaction. Since this comfort and satisfaction is a basic social aspect and pillar of sustainability. The planning of resources of the healthcare system is extensively reported in the literature [9–13].

Zeinali, et al. [14] addressed the resource planning model in the emergency department. The resources included were nurses, receptionists, and beds. However, they did not consider the physician, which are a core requirement of any healthcare system. In addition, they used a simulation model and performed a what-if analysis for resource management. However, the proposed model addresses the management of resources with the consideration of their absenteeism probability. Yousefi, et al. [15] focused the resource planning in the emergency department using a genetic algorithm. However, the outdoor patient department is not well focused. Feng, et al. [16] studied medical resource allocation in a healthcare system. They used a simulation integrated multi-objective algorithm for the allocation of resources such as beds, nurses, and other medical staff. The processes in healthcare centers are both discrete and continuous. However, simulation models only discrete events and system dynamic models focus on the continuous system. The proposed research integrates the Petri net with mixed integer nonlinear programming for modeling both discrete and continuous states in the outdoor patient department, which has not been well addressed in previous literature.

Most of the researchers have presented some static mathematical models for the planning of healthcare systems and optimized linear programming or heuristics [17–19]. However, these models might not be suitable for dynamic systems due to randomness in patient arrival, the nature of the disease, and treatment time of patients. Hence, healthcare systems should be efficient, sustainable, and flexible to incorporate the staff absenteeism factor. The dynamics of the healthcare system is modeled with either system dynamic (SD) approaches or discrete event simulation (DES). The objects in DES are different entities and discrete, while, in DS, are merged in a population to form a continuous quantity with a continuous state change [20].

A healthcare system has both characteristics DES and SD such as a number of patients in a system are discrete. However, the state change in a resource is a system dynamic. For example, a physician takes different treatment times for different patients, which is a continuous variable. SD methods are good for dynamic complexity, which makes it useful for strategic decision making and DES focuses on the detailed complexity of operational decision-making [21]. The Petri nets introduced by Adam Carl Petri (1962) combines the qualities of DES and SD. In Petri net modeling, the events, event calendar, entities, workflow, and constraints (decisional, operational) are presented in a DES pattern with a strong mathematical background [22]. The discrete event simulation (DES) has been widely used for the performance evaluation of healthcare systems such as utilization of resources, waiting time for patients, patient flow, and patient safety [23–27].

There are two types of resources in the healthcare system. One is an infrastructure of healthcare systems such as a number of beds, statures, and surgery equipment and other are staff members, doctor, nurses, and clerks. This research considers the human resource management, even though some researchers have considered the number of staff personnel required for patient's satisfaction [14,15,28]. Yalçındağ, et al. [29] addressed the planning of surgeons in the healthcare department but did not consider the absenteeism factor. Similarly, En-nahli, et al. [30] optimized the resources in the emergency department without staff absenteeism. However, this research focuses the resource management on the outdoor patient department for patient queue reduction and satisfaction. It was found that the required staff might not be available or on duty to meet the patient service. The absenteeism of staff may result in a long queue for the patient.

The concept of resource planning with absenteeism of staff has not received proper attention. Thus, the viability of such a healthcare system is necessary for sustainability and growth. The mixed integer nonlinear programming-based Petri net have been introduced to model this situation because of its discrete and continuous nature. Petri net itself combine the properties of simulation and mathematical models and it is possible to do what-if analysis. It is a time-consuming process and does not guarantee an optimal solution. However, in this research integration of Petri net with mixed integer nonlinear programming, reduces the computational time and iterations and gives optimal results. The major advantage of using mixed integer nonlinear programming Petri net model is reduced computational time, improved visualization of the flow of entities (patients), and its capability to find the best solution. The proposed mathematical solution and model approaches are useful for the sustainability of healthcare systems. The incorporation of Stochastic-Petri net approach with MINLP for real-time optimization in the scheduling model will enable the attainment of sustainable operational efficiency. The objective of the proposed model is to reduce the cycle time for the patient's satisfaction under the factor of absenteeism. In order to minimize cycle time and improve customer satisfaction, patient queues and resources should be minimized. The rest of the paper is organized as follows. The detailed literature review is in the second section. The third section includes the development of the model. The solution methodology of the research is represented in detail form as in the fourth section. A numerical example of a healthcare center is presented in the fifth section as an application of the proposed model.

2. Literature Review

The two major categories of the industries can be divided into manufacturing and service industry. Researchers are working to enhance the production, quality, and cycle time of the manufacturing firm to support the manufacturers, customers and government [31–33]. However, healthcare systems are essential for the population to provide services. Therefore, management and control is inevitable. Many operational and managerial strategies have been introduced to plan and control the healthcare systems. The planning of the resources especially human resources is critical because of the direct interaction with the patients. Healthcare models developed can be classified into four types including static models, dynamic models, simulation models, and Petri net models.

2.1. Static System Models

Bachouch, Guinet and Hajri-Gabouj [17] addressed the static strategic enterprise-resource-planning model for the healthcare systems and developed a multi-objective goal-programming model with finance, labor, revenue, capacity, and admission resources. Hans, Van Houdenhoven and Hulshof [11] proposed the scheduling model for healthcare systems for minimizing idle time of nurses in a week. Ghazalbash, et al. [34] developed a mathematical model for the scheduling of the operating rooms of teaching hospitals with the objectives of minimizing operation time and operating room idle time. Nasir and Dang [35] solved a flexible home health care routing and scheduling problem with the joint patient and nursing staff selection. Tan, et al. [36] presented a two-stage multi-objective mathematical model for the planning and scheduling of operating rooms considering the patient flow. This is clear from the literature that the static models have mostly focused the operating rooms, nurses, and operating equipment while neglecting the major functional department of healthcare systems such as an outdoor patient ward, where many patients visit on a daily basis. There are various reasons for modeling this department and the most prominent is the daily arrival of patients in healthcare systems. The static mathematical model does not provide the clarity of the dynamic behavior of the systems.

2.2. System Dynamic Models

Since the behavior of healthcare systems is dynamic and this factor should be taken into account while modeling the healthcare systems. Lane, Monefeldt and Rosenhead [13] provided the dynamic model for the accident and emergency department and planned the resources such as beds and the hospital process by minimizing the delay time of the patients. Carayon, et al. [37] improved the quality and patient safety by introducing the system engineering initiative for the patient safety (SEIPS) model, which analyzes the behavior of workers in human factor contexts. Atun [38] discussed the dynamic complexity of healthcare systems by considering the dynamic interactions between innovation and institutions. Fraher, et al. [39] developed the projection model for the forecasting of surgeons for healthcare systems for 2009 to 2028 with consideration of specialty, sex, and age to tackle the dynamic change in surgery complexity. There is plenty of literature on the system dynamic behavior of the healthcare system. The shorter review of dynamic modeling in healthcare is available in References [10,40]. The system dynamic models are helpful when a change in the state or occurrence of the event is continuous. However, it is not necessary that events are continuous in nature. The nature of events might be discrete, too. This limitation of dynamic models is not to consider the discrete change.

2.3. Discrete Event Simulation Models

A simulation is a tool for performance evaluation, planning, and scheduling of the healthcare systems because of its ability to model the discrete events. Rohleder, Lewkonia, Bischak, Duffy and Hendijani [27] reported the DES model for the performance improvement of the orthopedic outpatient clinic. The objective of their research was to minimize the waiting time of patients. Robinson, Radnor, Burgess and Worthington [26] introduced the concept of SimLean in healthcare, which is the process of performance improvement. It is a helpful tool for the planning and scheduling of healthcare systems.

Abo-Hamad and Arisha [41] presented an interactive program with simulation-based decision support systems for the optimization of the healthcare systems for optimum resource utilization. Zeltyn, et al. [42] focused the workforce staffing problems in the emergency department and developed the simulation model to evaluate different alternatives of the emergency department for customer satisfaction. The use of simulation models for the scheduling or planning of the healthcare systems is reported in the literature and some details can be found in the models of References [43–46].

2.4. Petri-Net Models

The Petri nets are a combination of dynamic modeling and simulation incorporating visualization with the help of graphs. These models deal with both discrete and continuous quantities. The planning and scheduling of the healthcare system involve both discrete and continuous variables. The number of patients, the number of physicians, and a number of nurses have a discrete nature. The examination time, medication time, and waiting time are the example of continuous variables. However, some mathematical models have been employed for scheduling they did not consider the dynamics of both discrete and continuous variables [47]. The processes within a healthcare center are random or stochastic. Therefore, in this research, stochastic Petri net modeling is chosen for realistic results.

There are very few articles on healthcare planning and scheduling using Petri nets. Mahulea, et al. [48] modeled the healthcare system by using Petri nets and evaluated the performance measure, but they did not optimize this process for customer satisfaction. Bertolini, et al. [49] analyzed the blood transfusion with component-based timed-arc Petri nets and modeled the healthcare workflow. Hicheur, et al. [50] modeled the flexible healthcare systems by using recursive and workflow Petri nets and evaluated the performance of the healthcare system by analyzing the medical properties of healthcare processes. In all of the above cases, Petri net have been used for flow modeling and process visualization and do not give us any information when the maximum number of patients will be served. However, proposed research not only models the healthcare system using a Petri net but also optimizes the process. In this regard, Petri nets have been integrated with a mixed integer nonlinear programming model that has been utilized to maximize customer satisfaction by reducing the cycle time.

Wang [51] modeled the emergency response in the emergency department by using Petri nets. The proposed methodology considered the stochastic Petri net modeling and optimization for outdoor patients in healthcare systems and the aim is to satisfy each patient with minimum waiting time while considering the staff (doctors, nurses, or cleric) unavailability. The concept of planning and scheduling the resources with consideration of staff absenteeism has not been paid proper attention.

3. Petri Nets Modeling of the Healthcare System

The Petri nets models are the graphical, simulation, and mathematical representation of the system [52]. A typical Petri net model is a bipartite directed graph, which consists of four elements called places, transitions, arcs, and token. Places and transitions are connected with each other with the help of the directed arc [53]. Figure 1 shows a basic Petri net graph and definition of each element in the Petri net is given below.

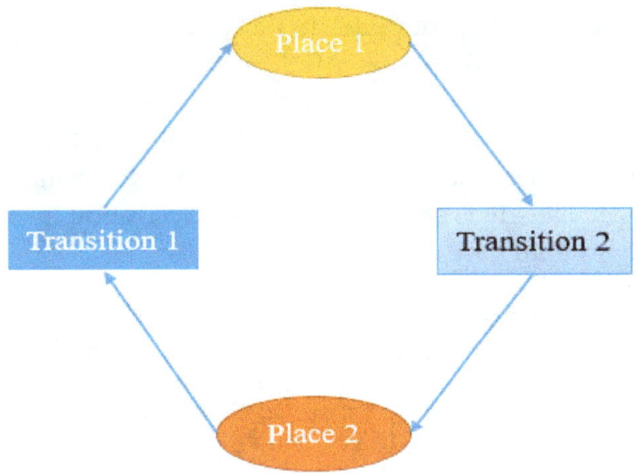

Figure 1. A Petri net graph.

a. Place

A place in the Petri net is a position with two states. A token is a process or event that occurs in place, so a place has a token or not. In a healthcare example, a physician is in a place with two states, either he/she sits idle or examines the patient. Places in Petri nets define discrete variables.

b. Transition

Transition shows the activity or process and its duration within the model. A transition represents continuous variables within the process. For example, patient examination time of the physician is a transition.

c. Arcs

Arcs are used to direct and connect the transitions with places. The sequence of a visit from registration to the medication is an example of art that shows the process plan.

d. Tokens

Tokens are the entities moving along arcs within transitions and places in a Petri net graph. The number of patients within the healthcare center are the example of tokens. The event in Petri net can be defined by introducing a transition between input and output places. The dynamic behavior analysis of Petri net involves availability or unavailability of tokens in the input or output places. If the token is available, then the condition associated with the place is true. Otherwise, it is false.

Definition 1. *A Petri net can be defined with five-tuple* $M = (\rho, \tau, I, O, \mu,)$, *where*

(1) $\rho = \{\sigma_1, \sigma_2, \ldots\ldots.\sigma_m\}$ *is the finite set of places*
(2) $\tau = \{t_1, t_2, \ldots\ldots t_n\}$ *is the finite set of transitions* $\rho \cup \tau \neq \varnothing$, *and* $\rho \cap \tau = \varnothing$
(3) $I : \rho \times \tau \rightarrow M$ *is a function, which acts as an input for defining a directed arc between places and transitions and "M" is set to positive integers.*
(4) $O : \tau \times \rho \rightarrow M$ *is a function, which acts as an output for defining a directed arc between places.*

(5) $\mu : \rho \rightarrow M$ *is called an initial marking.*

The assignment of the token to the places in Petri net is called marking. The position of token keeps on changing, according to the input and output directing arcs in Petri net modeling [54].

3.1. Petri Net Graphical Representation

The Petri net models can be explained well with the help of the graph, which is a bipartite directed multigraph. There are two types of nodes in Petri net graphs. A place node, which is usually represented by a circle and transition is represented by a bar or box. To understand the graphical illustration of the Petri net considers a healthcare center. There are two types of outdoor customers or patients. The patients arrive in the healthcare system, register themselves for treatment at the registration desk, and move to the next station where physicians examine them. The doctor completes the examination of the patient. He/she is sent to a dispenser for medicines. Table 1 shows the process plan for patient treatment in a healthcare center.

Table 1. Outdoor patient treatment process plan.

	Registration	Examination	Medication
$i = 1$	t_{11}	t_{12}	t_{13}
$i = 2$	t_{22}	t_{22}	t_{23}

The bipartite directed multigraph for the Petri net model can be developed using the process plan of patient treatment, as shown in Figure 2, which is built using the concept of Figure 1 and process plan in Table 1. It is clear from Figure 2 that there are two types of overlapping circuits called the patient circuit and sequencing circuits. The distribution and number of tokens in net control of its execution. The number of tokens in the patient circuit is equal to the number of patients waiting for service. The sequencing circuit is used for the representation of servers. Figure 2 is the graphical representation of the flow of entities such as patients in a healthcare system. Rectangles in a blue color show the transitions that are used for assigning times to processes. However, places in red circles are used to represent resources. Green-colored and orange-colored tokens represent two different types of patients.

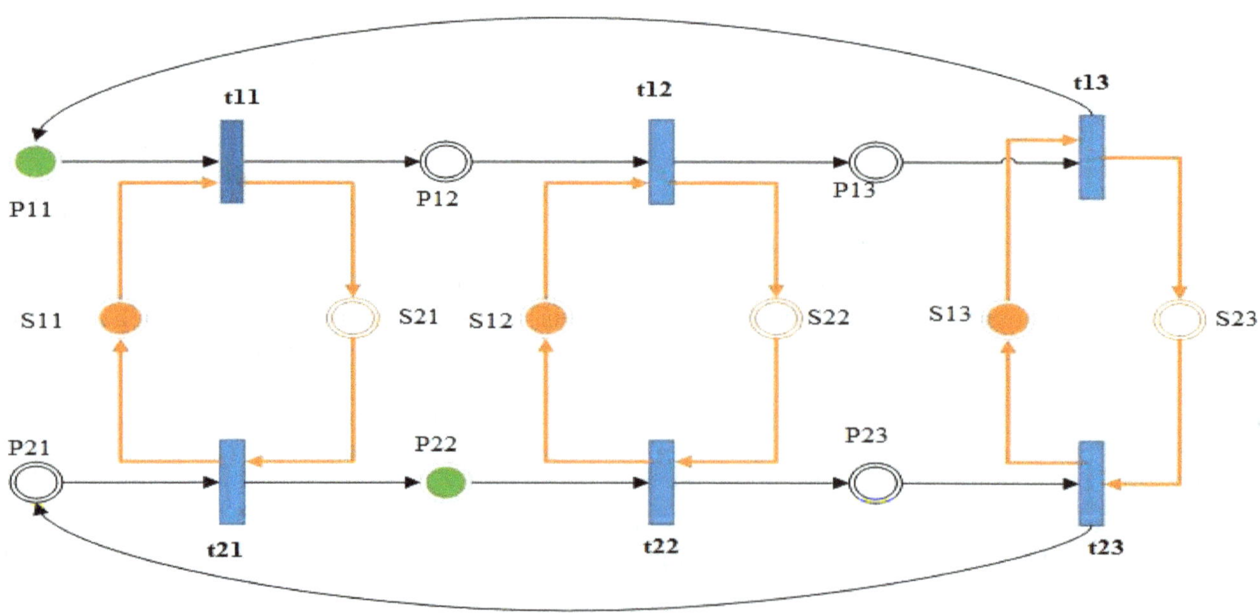

Figure 2. Petri net model of the healthcare system with a process plan.

3.1.1. Model Assumptions

To develop a mathematical model, the following assumptions must be taken into account.

1. The service times are stochastic.
2. One server can serve only one patient at a time.
3. Transitions and places cannot be together so, during any loop, this rule should not be violated.
4. Server interruption by another patient is not allowed when a server is in service of another patient.
5. The patient keeps arriving at the healthcare system randomly.

3.1.2. Mathematical Model

In the sequence loop, the number of tokens is invariants. The total transition time $\tau(\zeta)$ is equal to the sum of transition firing times in a sequence circuit. t_{ij} is the time taken by the server j to serve patient i.

$$\tau(\zeta) = \sum_{i=1}^{I}\sum_{j=1}^{J} t_{ij} \tag{1}$$

The total sum of tokens in a sequential loop or server loop $M(\zeta)$ is in Equation (2). In this case, S_{ij} shows a place of server j for serving patient i.

$$M(\zeta) = \sum_{j=1}^{J} S_{ij}\ \forall_i \tag{2}$$

Aziz, et al. [55] computed cycle time of each sequential circuit $C(\zeta)$ using Equation (3).

$$C(\zeta) = \frac{\tau(\zeta)}{M(\zeta)} \tag{3}$$

However, this cycle time does not consider the absenteeism of the server from the service stations. If there is some probability of absenteeism, then Equation (3) becomes PA, which represents the probability of absenteeism.

$$C(\zeta) = \frac{\tau(\zeta)}{PA \times M(\zeta)} \tag{4}$$

$\tau(\zeta_s)$ is the total service time in one service cycle by only one server to provide the service to a single patient. If the number of the server at each service station is known, then the service rate SR can be calculated using Equation (5).

$$SR = \frac{1}{C(\zeta)} \tag{5}$$

$$SR \times \tau(\zeta_s) = \frac{\tau(\zeta_s)}{C(\zeta_s)} \text{ and } M(\zeta_s) \geq \frac{\tau(\zeta_s)}{C(\zeta_s)} \tag{6}$$

After a finite transition, the system starts to operate at a steady state. As the system achieves the steady state, the service cycle time becomes equal to the maximum cycle time of the servers. The service cycle time reduces with the introduction of more tokens. The number of tokens in the server circuit is normally kept fixed per the capacity of healthcare centers. However, the number of patient tokens depends upon the arrival of patients and planning horizon. Equation (7) shows the required cycle time of the patient circuit $C(R\zeta_p)$,

$$C(R\zeta_p) = \frac{Planning\ horizon}{Number\ of\ patients} = \frac{PH}{NP}, \tag{7}$$

while

$$M(\zeta_s) \geq \frac{\tau(\zeta_s)}{C(R\zeta_p)} \quad (8)$$

Bottleneck server cycle time $C(B\zeta_s)$ and $C(R\zeta_p)$ have two possible cases as follows:

1. $C(B\zeta_s) \geq C(R\zeta_p)$
2. $C(B\zeta_s) < C(R\zeta_p)$

The performance of the system will be influenced because of the maximum number of bottleneck server cycle times $C(B\zeta_s)$ and required cycle times for the patient circuit $C(R\zeta_p)$ [56].

$$C(\zeta_{sMAX}) = Max\left\{C(B\zeta_s), C(R\zeta_p)\right\} \quad (9)$$

By considering the server bottleneck circuit as well as cycle time as the maximum cycle time, then the Petri net model is formulated as a mixed integer nonlinear programming problem [57].

$$Minimize \sum_{i=1}^{I}\sum_{j=1}^{j} P_{ij} \quad (10)$$

Subject to

$$C(\zeta) \leq C(\zeta_{sMAX})\ with\ C(\zeta) > C(R\zeta_p) \quad (11)$$

Equation (12) does not consider the absenteeism while Equation (13) considers it.

$$\tau(\zeta) = C(\zeta_{sMAX}) \times M(\zeta) \quad (12)$$

$$\tau(\zeta) = C(\zeta_{sMAX}) \times PA \times M(\zeta) \quad (13)$$

$$C(\zeta_p) \leq C(R\zeta_p)\ \forall\ \zeta_p \quad \text{with } C(\zeta_p) > C(R\zeta_p) \quad (14)$$

$$\forall\ P_{ij},\ S_{ij} \quad \text{Integer } \geq 0$$

The objective function of this model is to minimize the service cycle time. Equation (10) is the objective function. The constraint in Equation (11) ensures that the cycle must be within critical limits defined by the bottleneck cycle time or required cycle time for the patient loop. Equation (12) considers the effect of critical or maximum cycle time on total transition time without considering the absenteeism of the server. However, Equation (13) takes this factor into account. Equation (14) shows patient service achievement.

4. Solution Methodology

The method explained in Section 3 has been explained with the help of a numerical example of a hypothetical healthcare center.

4.1. Numerical Example

The numerical example of the proposed model is demonstrated as follows.

4.1.1. Problem Statement

Consider a healthcare center that provides services to the patients. Patients visiting the healthcare center have to process three steps including registration, medical examination, and medication. There are two servers at the registration boot, three physicians for the treatment of patients, and two servers in the dispensary. There are three possibilities due to the absenteeism of servers.

1. The server at the registration booth may be absent and this situation results in the long queue of patients at the registration booth.

2. The absenteeism of physicians may result in the long queue of patients and the station becomes a bottleneck.

3. If any of the servers from the last station is absent, then it also creates a longer wait time at the last station, which is also the reason for the bottleneck.

Table 2 shows the probability of absenteeism at registration, medical examination, and dispensary. The probability of absenteeism is computed using past data. The data has been collected through a brief interview and brainstorming session with healthcare managers. He suggested the following probability of absenteeism for each resource. Following one-week probability can be projected for one year.

Table 2. Absenteeism probability of servers at each step in the healthcare center.

	Absenteeism Probability		
Days	**Registration**	**Medical Examination**	**Medication**
Monday	~0.03	~0.07	~0.04
Tuesday	~0.05	~0.07	~0.05
Wednesday	~0.05	~0.08	~0.05
Thursday	~0.05	~0.10	~0.05
Friday	~0.12	~0.14	~0.10
	Average = 0.060	Average = 0.092	Average = 0.058

There are two types of patients. Type 1 patients have severe diseases and type 2 patients have less severe diseases. The processing time at all stations is stochastic, which can be modeled with the help of some distribution. Processing time of each server at each station is measured and data is analyzed to find the best fit distribution. In distribution analysis, data was analyzed on a set of distributions such as uniform, lognormal, binomial, and normal distribution. In this data analysis, uniform distribution was selected because it had the highest value of Log-likelihood and minimum value of Akaike information criterion. Table 3 shows the processing time of each process at each step in the healthcare center. Here, *U* represents un-firm distribution, in the case of medical examination *U* (3,7), which means that a physician takes three to seven minutes to examine a patient. Data analysis also showed that the number of patients visiting the hospital follows the Poisson distribution. The number of patients visiting in the next month has been generated randomly using a convolution theorem. Table 4 shows the number of patients visiting each month.

Table 3. Outdoor patient treatment process plan.

	Registration	Examination	Medication
$i = 1$	*U* (1,5)	*U* (3,7)	*U* (1,2)
$i = 2$	*U* (1,5)	*U* (4,6)	*U* (1,2)

Table 4. Number of patients visiting the healthcare center each month.

January	February	March	April	May	June	July	August	September	October	November	December
2979	2726	1798	2630	2809	1736	2208	1629	1750	1643	2334	2841

The hospital is interested in the planning of the outdoor patient department (OPD) for minimizing the number of patients waiting in the queue at all servers by determining the number of servers considering their absenteeism.

4.1.2. Problem Solution

The station in the system becomes a bottleneck when there is physician absenteeism at the medical examination center. Considering the server capacity at each station, the service cycle time can be calculated using the equation below.

$$Service\ Cycle\ time = \frac{total\ proces\sin g\ time}{Number\ of\ servers} \quad (15)$$

Service rate is computed using equation as follows.

$$Service\ rate = \frac{1}{Service\ cycle\ time} \quad (16)$$

In this case, the processing time is uniformly distributed. Therefore, the cycle time of each station will also be uniformly distributed. The planning horizon of the hospital is one month and there are 26 working days and, there are 12 operating hours. Therefore, the available time is 18,720 min. Considering the number of patients visiting the hospital from Table 4 and available time in each month, the required cycle time for the patient circuit can be computed using Equation (7). Table 5 shows the required cycle time of a patient circuit.

Table 5. Required patient circuit cycle time for each month.

Month#	January	February	March	April	May	June	July	August	September	October	November	December
$C(R\zeta_p)$	6.3	6.9	10.4	7.1	6.7	10.8	8.5	11.5	10.7	11.4	8.0	6.6

This actual cycle time is not according to the required cycle time. Therefore, the system should be redesigned in such a way that there should be a minimum queue of patients at the station. Mixed integer nonlinear programming model is solved for each required cycle time without considering the absenteeism.

4.1.3. Case for a Numerical Example

Two cases for the numerical example have been presented. The first case is monthly planning without absenteeism and the second case considers the absenteeism.

Case 1: Human Resource Planning without Absenteeism

In this case, the Petri net have been used to optimize and compute the number of servers and queues using a mathematical model consisting of Equations (10)–(12) and Equation (14). The processing time has been generated using Table 3. The required cycle time of each month shown in Table 6 is used for cycle time constraint in the Petri net model.

Table 6. Required resources and Queue length with absenteeism.

Month	$C(R\zeta_p)$	Patients in Queue			Number of Servers Required		
		Station 1	Station 2	Station 3	Station 1	Station 2	Station 3
January	6.3	2	2	2	1	2	2
February	6.9	1	2	1	1	1	2
March	10.4	1	1	1	0	1	1
April	6.71	1	2	1	1	2	2
May	6.70	1	2	1	1	2	2
June	10.8	1	1	1	0	2	1
July	8.50	1	1	1	1	2	1
August	11.5	1	1	1	1	1	0
September	10.7	1	1	1	1	1	0
October	11.4	1	1	1	1	1	1
November	8.0	1	2	1	1	1	1
December	6.60	2	2	2	1	2	2
Average		1.2	1.5	1.2	0.8	1.1	1.3

Case 2: Human Resource Planning with Absenteeism

In this case, the factor of absenteeism is considered for the planning of human resources. Equations (10), (11), (13) and (14) have been used in the model. The processing times are the same in either case.

5. Results and Discussion

Considering Figure 1 and the mathematical model in Section 3.1.2, an excel spreadsheet program is generated and solved using a personal computer with 4 GB RAM and 2.67 GHz processor.

5.1. *Results of Case 1*

It can be seen from Table 1 that there are two types of patients and three types of servers and service processes consist of three processes so the total number of places is 12 and the total number of transitions is six because there are three processes and two types of patients. There are three types of loops in the Petri net model, which are the patient loop, a sequential loop, and a mixed loop. There are three sequential loops because the number of processes of service is three. The number of patient loops is equal to the type of patients who visit the hospital. In each elementary circuit, the availability and unavailability of each token (server, patient) are represented by binary (0, 1) digits. It is clear from Table 6 that the required cycle time for the month of January is 6.3 min per patient. The problem is solved by using MINLP for the required cycle time of January.

Number of patients waiting at registration $= P11 + P21 = 0 + 1 = 1$
Number of patients waiting at the medical examination center = $P12 + P22 = 0 + 1 = 1$
Number of patients waiting at the dispensary = $P13 + P23 = 2 + 0 = 2$
Similarly, the required number of the server at each station can be calculated using the following formula.
Number of servers at the registration booth = $S11 + S21 = 0 + 2 = 2$
Number of servers at the medical examination center = $S12 + S22 = 1 + 1 = 2$
Required servers at the dispensary = $S13 + S23 = 1 + 0 = 1$
Similarly, the MINLP model is run for each cycle time in Table 5 for each month and the required number of servers and number of patients waiting in a queue obtained from Petri net-MINLP at each station are given in Table 7.

Table 7. Required resources and Queue length without absenteeism.

Month	$C(R\zeta_p)$	Patients in Queue			Number of Servers Required		
		Station 1	Station 2	Station 3	Station 1	Station 2	Station 3
January	6.3	1	2	2	3	2	1
February	6.9	1	2	2	2	1	2
March	10.4	1	1	1	1	0	1
April	7.1	1	2	2	1	1	2
May	6.70	1	2	2	2	2	1
June	10.8	1	1	1	1	1	1
July	8.50	1	1	1	1	1	1
August	11.5	1	1	1	1	1	1
September	10.7	1	1	1	1	1	1
October	11.4	1	1	1	1	1	1
November	8.0	1	2	1	1	1	1
December	6.60	1	2	2	2	2	1
Average		1	1.5	1.4	1.4	1.2	1.2

It is clear from Table 8 that the average queue length at station 1, station 2, and station 3 are 1, 1.5, and 1.4. However, the average number of servers at each station is 1.4, 1.2, and 1.2, respectively.

5.2. Results of Case 2

The aim of this research to design a healthcare center with a factor of resource absenteeism. The same numerical example is used for the planning of the healthcare center with the factor of absents. The objective is to determine the optimal number of resources at each station with a minimum queue length. To do so, the Petri net-mixed integer nonlinear programming (MINLP) model included Equation (13) instead of Equation (12). The Petri net-MINLP is run for each cycle for each month and the required number of resources and queue generated at the station is recorded. Table 6 shows that the average queue length of each station has been reduced while the required number of average resources has increased.

The results of resource planning with absenteeism are more robust and ensure patient satisfaction with reduced queue length. The Petri net-MINLP model used the process plan from Table 1 and employed Figure 2 for determining sequential, patient, and mixed loops. Since there are three stations, it contains three sequential loops. A number of patient loops are two because two types of patients visit the healthcare center. The mixed loops depend on the sequential and patient loops. These loops can be computed using the following Equation (17).

$$Mixed\ loops = Patient\ loops \times sequential\ loops \quad (17)$$

Table 8 shows markings of the circuit with cycle time for the Petri net model for the healthcare system. The binary digits show the availability and unavailability of the token in places, hiring, and firing from the transitions. Two types of cycle time have been used for this Petri net model. Circuit cycle time is known from the process plan and Petri net diagram. However, for the required cycle, the Petri net model is hybridized with MINLP to get the optimal number of servers and queue length.

Table 8. Marking of the circuit with cycle time for the Petri net model for the healthcare system in the month of January without absenteeism.

Transition Time	U (1,1.5)	U (3,7)	U (1,2)	U (1,1.5)	U (4,6)	U (1,2)	U (1,1.5)	U (1,1.5)	U (3,7)	U (4,6)	U (1,2)	U (1,2)				
Transition Name	t11	t12	t13	t21	t22	t23	t11	t21	t12	t22	t13	t23				
Place Name	P11	P12	P13	P21	P22	P23	S11	S21	S12	S22	S13	S23				
No.	1	2	3	4	5	6	7	8	9	10	11	12	$\tau(\zeta)$	L.T	$M(\zeta)lp$	$C(\zeta)lp$
1	0	0	0	0	0	0	1	1	0	0	0	0	8	S.C	1.27	6.30
2	0	0	0	0	0	0	0	0	1	1	0	0	12	S.C	1.90	6.30
3	0	0	0	0	0	0	0	0	0	0	1	1	11	S.C	1.75	6.30
4	1	1	1	0	0	0	0	0	0	0	0	0	15	P.C	2.38	6.30
5	0	0	0	1	1	1	0	0	0	0	0	0	16	P.C	2.54	6.30
6	1	0	0	0	1	1	0	1	0	0	1	0	27	MX	4.29	6.30
7	0	1	1	1	0	0	1	0	0	0	0	1	23	MX	3.65	6.30
8	1	1	0	0	0	1	0	0	0	1	1	0	24	MX	3.81	6.30
9	0	1	0	1	0	1	1	0	0	1	0	0	21	MX	3.33	6.30
10	1	0	1	0	1	0	0	1	1	0	0	0	30	MX	4.76	6.30
11	0	0	1	1	1	0	0	0	1	0	0	1	30	MX	4.76	6.30
Marking	1	1	1	1	1	0	0	1	1	1	1	1	9.8			

L.T: Loop type. S.C: Sequencing circuit. P.C: Patient circuit. MX: Mixed circuit of sequencing and patient. $M(\zeta)lp$: Sum of tokens in circuits in the mixed integer nonlinear programming model (MINLP). $C(\zeta)lp$: Cycle time of the circuit for MI.

5.3. *Comparison of Results of Case 1 and Case 2 with the Existing System*

In this section, a detailed analysis of servers and queue is presented and results are compared with the existing system. Figure 3 shows a healthcare center and its process plan.

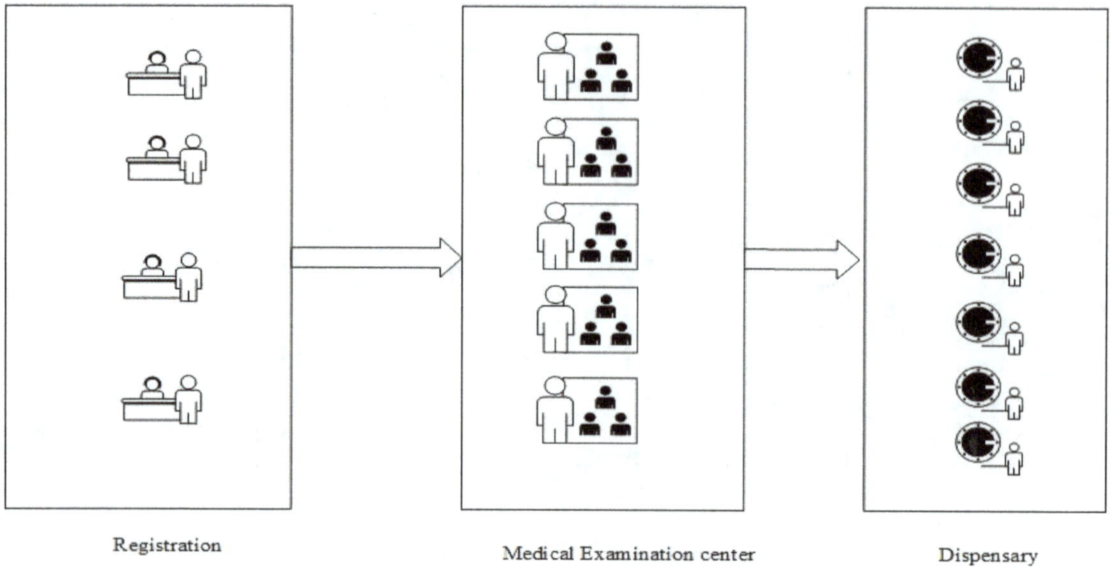

Figure 3. Process plan of a healthcare center.

5.3.1. Analysis of Servers

The healthcare center operated the overutilization of servers. The existing healthcare center contained two servers at station 1, three servers at station 2, and two servers at station 3. Figure 4 shows the number of servers required in both cases with and without the absenteeism case. It is clear from Figure 4 that the existing number of servers at the registration booth are in access. However, the required number of servers in both cases is reduced for the effective utilization of resources. Figure 5 shows the analysis of a number of servers at the medical examination center. It is shown that the required number of servers, in either case, is less than the existing servers.

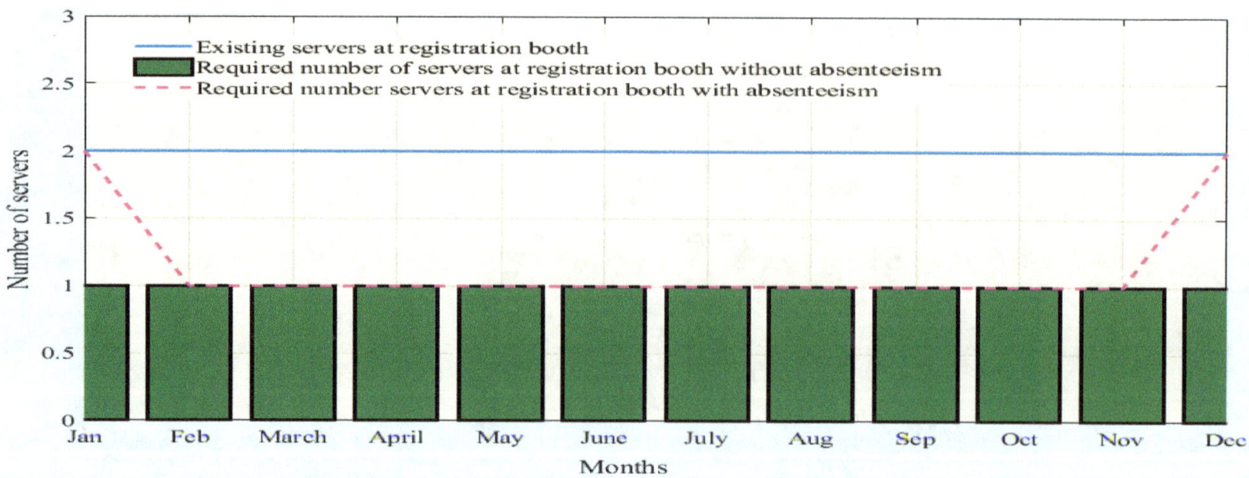

Figure 4. Number of servers at the registration booth during each month.

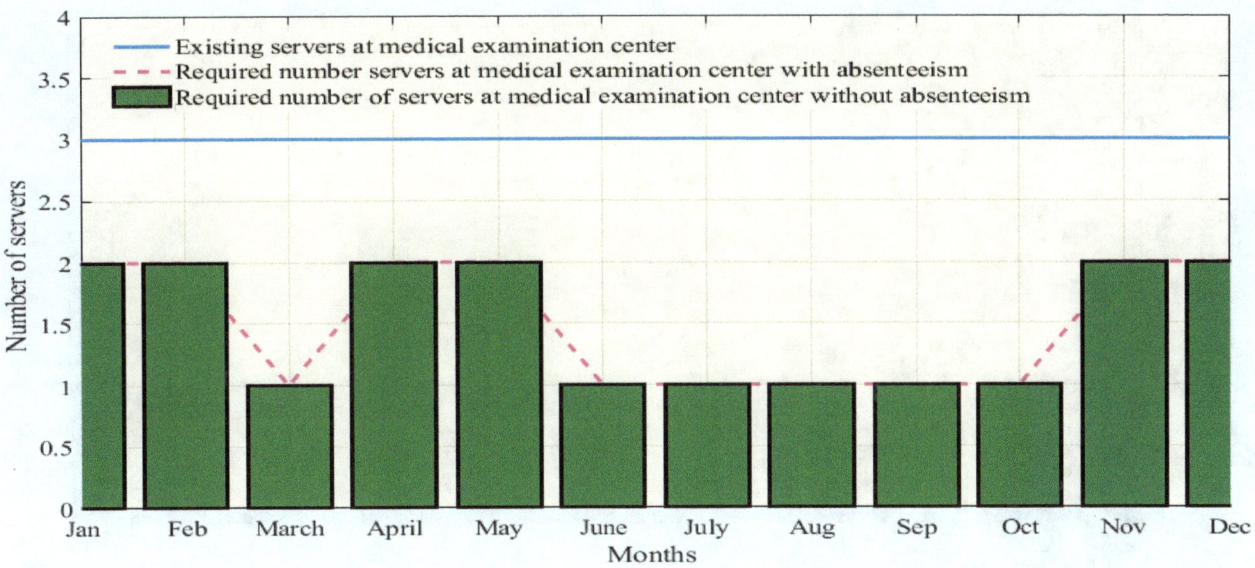

Figure 5. Number of servers at a medical examination center for each month.

Figure 6 shows the number of servers at the dispensary. It can be seen that a number of servers with an absenteeism case are higher from April to June.

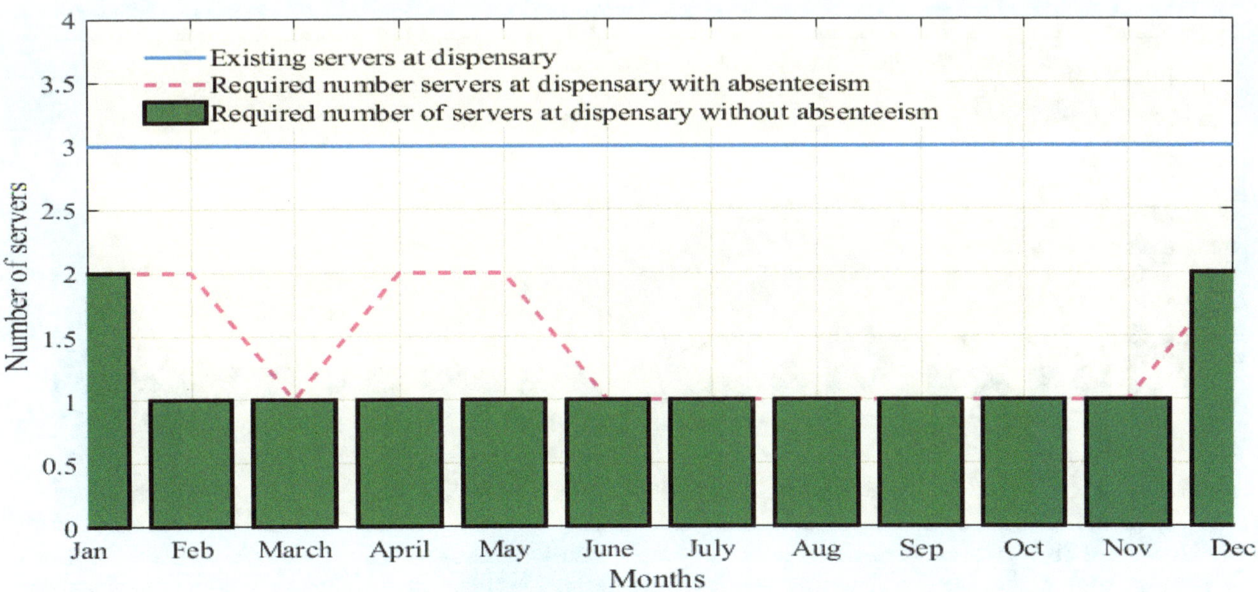

Figure 6. Number of servers at a dispensary for each month.

5.3.2. Analysis of Queue

Considering the case of server absenteeism reduced the queues of patients at all booths in a healthcare center. Figure 7 shows the comparison of queues for both cases of absenteeism and without absenteeism.

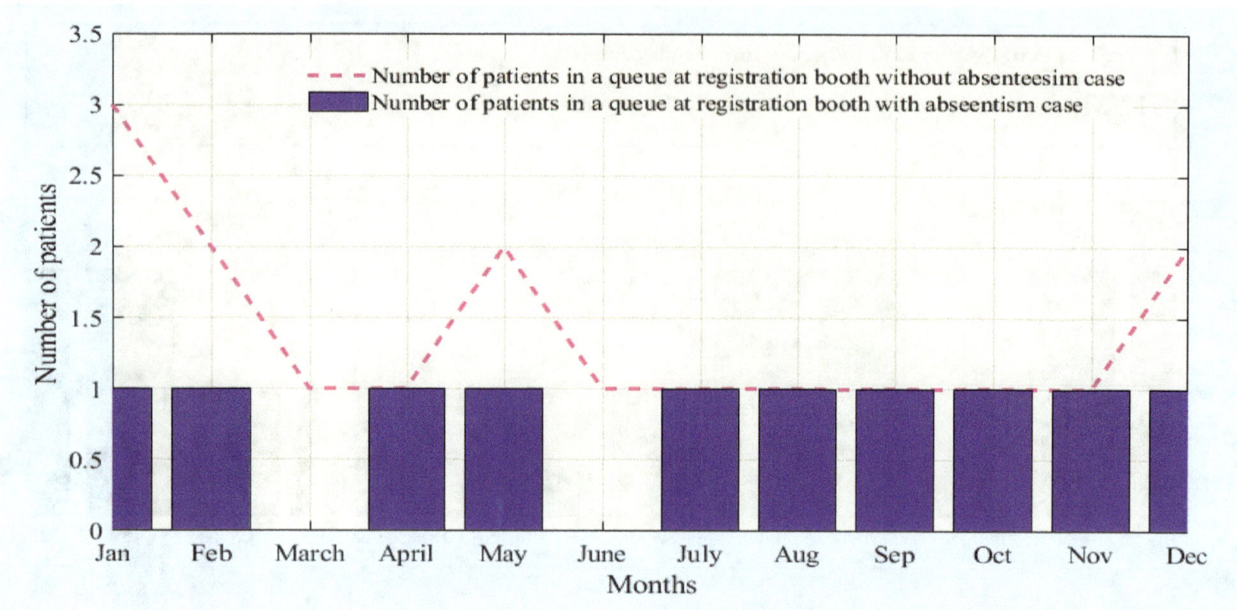

Figure 7. Number of patients waiting for service at the registration booth.

Figure 8 shows the number of patients waiting for medical treatment at the medical examination center. Planning resources as the factor of absenteeism reduced the queue length at the examination center.

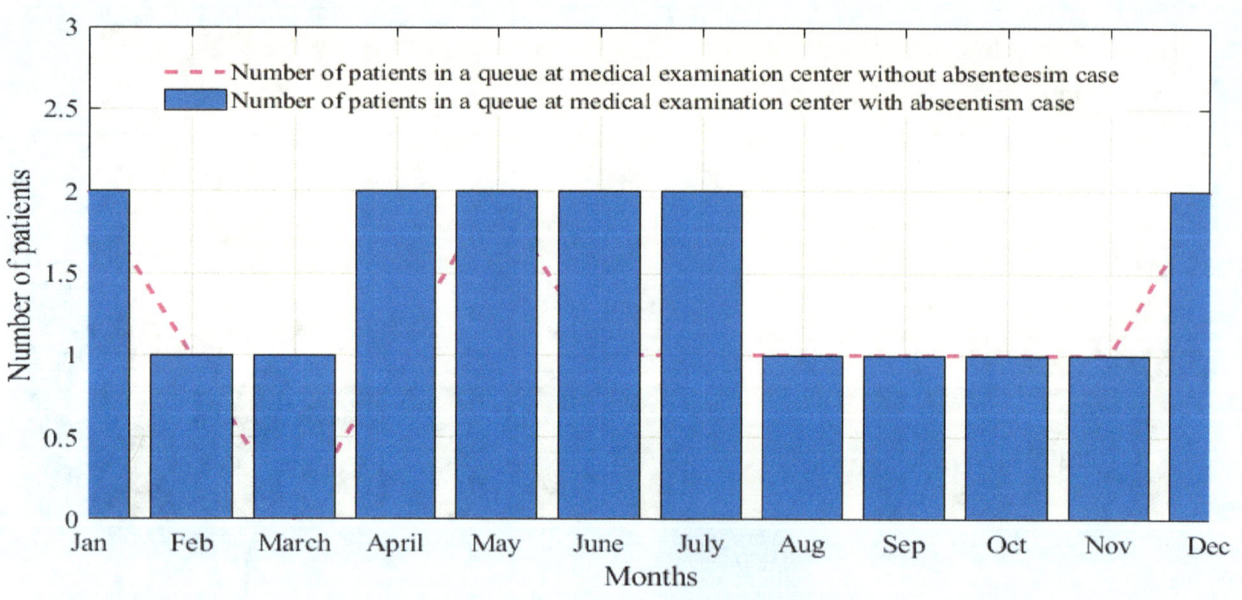

Figure 8. Number of patients waiting for service at the medical examination center.

It is prominent from Figures 7–9 that the length of queue reduced by determining the number of servers at each station. Therefore, this model provides the best approach for the planning of resources in a healthcare system. The consideration of server absenteeism in the planning of resources reduces queue length substantially, which increases resource utilization and patient satisfaction.

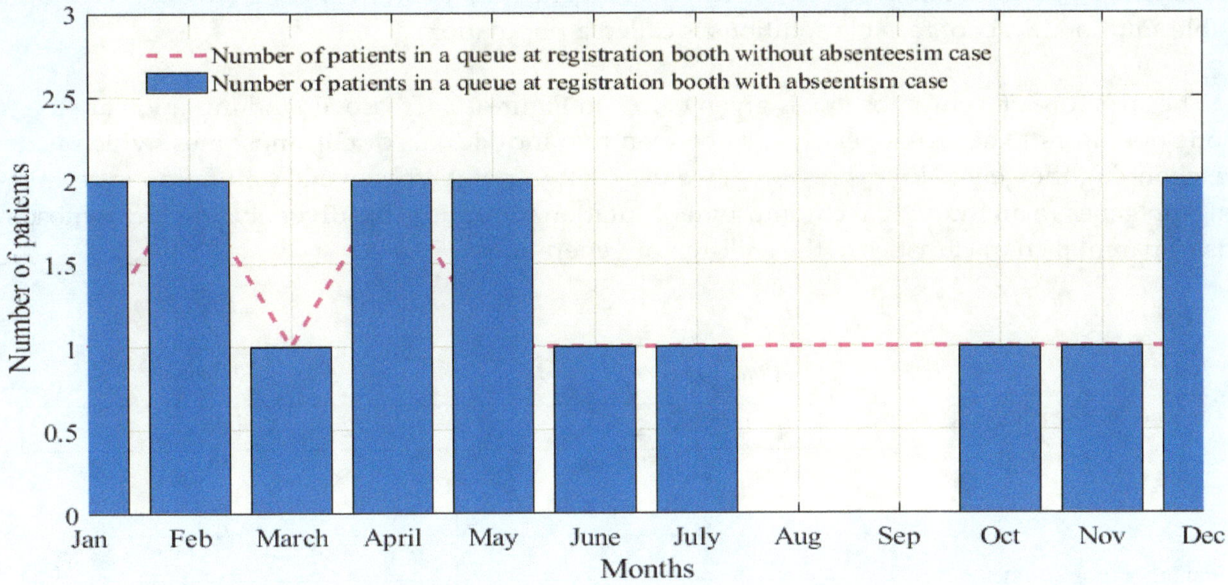

Figure 9. Number of patients waiting for service at the dispensary.

5.4. Petrinet Based Mixed Integer Nonlinear Programming (Petrinet-MINLP)

Most of the real-life problems have a nonlinear nature because of their complexity and involvement of many dependent decision variables. An optimization problem is said to be nonlinear if it contains a nonlinear equation either in the objective function or in a constraint. There are different software packages used for nonlinear optimization such as MATLAB, CPL EX, GAMS, and LINGO. However, in all these packages, an interior point algorithm is used for the optimization in mixed integer nonlinear programming. Mixed integer nonlinear programming is an exact method and guarantees the global optima in the optimization problem (Tawarmalani, Sahinidis, & Sahinidis, 2002). Traditionally, MINLP was used for mathematical optimization.

Petri net is a combination of simulation, mathematical, and graphical methods that are used to represent the real systems in both dynamic and static ways. Traditionally, Petri nets have been used for what-if analysis and could not optimize the complete process in a dynamic environment (Jensen, 2013). The optimization of the Petri net has not been well focused. In this research, we made an attempt to integrate the Petri net with mixed integer nonlinear programming for the optimization of real systems in a dynamic way. In this research, the Petri net-based mixed integer nonlinear programming is used to solve a healthcare resource planning problem. However, this proposed method can be applied to all business problems for their dynamic analysis and optimization.

5.5. Performance Evaluation of Petri Net-Based Mixed-Integer Nonlinear Programming

In this research study, a nonlinear mixed integer nonlinear programming model has been integrated with a Petri net for the dynamic behavior of patients flow in a healthcare center. In order to evaluate the performance of a proposed method, Petri nets have been integrated with genetic algorithm, a pattern search method, and case, which considers that the absenteeism has been solved using a Petri net integrated genetic algorithm and a patient integrated pattern search algorithm. Details of each method is given below.

5.5.1. Petri Net Based Genetic Algorithm

Genetic algorithm is a heuristic optimization method based on the theory of evaluation by Darwin. According to the theory of Darwin, only the fittest individuals survive in the next generation. In a genetic algorithm, an initial solution for an initial population is randomly generated. A randomly generated solution is called chromosome or individuals. The chromosome is composed of genes.

In optimization, a decision variable is treated as a gene and a chromosome is treated as an initial feasible solution. A set of feasible solutions is called a population.

The structure of a chromosome is given below in Figure 10. Crossover and mutation are the two genetic operators. The crossover occurs between two individuals or chromosomes, which is done by exchanging the genes. In this research, we used a two-point cross over. A mutation is because of change in genes in an individual chromosome. Mutation introduces the diversity while chromosomes transfer parental characteristics to the children or generations.

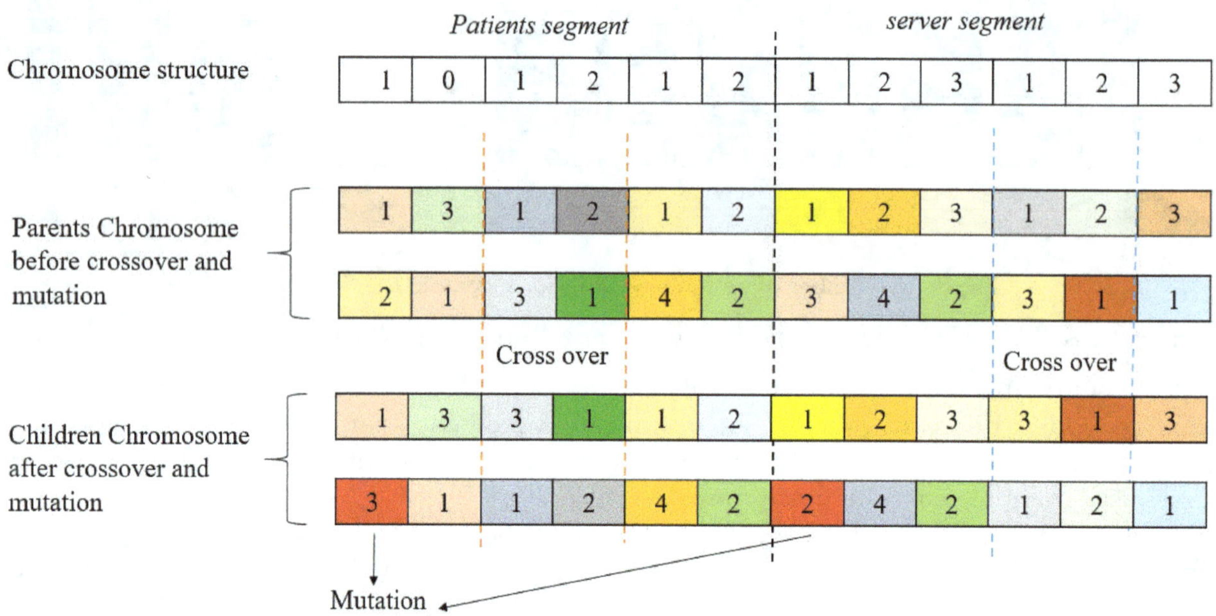

Figure 10. Chromosome structure and genetic operators.

The fittest chromosome in a population is called elite chromosome. Genetic algorithm is a heuristic method and it may or may not give the global optima [58]. In addition, it follows the random search method. Therefore, its computation time is higher than the mixed integer nonlinear programming. In solving the numerical example, the Petri net is integrated with a genetic algorithm [59]. In a Petri net-based genetic algorithm, the population consisted of 200 chromosomes. The mutation function was constraint-dependent. Elitism was $0.25 \times$ *population size* with 80% crossover probability, which was used for the diversity in next generations, and a stochastic uniform selection method was adopted for the selection of individuals or chromosomes for the next or upcoming generation. Termination criteria for the genetic algorithm were the repetition of the same elite in 20 successive generations.

5.5.2. Petri Net-Based Pattern Search Method

The pattern search is a derivative-free search method and its independent of a gradient. In this method, the initial solution is provided and it follows the basic procedure of other optimization methods. This method was introduced in 1961 and emerged in two domains such as the exploratory search and pattern move. The exploratory search follows the direction along the gradient and the pattern move keeps improving its search. The pattern search method involves some parameters that affect its performance for a different problem data set. Kang, Li, and Li (2013) explained the pattern move with the help of Figure 11.

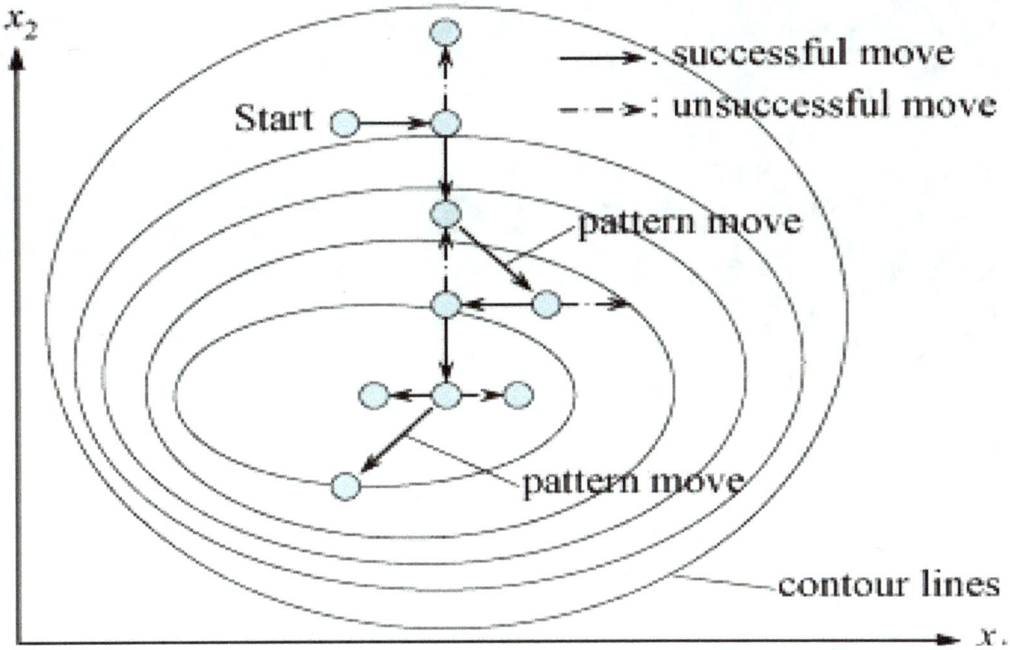

Figure 11. Pattern search illustration.

It is clear that the pattern search method initiates its move to find an appropriate direction for one variable at a time along with the coordinates of the individual from the base point. Once the exploratory search is over, the pattern move starts accelerating along the predetermined direction by the exploratory search. The expansion and contraction during the move are the most influencing parameters in the pattern search. In order to optimize the numerical example, the Petri net was integrated with a pattern search method [60]. In the patient-pattern search method, the same termination criteria were used as in the Petri net-genetic algorithm, expansion, and contraction factors, which were restricted to 2 and 0.5.

5.5.3. Comparison of Methods

The optimization problems can be classified into three categories based on the search method, which include the heuristic method, the direct search method, and the exact method. In this research study, the Petri net has been integrated with each kind of method. In order to evaluate the relative performance of these methods, the same numerical problem presented in Section 4.1.1 is solved with all methods. All methods have been coded in MATLAB (2017) Ra on a personal computer with 8 GB RAM and a 2.0 GHz Processor. The objective function in Equation (10) has been optimized subject to constraints in Equations (11)–(14). Table 9 shows the results of all the algorithms. In order to simplify the results, an average number of servers, and the average number of queues of all stations is considered.

The problem is solved for all required cycle times using both methods and an average number of servers and queues are given in Table 9 for all months of the relevant years. The computational power of methods has been measured in terms of a number of iterations and computation time. The relative performance of each method is computed using a gap analysis. Equation (18) has been used for the computation of the percentage gap.

$$\%Gap = \frac{Achieved\ value - Best\ value}{Achieved\ value} \times 100 \tag{18}$$

Table 9. Results of different algorithms for a case II with an absenteeism factor.

	Petri Net-Genetic Algorithm				Petri Net-Pattern Search				Petri Net-Mixed Integer Nonlinear Programming			
	Average Queue	Average Servers	Iterations	CPU Time (s)	Average Queue	Average Servers	Iterations	CPU Time (s)	Average Queue	Average Servers	Iterations	CPU Time (s)
January	3	4	59	7.12	4	3	4	5.34	2	2	254	2.33
February	3	4	55	6.35	4	4	5	4.99	1	1	258	2.78
March	4	4	416	10.12	5	4	5	4.54	1	1	268	3.1
April	3	3	92	7.87	4	3	5	4.46	1	2	253	2.35
May	3	3	75	7.23	4	4	5	4.25	1	2	265	2.78
Jun	3	3	69	7.15	4	4	5	4.21	1	1	259	3.32
July	3	3	74	7.31	3	3	5	4.29	1	1	270	3.42
August	3	4	89	7.67	3	3	5	4.23	1	1	257	2.45
September	3	3	52	6.45	4	3	5	4.19	1	1	267	2.43
October	4	3	68	7.78	4	2	5	4.89	1	1	268	2.35
November	4	4	67	7.53	5	3	5	4.56	1	1	255	2.34
December	3	4	57	7.23	4	3	5	4.45	2	2	254	2.31
Average	3	4	98	7.48	4	3	5	4.53	1	1	261	2.66

In the case of minimization/maximization problems, the best will be a minimum/maximum value of the specific performance measure. It can be seen from Table 9 that a number of iterations are greater in a Petri net mixed integer nonlinear program.

MINLP is an exact method and evaluates one solution in one iteration. Therefore, a number of iterations are evaluated in a short period of time. However, in the case of the Genetic algorithm and pattern search, it takes more time to evaluate one iteration. In one generation, multiple solutions are evaluated. Average queue length, the average number of servers at the station, the number of iterations, and computational time has been considered as performance measures of the proposed method. Table 10 shows the average performance measures computed from three different methods.

Table 10. Average performance measurement of algorithms.

	Monthly Average Queue	Monthly Average Servers	Average Iterations	Average CPU Time
Petri net-Genetic algorithm	3	4	98	7.5
Petri net-Pattern search	4	3	5	4.5
Petri net-Mixed-integer nonlinear programming	1	1	261	2.7

Using Equation (18) and Table 10, the relative percentage gap has been computed, which is given in Table 11. In order to understand the percentage gap, consider the monthly average queue in Table 10. The best value of the monthly queue is one patient. Therefore, using Equation (18), the percentage gap of the Petri net-genetic algorithm can be computed as follows. Similarly, performance measures of rest of the algorithms are given in Table 11.

$$\%\ Gap = \frac{3-1}{3} \times 100 = 66.67\% \tag{19}$$

Table 11. Relative percentage gap in different methods.

	Monthly Average Queue	Monthly Average Servers	Average Iterations	Average CPU Time	Sum
Petri net-Genetic algorithm	66.67	75.00	94.90	64.00	300.56
Petri net-Pattern Search	75.00	66.67	0.00	40.00	181.67
Petri net-MINLP	0.00	0.00	98.08	0.00	98.08

It is clear that the total percentage gap is minimum in case of Petri net-based mixed integer nonlinear programming as compared to the other methods. The Petri net-genetic algorithm and Petri net-paterren search are the heuristics and follow a random search pattern. However, in mixed integer non-linear programming, the branch and bound algorithm is used in the exact solution method.

5.6. Advantages and Disadvantages of Petri Net Integrated Mixed Integer Non-Linear Programming

The Petri net combines the characteristics of graphics, simulation, and mathematical modeling for the imitation of real systems. Petri nets are good tools for analyzing the dynamic behavior of systems with both continuous and discrete state changes. Integration of optimization with Petri nets provides the following benefits.

1. It enables real-time optimization of systems for more practical results. In traditional literature, a mathematical model is optimized and results are then used in a simulation model for the validation. However, in this research, simulation and optimization takes place at the same time. Therefore, the word real-time optimization is more suitable for this type of optimization.

2. It visualizes the patterns and their flow within the program and highlights the state changes in a system at a different process.
3. It combines the computational power of simulation and mathematics for more robust results.
4. Integration of Petri net with exact optimization reduces computational time and improves the performance.

In addition to the advantages, Petri nets have some disadvantages as well. Complex systems involving more than three processes and entities are hard to model. In Petri nets, model construction is a cumbersome process and needs computational and graphical efforts.

5.7. Managerial Insights

This research is useful for operation managers in healthcare for efficient planning of resources such as physicians and other medical staff. The consideration of the factor of absenteeism makes it more robust for long-term decision-making in determining the optimal number of resources. This planning model can be useful in determining the more accurate number of resources even in case of seasonal diseases or some disease caused by viral viruses or bacteria.

6. Conclusions

This paper presented a Petri net based mixed integer nonlinear programming model for the outdoor patient department (OPD) of the sustainable healthcare center. The integration of the Stochastic-Petri net approach with MINLP for real-time optimization in the planning model enabled the realization of sustainable operational utilization with a long-term competitive advantage. The objective of the research was to minimize the queue length of patients and the required number of servers for a sustainable future. The reason for the queue is the limited resources including unavailability or absenteeism of servers at their stations. The introduction of absenteeism in sustainable planning made this research a novel. A numerical example of a sustainable healthcare center is presented as a case study. The example provided resource planning for the period of one year. The forecast of a number of patients visiting the healthcare system is used to determine the required cycle time for each month. The Petri net mixed integer nonlinear programming is used to solve the model for determining an optimal number of servers at each station of OPD for two cases. The first case considered the determination of resources without absenteeism and the second case considered the determination of resources with absenteeism. It was observed that long queues (two to three patients) are generated if we plan the resource without the absenteeism factor and queue is substantially reduced to one patient only when absenteeism factors are considered. The reduced queue has various benefits associated with it such as maximum resource utilization and improved efficiency of the system with the highest degree of patient satisfaction. In order to evaluate the performance of a Petri net-based mixed integer nonlinear programming model, the same numerical example is solved with other methods such as the Petri net-genetic algorithm and the Petri net-pattern search method. The minimum gap of 98.08% showed that the performance of Petri net-MINLP is better in terms of computation time and function value. This research is useful for the operation manager for resource planning in a healthcare center. This is a planning model for the period of one year and estimates the required number of servers at each station. However, it does not provide any information for controlling the mechanism of servers if the server suddenly leaves the system without serving the patients on a daily basis. The sudden absenteeism control of servers is included in future research.

Author Contributions: Conceptualization, B.S.; methodology and writing original draft preparation, M.I.; writing review and editing, M.O.; formal analysis, W.A.; data curation, M.U.; supervision, C.W.K.

Abbreviations

The notations and abbreviations used in this model are given below.

Indices:

i	patient type $i = 1, 2 \ldots, I$
j	server type $j = 1, 2 \ldots, J$
Legends:	
$\rightarrow$	directed arc shows the flow of token between transition and place
$\circ$	token (patient)
Parameters:	
t_{ij}	attendance time of patient "i" at server "j".
ρ_{ij}	a place for waiting of patient "i" to be served by server "j".
S_{ij}	place of server "j" for serving patient "i".
ζ_p	overlapping circuit for each patient
ζ_s	overlapping sequencing circuits
$\tau(\zeta)$	total transition time
$C(\zeta)$	the cycle time of each sequential loop
PA_j	the probability of absenteeism of server "j"
SR	service rate
PH	planning horizon
NP	the expected number of patients
$C(\zeta_{sMAX})$	the maximum cycle time of the bottleneck server or required cycle time for a patient
Decision Variables:	
NQ_{ij}	Number of patients " " waiting for service at server "j"
NS_{ij}	Number of servers "j" serving the patient "i"

References

1. Brundtland, G.H.; Khalid, M.; Agnelli, S.; Al-Athel, S. *Our Common Future*; Oxford University Press: Oxford, UK, 1987.
2. Omair, M.; Sarkar, B.; Cárdenas-Barrón, L.E. Minimum Quantity Lubrication and Carbon Footprint: A Step towards Sustainability. *Sustainability* **2017**, *9*, 714. [CrossRef]
3. Habib, M.S.; Sarkar, B.; Tayyab, M.; Saleem, M.W.; Hussain, A.; Ullah, M.; Omair, M.; Iqbal, M.W. Large-scale disaster waste management under uncertain environment. *J. Clean. Prod.* **2019**, *212*, 200–222. [CrossRef]
4. Omair, M.; Noor, S.; Hussain, I.; Maqsood, S.; Khattak, S.; Akhtar, R.; Haq, I.U. Sustainable development tool for Khyber Pakhtunkhwa's dimension stone industry. *Technol. J.* **2015**, *20*, 160–165.
5. Sarkar, B.; Omair, M.; Choi, S.-B. A Multi-Objective Optimization of Energy, Economic, and Carbon Emission in a Production Model under Sustainable Supply Chain Management. *Appl. Sci.* **2018**, *8*, 1744. [CrossRef]
6. Tayyab, M.; Sarkar, B.; Ullah, M. Sustainable Lot Size in a Multistage Lean-Green Manufacturing Process under Uncertainty. *Mathematics* **2018**, *7*, 20. [CrossRef]
7. Carnero, M.C. Assessment of environmental sustainability in health care organizations. *Sustainability* **2015**, *7*, 8270–8291. [CrossRef]
8. Liou, J.J.; Lu, M.-T.; Hu, S.-K.; Cheng, C.-H.; Chuang, Y.-C. A Hybrid MCDM Model for Improving the Electronic Health Record to Better Serve Client Needs. *Sustainability* **2017**, *9*, 1819. [CrossRef]
9. Carpman, J.R.; Grant, M.A. *Design That Cares: Planning Health Facilities for Patients and Visitors*; John Wiley & Sons: Hoboken, NJ, USA, 2016; Volume 142.
10. Dangerfield, B. System Dynamics Applications to European Healthcare Issues. In *Operational Research for Emergency Planning in Healthcare*; Springer: Berlin, Germany, 2016; Volume 2, pp. 296–315.
11. Hans, E.W.; Van Houdenhoven, M.; Hulshof, P.J. A Framework for Healthcare Planning and Control. In *Handbook of Healthcare System Scheduling*; Springer: Berlin, Germany, 2012; pp. 303–320.
12. Hulshof, P.J.; Kortbeek, N.; Boucherie, R.J.; Hans, E.W.; Bakker, P.J. Taxonomic classification of planning decisions in health care: A structured review of the state of the art in OR/MS. *Health Syst.* **2012**, *1*, 129–175. [CrossRef]

13. Lane, D.C.; Monefeldt, C.; Rosenhead, J. Looking in the wrong place for healthcare improvements: A system dynamics study of an accident and emergency department. In *Operational Research for Emergency Planning in Healthcare*; Springer: Berlin, Germany, 2016; Volume 2, pp. 92–121.
14. Zeinali, F.; Mahootchi, M.; Sepehri, M.M. Resource planning in the emergency departments: A simulation-based metamodeling approach. *Simul. Model. Pract. Theory* **2015**, *53*, 123–138. [CrossRef]
15. Yousefi, M.; Yousefi, M.; Ferreira, R.P.M.; Kim, J.H.; Fogliatto, F.S. Chaotic genetic algorithm and Adaboost ensemble metamodeling approach for optimum resource planning in emergency departments. *Artif. Intell. Med.* **2018**, *84*, 23–33. [CrossRef]
16. Feng, Y.-Y.; Wu, I.-C.; Chen, T.-L. Stochastic resource allocation in emergency departments with a multi-objective simulation optimization algorithm. *Health Care Manag. Sci.* **2017**, *20*, 55–75. [CrossRef] [PubMed]
17. Bachouch, R.B.; Guinet, A.; Hajri-Gabouj, S. An integer linear model for hospital bed planning. *Int. J. Prod. Econ.* **2012**, *140*, 833–843. [CrossRef]
18. Barado, J.; Guergué, J.M.; Esparza, L.; Azcárate, C.; Mallor, F.; Ochoa, S. A mathematical model for simulating daily bed occupancy in an intensive care unit. *Crit. Care Med.* **2012**, *40*, 1098–1104. [CrossRef] [PubMed]
19. Ma, G.; Demeulemeester, E. A multilevel integrative approach to hospital case mix and capacity planning. *Comput. Oper. Res.* **2013**, *40*, 2198–2207. [CrossRef]
20. Brailsford, S.; Hilton, N. A Comparison of Discrete Event Simulation and System Dynamics for Modelling Health Care Systems. In *Planning for the Future: Health Service Quality and Emergency Accessibility*; Riley, J., Ed.; University of Southampton: Southampton, UK, 2001.
21. Morecroft, J.; Robinson, S. Explaining Puzzling Dynamics: Comparing the Use of System Dynamics and Discrete-Event Simulation. In Proceedings of the 23rd International Conference of the System Dynamics Society, Boston, MA, USA, 17–21 July 2005; pp. 17–21.
22. Duggan, J. A Comparison of Petri Net and System Dynamics Approaches for Modelling Dynamic Feedback Systems. In Proceedings of the 24th International Conference of the Systems Dynamics Society, Nijmegen, The Netherlands, 23–27 July 2006.
23. Eppich, W.; Howard, V.; Vozenilek, J.; Curran, I. Simulation-based team training in healthcare. *Simul. Healthc.* **2011**, *6*, 14–19. [CrossRef] [PubMed]
24. Hall, R. Patient flow. *AMC* **2014**, *10*, 12.
25. Meiller, Y.; Bureau, S.; Zhou, W.; Piramuthu, S. Adaptive knowledge-based system for health care applications with RFID-generated information. *Decis. Support. Syst.* **2011**, *51*, 198–207. [CrossRef]
26. Robinson, S.; Radnor, Z.J.; Burgess, N.; Worthington, C. SimLean: Utilising simulation in the implementation of lean in healthcare. *Eur. J. Oper. Res.* **2012**, *219*, 188–197. [CrossRef]
27. Rohleder, T.R.; Lewkonia, P.; Bischak, D.P.; Duffy, P.; Hendijani, R. Using simulation modeling to improve patient flow at an outpatient orthopedic clinic. *Health Care Manag. Sci.* **2011**, *14*, 135–145. [CrossRef]
28. Du, G.; Liang, X.; Sun, C.J.S. Scheduling optimization of home health care service considering patients' priorities and time windows. *Sustainability* **2017**, *9*, 253. [CrossRef]
29. Yalçındağ, S.; Cappanera, P.; Scutellà, M.G.; Şahin, E.; Matta, A.J.C.; Research, O. Pattern-based decompositions for human resource planning in home health care services. *Comput. Oper. Res.* **2016**, *73*, 12–26. [CrossRef]
30. En-nahli, L.; Allaoui, H.; Nouaouri, I.J.I.-P. A multi-objective modelling to human resource assignment and routing problem for home health care services. *Sci. Direct* **2015**, *48*, 698–703. [CrossRef]
31. Ganguly, B.; Pareek, S.; Sarkar, B.; Sarkar, M.; Omair, M. Influence of controllable lead time, premium price, and unequal shipments under environmental effects in a supply chain management. *RAIRO-Oper. Res.* **2018**. [CrossRef]
32. Wook Kang, C.; Ullah, M.; Sarkar, M.; Omair, M.; Sarkar, B. A Single-Stage Manufacturing Model with Imperfect Items, Inspections, Rework, and Planned Backorders. *Mathematics* **2019**, *7*, 446. [CrossRef]
33. Omair, M. Effect of Workplace Stress in Short Term and Long Term Controllable Production System. Ph.D. Thesis, Graduate School of Hanyang University, Seoul, Korea, 2019. Available online: http://dcollection.hanyang.ac.kr/common/orgView/000000107873.
34. Ghazalbash, S.; Sepehri, M.M.; Shadpour, P.; Atighehchian, A. Operating room scheduling in teaching hospitals. *Adv. Oper. Res.* **2012**, *2012*, 16. [CrossRef]

35. Nasir, J.A.; Dang, C. Solving a More Flexible Home Health Care Scheduling and Routing Problem with Joint Patient and Nursing Staff Selection. *Sustainability* **2018**, *10*, 148. [CrossRef]
36. Tan, Y.; El Mekkawy, T.; Peng, Q.; Oppenheimer, L. Mathematical programming for the scheduling of elective patients in the operating room department. *Proc. Can. Eng. Educ. Assoc.* **2011**. [CrossRef]
37. Carayon, P.; Wetterneck, T.B.; Rivera-Rodriguez, A.J.; Hundt, A.S.; Hoonakker, P.; Holden, R.; Gurses, A.P. Human factors systems approach to healthcare quality and patient safety. *Appl. Ergon.* **2014**, *45*, 14–25. [CrossRef]
38. Atun, R. Health systems, systems thinking and innovation. *Health Policy Plan.* **2012**, *27*, 4–8. [CrossRef]
39. Fraher, E.P.; Knapton, A.; Sheldon, G.F.; Meyer, A.; Ricketts, T.C. Projecting surgeon supply using a dynamic model. *Ann. Surg.* **2013**, *257*, 867–872. [CrossRef]
40. Lee, C.W.; Kwak, N. Strategic enterprise resource planning in a health-care system using a multicriteria decision-making model. *J. Med. Syst.* **2011**, *35*, 265–275. [CrossRef] [PubMed]
41. Abo-Hamad, W.; Arisha, A. Simulation-based framework to improve patient experience in an emergency department. *Eur. J. Oper. Res.* **2013**, *224*, 154–166. [CrossRef]
42. Zeltyn, S.; Marmor, Y.N.; Mandelbaum, A.; Carmeli, B.; Greenshpan, O.; Mesika, Y.; Wasserkrug, S.; Vortman, P.; Shtub, A.; Lauterman, T. Simulation-based models of emergency departments: Operational, tactical, and strategic staffing. *ACM Trans. Modeling Comput. Simul. (TOMACS)* **2011**, *21*, 24. [CrossRef]
43. Hamrock, E.; Parks, J.; Scheulen, J.; Bradbury, F.J. Discrete event simulation for healthcare organizations: A tool for decision making. *J. Healthc. Manag.* **2013**, *58*, 110. [CrossRef] [PubMed]
44. Jahangirian, M.; Naseer, A.; Stergioulas, L.; Young, T.; Eldabi, T.; Brailsford, S.; Patel, B.; Harper, P. Simulation in health-care: Lessons from other sectors. *Oper. Res.* **2012**, *12*, 45–55. [CrossRef]
45. Zhang, Y.; Puterman, M.L.; Nelson, M.; Atkins, D. A simulation optimization approach to long-term care capacity planning. *Oper. Res.* **2012**, *60*, 249–261. [CrossRef]
46. Ozcan, Y.A.; Tànfani, E.; Testi, A. A Simulation-Based Modeling Framework to Deal with Clinical Pathways. In Proceedings of the Winter Simulation Conference, Phoenix, AZ, USA, 11–14 December 2001; pp. 1190–1201.
47. Granja, C.; Almada-Lobo, B.; Janela, F.; Seabra, J.; Mendes, A. An optimization based on simulation approach to the patient admission scheduling problem using a linear programing algorithm. *J. Biomed. Inform.* **2014**, *52*, 427–437. [CrossRef] [PubMed]
48. Mahulea, C.; Mahulea, L.; García-Soriano, J.-M.; Colom, J.-M. Petri Nets with Resources for Modeling Primary Healthcare Systems. In Proceedings of the 18th International Conference System Theory, Control and Computing (ICSTCC), Sinaia, Romania, 17–19 October 2014; pp. 639–644.
49. Bertolini, C.; Liu, Z.; Srba, J. Verification of Timed Healthcare Workflows Using Component Timed-Arc Petri Nets. In Proceedings of the International Symposium on Foundations of Health Informatics Engineering and Systems, Paris, France, 27–28 August 2012; pp. 19–36.
50. Hicheur, A.; Dhieb, A.B.; Barkaoui, K. Modelling and Analysis of Flexible Healthcare Processes Based on Algebraic and Recursive Petri Nets. In Proceedings of the International Symposium on Foundations of Health Informatics Engineering and Systems, Paris, France, 27–28 August 2012; pp. 1–18.
51. Wang, J. Emergency healthcare workflow modeling and timeliness analysis. *IEEE Trans. Syst. Man Cybern. Part. A Syst. Hum.* **2012**, *42*, 1323–1331. [CrossRef]
52. Zeng, Q.; Lu, F.; Liu, C.; Duan, H.; Zhou, C. Modeling and verification for cross-department collaborative business processes using extended Petri nets. *IEEE Trans. Syst. Man Cybern. Syst.* **2015**, *45*, 349–362. [CrossRef]
53. Mahulea, C.; Mahulea, L.; Soriano, J.M.G.; Colom, J.M. Modular Petri net modeling of healthcare systems. *Flex. Serv. Manuf. J.* **2018**, *30*, 329–357. [CrossRef]
54. Wang, J. Petri Nets for Dynamic Event-Driven System Modeling. *Handb. Dyn. Syst. Modeling* **2007**, *1*.
55. Aziz, M.; Bohez, E.L.; Pisuchpen, R.; Parnichkun, M. Petri Net model of repetitive push manufacturing with Polca to minimise value-added WIP. *Int. J. Prod. Res.* **2013**, *51*, 4464–4483. [CrossRef]
56. Mukhlash, I.; Rumana, W.N.; Adzkiya, D.; Sarno, R. Business Process Improvement of Production Systems Using Coloured Petri Nets. *Bull. Electr. Eng. Inform.* **2018**, *7*, 102–112.
57. Premchaiswadi, W.; Porouhan, P. Process modeling and bottleneck mining in online peer-review systems. *SpringerPlus* **2015**, *4*, 441. [CrossRef] [PubMed]
58. Ren, Y.; Zhang, C.; Zhao, F.; Xiao, H.; Tian, G. An asynchronous parallel disassembly planning based on genetic algorithm. *Eur. J. Oper. Res.* **2018**, *269*, 647–660. [CrossRef]

59. Si, Y.-W.; Chan, V.-I.; Dumas, M.; Zhang, D. A Petri Nets based Generic Genetic Algorithm framework for resource optimization in business processes. *Simul. Model. Pract. Theory* **2018**, *86*, 72–101. [CrossRef]
60. Tax, N.; Sidorova, N.; van der Aalst, W.M.; Haakma, R. LocalProcessModelDiscovery: Bringing Petri Nets to the Pattern Mining World. In Proceedings of the 39th International Conference on Applications and Theory of Petri Nets and Concurrency, Bratislava, Slovakia, 24–29 June 2018; pp. 374–384.

Permissions

All chapters in this book were first published in MDPI; hereby published with permission under the Creative Commons Attribution License or equivalent. Every chapter published in this book has been scrutinized by our experts. Their significance has been extensively debated. The topics covered herein carry significant findings which will fuel the growth of the discipline. They may even be implemented as practical applications or may be referred to as a beginning point for another development.

The contributors of this book come from diverse backgrounds, making this book a truly international effort. This book will bring forth new frontiers with its revolutionizing research information and detailed analysis of the nascent developments around the world.

We would like to thank all the contributing authors for lending their expertise to make the book truly unique. They have played a crucial role in the development of this book. Without their invaluable contributions this book wouldn't have been possible. They have made vital efforts to compile up to date information on the varied aspects of this subject to make this book a valuable addition to the collection of many professionals and students.

This book was conceptualized with the vision of imparting up-to-date information and advanced data in this field. To ensure the same, a matchless editorial board was set up. Every individual on the board went through rigorous rounds of assessment to prove their worth. After which they invested a large part of their time researching and compiling the most relevant data for our readers.

The editorial board has been involved in producing this book since its inception. They have spent rigorous hours researching and exploring the diverse topics which have resulted in the successful publishing of this book. They have passed on their knowledge of decades through this book. To expedite this challenging task, the publisher supported the team at every step. A small team of assistant editors was also appointed to further simplify the editing procedure and attain best results for the readers.

Apart from the editorial board, the designing team has also invested a significant amount of their time in understanding the subject and creating the most relevant covers. They scrutinized every image to scout for the most suitable representation of the subject and create an appropriate cover for the book.

The publishing team has been an ardent support to the editorial, designing and production team. Their endless efforts to recruit the best for this project, has resulted in the accomplishment of this book. They are a veteran in the field of academics and their pool of knowledge is as vast as their experience in printing. Their expertise and guidance has proved useful at every step. Their uncompromising quality standards have made this book an exceptional effort. Their encouragement from time to time has been an inspiration for everyone.

The publisher and the editorial board hope that this book will prove to be a valuable piece of knowledge for researchers, students, practitioners and scholars across the globe.

List of Contributors

M. Ali Ülkü
Rowe School of Business, Dalhousie University, Halifax, NS B3H 4R2, Canada
Centre for Research in Sustainable Supply Chain Analytics, Dalhousie University, Halifax, NS B3H 4R2, Canada
Department of Industrial Engineering, Dalhousie University, Halifax, NS B3H 4R2, Canada

Uday Venkatadri, Claver Diallo and Simranjeet S. Chadha
Centre for Research in Sustainable Supply Chain Analytics, Dalhousie University, Halifax, NS B3H 4R2, Canada
Department of Industrial Engineering, Dalhousie University, Halifax, NS B3H 4R2, Canada

Melek Akgün
Sakarya Business School, Sakarya University, Sakarya 54050, Turkey

Sidharth Sankhye and Guiping Hu
Department of Industrial and Manufacturing Systems Engineering, Iowa State University, Ames, IA 50011, USA

Ata Allah Taleizadeh
School of Industrial Engineering, College of Engineering, University of Tehran, Tehran 456311155, Iran

Masoumeh Sadat Babaei
Department of Industrial Engineering, Islamic Azad University, South Tehran Branch, Tehran 1584743311, Iran

Shib Sankar Sana
Principal, Kishore Bharati Bhagini Nivedita College, Ramkrishna Sarani, Behala, Kolkata 700060, India

Biswajit Sarkar
Department of Industrial & Management Engineering, Hanyang University, Ansan, Gyeonggi-do 15588, Korea

Sonia Irshad Mari and Muhammad Saad Memon
Department of Industrial Engineering and Management, Mehran University of Engineering and Technology, Jamshoro, Sindh 76062, Pakistan

Muhammad Babar Ramzan
Department of Garment Manufacturing, National Textile University, Faisalabad 37610, Pakistan

Sheheryar Mohsin Qureshi
Department of Industrial and Manufacturing Engineering, NED University of Engineering and Technology, Karachi 75270, Pakistan

Muhammad Waqas Iqbal
Department of Industrial Engineering, Hongik University, Seoul 04066, Korea

Sharmistha Halder Jana
Department of Mathematics, Midnapore College (Autonomous), Midnapore 721101, India

Biswapati Jana
Department of Computer Science, Vidyasagar University, Midnapore 721102, India

Barun Das
Department of Mathematics, Sidho Kanho Birsha University, Purulia 723104, India

Goutam Panigrahi
Department of Mathematics, National Institute of Technology, Durgapur 713209, India

Manoranjan Maiti
Department of Mathematics, Vidyasagar University, Midnapore 721102, India

Muhammad Tayyab, Biswajit Sarkar and Mehran Ullah
Department of Industrial & Management Engineering, Hanyang University, Ansan Gyeonggi-do 155 88, Korea

Bernardo Nugroho Yahya
Industrial and Management Engineering Department, Hankuk University of Foreign Studies, 81, Oedae-ro, Mohyeon-myeon, Cheoin-gu, Yongin-si, Gyeonggi-do 17035, Korea

Mohammad A. M. Abdel-Aal and Shokri Z. Selim
Systems Engineering Department, King Fahd University of Petroleum and Minerals, Dhahran 31261, Saudi Arabia

Seok-Beom Choi
Department of International Business, Cheju Halla University, Jeju-do 63092, Korea

Juhyun Lee
Department of Industrial and Management Engineering, Hanyang University, Seoul 04763, Korea

Byunghoon Kim and Suneung Ahn
Department of Industrial and Management Engineering, Hanyang University ERICA, Ansan 15588, Korea

Tien-Muoi Le
Department of Industrial Engineering and Management, National Kaohsiung University of Sciences and Technology, Chien Kung Campus, 415 Chien Kung Road, Kaohsiung 807, Taiwan

Chia-Nan Wang
Department of Industrial Engineering and Management, National Kaohsiung University of Sciences and Technology, Chien Kung Campus, 415 Chien Kung Road, Kaohsiung 807, Taiwan
Department of Industrial Engineering and Management, Fortune Institute of Technology, Kaohsiung 83160, Taiwan

Han-Khanh Nguyen
Faculty of Economics, Thu Dau Mot University, Number 6, Tran Van On Street, Phu Hoa Ward, Thu Dau Mot City 820000, Binh Duong Province, Vietnam

Chang Wook Kang, Mitali Sarkar and Biswajit Sarkar
Department of Industrial and Management Engineering, ERICA Campus, Hanyang University, Ansan, Gyeonggi-do 15588, Korea

Misbah Ullah
Department of Industrial Engineering, University of Engineering and Technology, Peshawar 25000, Pakistan

Muhammad Omair
Department of Industrial Engineering, Jalozai Campus, University of Engineering and Technology, Peshawar 25000, Pakistan

Muhammad Imran and Waqas Ahmed
Department of Management & HR', NUST Business School, National University of Sciences & Technology, Islamabad 44000, Pakistan

Index

R

S

T

www.ingramcontent.com/pod-product-compliance
Lightning Source LLC
LaVergne TN
LVHW081019210926
848485LV00004B/905